VORTEX PROCESSES AND SOLID BODY DYNAMICS

FLUID MECHANICS AND ITS APPLICATIONS
Volume 25

Series Editor: R. MOREAU
MADYLAM
Ecole Nationale Supérieure d'Hydraulique de Grenoble
Boîte Postale 95
38402 Saint Martin d'Hères Cedex, France

Aims and Scope of the Series

The purpose of this series is to focus on subjects in which fluid mechanics plays a fundamental role.

As well as the more traditional applications of aeronautics, hydraulics, heat and mass transfer etc., books will be published dealing with topics which are currently in a state of rapid development, such as turbulence, suspensions and multiphase fluids, super and hypersonic flows and numerical modelling techniques.

It is a widely held view that it is the interdisciplinary subjects that will receive intense scientific attention, bringing them to the forefront of technological advancement. Fluids have the ability to transport matter and its properties as well as transmit force, therefore fluid mechanics is a subject that is particulary open to cross fertilisation with other sciences and disciplines of engineering. The subject of fluid mechanics will be highly relevant in domains such as chemical, metallurgical, biological and ecological engineering. This series is particularly open to such new multidisciplinary domains.

The median level of presentation is the first year graduate student. Some texts are monographs defining the current state of a field; others are accessible to final year undergraduates; but essentially the emphasis is on readability and clarity.

For a list of related mechanics titles, see final pages.

Vortex Processes and Solid Body Dynamics

The Dynamic Problems of Spacecrafts and Magnetic Levitation Systems

by

BORIS I. RABINOVICH

Moscow Institute for Control Devices Design,
Moscow, Russia

VALERYI G. LEBEDEV

Research and Design Institute,
Moscow, Russia

and

ALEXANDER I. MYTAREV

Research and Design Institute,
Moscow, Russia

SPRINGER-SCIENCE+BUSINESS, MEDIA, B.V.

Library of Congress Cataloging-in-Publication Data

Rabinovich, B. I. (Boris I.)
Vortex processes and solid body dynamics : spacecrafts and magnetic levitation systems dynamics problems / by Boris I. Rabinovich, Valeryi G. Lebedev, Alexander I. Mytarev.
p. cm. -- (Fluid mechanics and its applications v. 25)
Includes bibliographical references and index.
ISBN 978-94-010-4452-3 ISBN 978-94-011-1038-9 (eBook)
DOI 10.1007/978-94-011-1038-9
1. Vortex-motion--Mathematical models. 2. Space vehicles--Dynamics--Mathematical models. 3. Magnetic suspension--Mathematical models. 4. Magnetohydrodynamics--Mathematics. I. Lebedev, V. G. (Valeriĭ G.) II. Mytarev, Alexander I. III. Title. IV. Series.
TL574.V6R33 1994
629.132'3--dc20 94-30833

ISBN 978-94-010-4452-3

Translated from the Russian
by A. S. Leviant

Printed on acid-free paper

Contents

Preface

...a wise man knows all things in a manner in which this is possible, not, however, knowing them individually.

Aristotle. *Metaphysics**

The problem of consideration of vortex fields' influence on solid body dynamics has a long history. One constantly comes upon it in flight dynamics of airplanes, helicopters, and other flying vehicles (FV) moving in the atmosphere, in dynamics of ships with hydrofoils, and in dynamics of rocket carriers (RC) and spacecrafts (SC) with liquid-propellant rocket engines (LPRE), that are equipped with special damping devices and other structural elements inside fluid tanks.

Similar problems occur when solving problems related to attitude control and stabilization of artificial Earth satellites (AES) and spacecrafts with magnetic (electromagnetic) systems, in conducting elements of which eddy currents are induced while control of those vehicles' angular position. It is also true with special test facilities for dynamic testing of space vehicles and their systems, with modern high-speed magnetic suspension transport systems (those based on the phenomenon of 'magnetic levitation'), with generators having rotors carried in 'magnetic bearings', and so on.

At certain stages of research it was possible to assume, that vortex fields were localized within a thin boundary layer, and to consider them a by-effect. Recently, however, due to rapid complication of both vehicles themselves and their control systems (systems of stabilization and attitude control), still sharper has become the problem of inclusion into the vehicle's mathematical model of one or another model of 'non-small' vortex fields, the kinetic energy of which is comparable to potential fields kinetic energy. It is true for low-viscous liquid vortex motions caused by various stalling effects, as well as for magnetic fields induced by eddy currents in cores having comparatively high electrical conductivity. The problem of control of those fields moves to the foreground.

Complete mathematical description of unsteady vortex fields requires the introduction of a number of hardly formalizable physical factors and solution of complicated boundary-value problems (especially with stalled flow). These questions are currently covered with the most completeness in the literature on winged vehicle dynamics and on aeroelasticity, with respect to the problems of unsteady flow around wings, tail units, and the flying vehicle as a whole (external boundary-value problems).

Literature devoted to dynamics of rocket carriers and spacecrafts with liquid-propellant rocket engines having tanks equipped with liquid sloshing dampers and

* Transl. with comment. and glossary by Hippocrates G. Apostle. Bloomington—London: Indiana University Press, Cop. 1966, (p. 14)

literature on eddy currents in cores reduce analysis of this problem to investigation of the influence of purely dissipative effects (these are mainly internal boundary-value problems), that, naturally, narrows the statement and the investigation of the problem.

The present book is devoted to a more complete investigation of that class of problems, including not only dissipation effects, but also the effect of liquid vortex motion kinetic energy. Dynamics of objects, for which vortex fields may play a dominant role, is described based on a common phenomenologic model of unsteady eddy currents and vortex motion of liquid.

The appropriate mathematical models describing the considered objects' motion are sets of singular integrodifferential equations allowing fruitful investigation with modern analytical and numerical methods. Phenomenologic models of that kind let us obtain information of the system's behaviour as a whole ("know all things in a manner in which this is possible") and reveal a number of new delicate dynamic effects, not knowing the details of eddies distribution ("not, however, knowing them individually").

The book consists of eight chapters. Chapters 1 and 2 are devoted to the synthesis of phenomenologic mathematical models of objects, for which the consideration of vortex fields plays a dominant role in the formulation of those models.

Success of those models application to the analysis of complex controlled systems dynamics is determined by the possibility of creating, based on the models, efficient analytical and numerical algorithms for perturbed motion investigation and for solution of the system 'object – regulator' stability problems. Note, that though obtained mathematical models seem simple, the general solution of the sets of integrodifferential equations, that the models yield, is by no means trivial. Chapter 3 deals with this question.

Chapter 4 is devoted to the experimental checking of the introduced models and to estimation of their applicability limits.

The first four chapters form thus a basis for an unbiassed estimation of the influence of new physical factors, reflected in the suggested mathematical models, on dynamics of the stabilization and attitude control systems elements and on dynamics of the closed system 'object – regulator'.

Chapter 5 analyzes influence on the electromagnetic levitation system stability of eddy currents in ferromagnetic core elements having comparably high electrical conductivity (HECF *elements*). Electromagnets with controlled winding currents are the actuators of that system. Chapter 6 presents an analysis of the influence on spacecrafts dynamics and stability of vortex motions of the low-viscous liquid (LV *liquid*) partially filling the vehicle's tanks and characterized by high vorticity caused by damping ribs.

Those two chapters successively consider mathematical models of various degrees of completeness, including traditional models, that take into account only dissipative components of the adjacent vortex fields. Some model problems are considered, which solutions enable an estimation of those factors' influence on the 'object – regulator' closed system dynamics. Examples within the chapters are intended to illustrate the possibility of proceeding from "knowing individual things" to "the general view on similar things" and the possibility to allow for different physical phenomena in a common formalism. On the other side, those examples let you judge, in which cases and to what extent it is necessary to use the suggested more exact mathematical models, and when the simplified ones are sufficient.

Chapter 7 presents examples of synthesizing a control law based on one of the models of Chapter 1 for the air gap stabilization of the magnetic levitation system of the test facility.

Somewhat separate is Chapter 8 completing the book. It deals with the synthesis and investigation of a general mathematical model of the system 'solid – low-viscous electrically conductive ferromagnetic liquid (LVECF *liquid*)', which is based on the classical equations of magnetohydrodynamics [11–13]. In particular cases low-viscous conducting liquid (LVEC *liquid*) may be an element of that system, the magnetic field being implemented with a special coil in a body related coordinate system. The mathematical models of the first two chapters may be obtained from that model as particular cases.

All the problems considered in those eight chapters may be related to non-classical problems of solid body dynamics, where vortex fields of one or another physical nature play a dominant role.

Chapters 1 and 2 are written by Boris Rabinovich; Chapter 3 by Valery Lebedev, Alexander Mytarev, and Boris Rabinovich together; Chapters 4 and 6 by Valery Lebedev together with Alexander Mytarev; Chapters 5 and 7 by Alexander Mytarev, Chapter 8 by Boris Rabinovich together with Alexander Mytarev.

The authors would like to express their acknowledgement to Professor V. Beletsky for a number of useful advices when reviewing the manuscript and to A.S.Leviant, Candidate of Science in Physics and Mathematics, for translating the book into English.

The authors.

Preface to the English Edition

The present edition is an English translation of the authors' book having the same title (published in Moscow by Nauka Publishers, the Main Editorial Board for Physical and Mathematical Literature, in 1992).

This book differs from the mentioned one in essential extension of Chapter 8, to which three new articles are added. Those articles are devoted to revealing dynamic features of the system 'solid body – electrically conductive liquid moving in magnetic field related to the body.'

The list of bibliography includes, where available, references to the English language editions of papers and books published originally in Russian. Besides that, some denominations are changed in order to make them closer to those traditional for the English language scientific literature. The noticed misprints were corrected.

The authors

List of Main Symbols

1. Constants and Functions of One or Several Arguments

A, Ψ, a, β – scalars; $\mathbf{R}$, $\vec{\Phi}$, r, $\vec{\alpha}$ – vectors; J, A – tensors and matrices; $\mathfrak{J}$ – operators;
Sh – dimensionless congruence criteria.

2. Superscripts

$\dot{}$ – time derivative;
$'$, $''$ – coordinate derivatives; quantities associated with the fixed cover model of liquid motions; in particular places of the book they denote some coefficients at the first and second derivatives and some auxiliary coeficients;
$\bar{}$ – conjugated quantity (a tensor or a complex quantity);
$\sim$ – normalized quantity, particularly a dimensionless one;
$\wedge$ – boundary layer type function; a temporary symbol in intermediate transformations;
$\circ$ – quantity related to unperturbed motion or state (in the latter case it means absence of perturbations associated with air gap, liquid fluidity, presence of damping ribs inside a tank, etc.);
(0) – quantity related to solidified liquid;
$*$ – quantity directly associated with vortex motions of liquid or with eddy currents;
e – external domain;
i – internal domain;
a – self-sustained oscillation processes in the roll channel.

3. Subscripts

F – quantity characterizing liquid motion;
M – quantity associated with electromagnetic field;
E – experimental value;
O, G – the point, the given quantity is referred to, in particular, it is the system's centre of mass;
0 – initial value in Cauchy problem or initial estimate;
a – absolute motion of liquid; self-sustained oscillations.

4. Adopted Abbreviations

HV liquid – high-viscous liquid;
LV liquid – low-viscous liquid;
LVEC liquid – low-viscous electrically conductive liquid;
LVECF liquid – low-viscous electrically conductive ferromagnetic liquid;
HECF element – ferromagnetic element with high electrical conductivity;
LECF element – ferromagnetic element with low electrical conductivity;
LPR engine – liquid-propellant rocket engine.

Introduction

Now art comes into being when out of many notions from experience we for one universal belief concerning similar facts.

<...> The cause of this is that experience is knowledge of individuals but art is universal knowledge <...>

Aristotle. *Metaphysics* (p. 12, 13) *

The book deals with vehicles represented by solid bodies, having a certain space orientation and stabilized with respect to one or another coordinate system. Particular cases are artificial Earth satellites (AES), rocket carriers (RC), and vehicles with magnetic levitation systems (see V.I.Bocharov and V.D.Nagorsky (Ed.) [9], T.I.Katsan, V.G.Lebedev, and A.I.Mytarev [35], Yu.A.Nikitenko *et al.* [71], K.H.Brock *et al.* [14], E.Gottzein [25], E.Gottzein, L.Miller, and R.Meisinger [27], P.K.Sinha [92], R.D,Thornton [96]).

A specific feature of all those vehicles is presence of control systems (performing attitude control, stabilization, levitation control) of various degrees of complexity determined by the considered object's purpose. Some elements of those systems can, in their turn, be considered as stabilized objects. Some problems of dynamics are considered, that makes it possible to reveal the effect of structure and parameters of the object's mathematical model on the choice of parameters for the control system affording motion stability.

The problem by itself is not new, and many aspects of it are concerned in numerous publications, for instance, consideration of liquid fuel (partially filling fuel tanks) components sloshing and airframe elasticity effects on stability of flying vehicles with LPR engines, and synthesis of appropriate mathematical models (see, e.g. V.D.Arens, S.M.Fedorov, and M.S.Khitrik [4], G.N.Mikishev and B.I.Rabinovich [63, 64], G.S.Narimanov [70], B.I.Rabinovich [75, 76], V.M.Rogovoy and S.V.Cheremnykh [86], I.M.Sidorov, L.Ye.Goncharova, and V.G.Lebedev [90], H.N.Abramson (Ed.) [1]).

To a lesser extent this is true for AES stabilization and attitude control system ferromagnetic elements interaction with the Earth magnetic field and with interior electromagnetic fields, which is also dealt with in a number of papers (e.g. V.V.Beletsky and A.A.Khentov [7], D.M.Veinberg *et al.* [100]).

The listed objects and problems associated with them have a number of individual features. It is possible, however, to pick out a narrower class of problems characterized

---

* Aristotle's Metaphysics. Transl. with comm. and glossary by H.G.Apostle. Bloomington – London. Indiana University Press, 1966.

by common structure of mathematical models. The methodological base of those models is the consideration of vortex fields, that allows a more complete reflection of the physical reality and hence a better description of perturbed motion of space vehicle as a whole.

This lead in dynamics of a solid stabilized in the space (RC, SC, AES, and other flying vehicles) is comparatively new. It is necessary to emphasize, that traditional models, playing the main role in applied research up to this day, pay the principal attention to considering dissipative components of vortex fields only. Inertial components are usually not considered, that is equivalent to neglecting vortex fields kinetic energy in comparison with potential fields energy. For instance, such an assumption, that comes from the concept of boundary layer, which contains the majority of vortexes, is the basis of conventionally used in practice mathematical models of RC and SC (see, e.g. G.N.Mikishev [62], B.I.Rabinovich [76], V.M.Rogovoy and S.V.Cheremnykh [86]).

A specific feature of this book is that it consistently follows another idea expressed by B.I.Rabinovich and V.M.Rogovoy [80, 81], B.I.Rabinovich [77], that LV liquid velocity fields and magnetic fields in ferromagnets with high electrical conductivity (HECF *elements*) are simulated with high-vorticity fields. One of the models, considered in Chapter 1, describes eddy currents in stabilization and orientation systems ferromagnetic elements in the integral sense; the other one, considered in Chapter 2, describes vortex motions of liquid in tanks equipped with interior damping ribs.

A generalization of those models is the mathematical model of a solid with a cavity containing electrically conducting ferromagnetic liquid (LVECF liquid) that moves in magnetic field. This model is discussed in Chapter 8.

A striking feature of all the models, that has already been mentioned, is common character of their structure, particularly integral operators with identical singular kernels of the type met with in viscous liquid hydrodynamics classical problems and in heat transfer problems (see, e.g. H.Lamb [40], L.D.Landau and E.M.Lifshits [47], A.I.Lurie [58]). These operators are a particular case of hose met with in mathematical physics problems taking into account aftereffects, for instance, theory of creeping (see N.Kh.Arutiunian and V.B.Kolmanovsky [5], V.B.Kolmanovsky and V.R.Nosov [41], Yu.N.Rabotnov [85]).

Chapter 1

Mathematical Models of High Electrical Conductivity Ferromagnetic Elements with Eddy Currents

The study of nature with a view to works is engaged in by the mechanic, the mathematician, the physician, the alchemist, and the magician; but all (as things now are) with slight endeavour and scanty success.

Francis Bacon. *Novum Organum.* V.*

The problem of considering eddy currents in elements of electrical machines and apparatus, especially materials with comparatively high electrical conductivity (skin effect), is a traditional one for electrical engineering and is well studied. Practical aspects of this problem has determined two specific features of its statement, which are completely described in literature (see, e.g. N.N.Fedorov [20]):

1. Consideration of steady-state harmonic oscillations.
2. Extraction and detailed consideration of the solution part determining energy dissipation due to eddy currents.

Appearance of still more and more complicated automatic control systems for moving vehicles (magnetic suspension high-speed transport moving stock and turbo-generator rotors, AES and SC equipped with magnetic attitude control and stabilization systems) has put new problems on the agenda, namely adequate description of eddy currents in electrically conductive ferromagnetic elements of those systems, considered as objects of automatic control.

Specific features of those problems are the following ones:

1. Essential non-stationarity of processes, and system's perturbed motion distinction from harmonic oscillations.
2. Necessity of complete description of the object's dynamic properties, including eddy currents effect, within the scope of a common mathematical model admitting both

---

* The Works of Francis Bacon, Baron of Verulam, Viscount St. Alban, and Lord High Chancellor of England. Collected and edited by James Spedding, M.A. of Trinity College, Cambridge; Robert Leslie Ellis, M.A., Late Fellow of Trinity College, Cambridge; Douglas Denon Heath, Barrister-at-Law, Late Fellow of Trinity College, Cambridge. Vol. IV. Translations of philosophical works, Vol. I, London, 1870, (p. 47-48).

an appropriate control system synthesis and analysis of the system 'object – regulator' perturbed motion unknown a priori.

The most important for practice is the class of comparatively 'slow' (quasi-steady) motions and, at the same time, of comparatively high electrical conductivity of the material (ferromagnet) in which eddy currents are induced. Although, fortunately, the state of affairs with synthesis of mathematical models describing phenomena of this class does not look so regrettable, as in the situation to which F.Bacon's words quoted in the epigraph refer, the results here are yet more modest than in the field of classical electrical engineering problems associated with skin effect.

Note that, according to B.I.Rabinovich and V.M.Rogovoy [80, 81], there is a deep analogy between problems of attitude control and stabilization of SC and AES, related to dynamics of conducting solid body in magnetic field, and dynamics of a solid body containing cavities with viscous incompressible liquid. In the limiting cases of low-viscous liquid and high-frequency magnetic field, as well as high-viscous liquid and slowly alternating external magnetic field, liquid velocity field and magnetic field construction is accomplished in the first-order approximation by L.D.Landau and E.M.Lifshits [47, 44] (the asymptotics of great and small Reynolds numbers respectively). The works also note the analogy of space and time distribution of fields.

In the first case of closed volume with impenetrable boundary liquid velocity field agrees with the case of ideal liquid and magnetic field agrees with that in a superconductor of the same shape. Difference is detected in a thin wall layer only, near to the body surface, where behaviour of the field component normal to the surface has no singularities, and the tangential component displays rapid change from the value at the body surface, that meets appropriate boundary conditions, to the values of the 'parent' solution (space boundary layer, skin effect).

In the second case the first order approximation is the solution of the adjacent steady problem (boundary layer with respect to time axis).

Further development of those results is concerned with construction of high-order approximations using asymptotical methods of singularly perturbed problems solution (see A.B.Vasilieva and V.F.Butuzov [99], I.B.Bogoriad *et al.* [10]), with investigation of internal and external dynamics problems of a solid body interacting with viscous liquid (see F.L.Chernousko [15], B.I.Rabinovich and V.M.Rogovoy [79], V.M.Rogovoy and S.V.Cheremnykh [86], B.I.Rabinovich [76]), and with dynamics of a solid body in magnetic field, a steady one or varying in a specified manner (see V.V.Beletsky and A.A.Khentov [7], A.I.Kobrin and Yu.G.Martynenko [37], R.V.Linkov and Yu.V.Urman [57], R.V.Linkov [55], Yu.G.Martynenko [61]).

When constructing attitude control and stabilization systems for vehicles, still more and more important role is played, as it has already been mentioned, by essentially unsteady problems, when the external magnetic field is controllable and is used in the control system actuators (see D.M.Veinberg *et al.* [100], E.Gottzein [25], E.R.Laithwaite [42], V.I.Bocharov and V.D.Nagorsky [9]). The resultant field is highly dependent on the control current as well as on secondary field raised by eddy currents induced in the conducting ferromagnet.

The present chapter, which contents is a further development of works by B.I.Rabinovich and V.M.Rogovoy [80] and V.G.Lebedev, A.I.Mytarev, and B.I.Rabinovich [51], is devoted to construction of an adequate mathematical model of eddy currents in conducting ferromagnets in the frequency band specific for AES and SC attitude control and stabilization problems and for various magnetic levitation problems, and to synthesis based on it of adjacent systems elements' mathematical models with external currents and mechanical degrees of freedom. The basis of those models is constructed by asymptotic solutions of field problems, in which both types of boundary layer are involved (time and space one), that is an important feature of those problems. Obtained mathematical models allow description of complicated transient processes, particularly oscillations with 'slowly' varying frequency, amplitude, and phase. The appropriate formalism is 'adjusted' to the possibility of closing the control loop of the object's mathematical model with one or another control equation.

1.1. Statement of the Problem

Let us consider a three-dimensional doubly connected body $G^{(i)}$, shaped in the general case as an irregular torus made of ferromagnetic material having electrical conductivity, which is 'not small' (the meaning of that term will be defined more exactly later on), i. e. HECF element of toroidal shape. Suppose the body $G^{(i)}$ (further on called schematically *the torus*) is dissected by a surface close to the diameter plane (coinciding with the latter in the case of a circular torus) into two simply connected parts (semi-torus) $G_j^{(i)}$ with end surfaces being the coasts of the section Σ_j ($j = 1, 2$), so that

$$G^{(i)} = G_1^{(i)} + G_2^{(i)}, \quad \Sigma = \Sigma_1 + \Sigma_2 .$$

Denote the characteristic size of the body $G^{(i)}$ by l, the characteristic size of its simply connected cross-section by b, area external to the torus by $G^{(e)}$. Semi-torus $G_1^{(i)}$ and $G_2^{(i)}$ may be additionally connected by elastoviscous ties.

Orient the torus $G^{(i)}$ in the space for the section plane to become horizontal (perpendicular to the field of mass forces of unperturbed motion related to absence of section) and attach its upper half $G_2^{(i)}$ to some 'absolute' (motionless) coordinate system $CXYZ$ (Fig. 1.1). CY axis of this coordinate system is directed vertically up (i.e. it is antiparallel to unperturbed motion mass forces field gradient), CX and CZ axes lie in the section plane, making a right triplet. Suppose the coordinate origin O is placed so, that the semi-torus $G_1^{(i)}$ mass centre C with zero relative displacement lies on CY axis. The distance from the point C to point O denote by y_O (normally $y_O > 0$).

We introduce, further, the coordinate system $Oxyz$ related to the lower semi-torus, the origin being coincident with its mass centre, and the axes being parallel to the coordinate system $CXYZ$ axes.

Let us consider small, compared to the characteristic size b, translational displacements of the semi-torus $G_1^{(i)}$ with respect to the motionless coordinate system $CXYZ$, that are determined by the vector

$$\mathbf{u}_O = -\mathbf{i}_2 y_O + \mathbf{u}, \quad \mathbf{u} = \mathbf{i}_1 \xi + \mathbf{i}_2 \eta + \mathbf{i}_3 \zeta = (\xi, \eta, \zeta), \tag{1.1}$$

where $\xi \equiv 0$, $\eta \equiv 0$, $\zeta \equiv 0$, when the halves $G_1^{(i)}$ and $G_2^{(i)}$ form a whole $G^{(i)}$. Thus the semi-torus $G_1^{(i)}$ is considered a system with three degrees of freedom (displacements along three relatively perpendicular axes CX, CY, CZ). For a small displacement along the negative CY axis a small gap $\eta = -s$, $s \geq 0$, arises between the cut coasts Σ_1 and Σ_2.

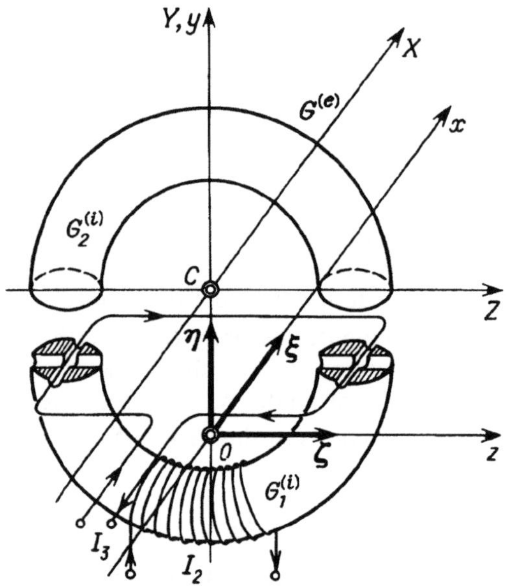

Fig. 1.1. The scheme of the electromagnetic actuator and the control coils generating forces along Oy and Oz axes (the air gap is shown approximately one order greater than it is in reality)

Suppose we have three systems of coils making closed loops rigidly connected with the torus $G_1^{(i)}$. External currents

$$\mathbf{j}_1(\mathbf{R}, t), \quad \mathbf{j}_2(\mathbf{R}, t), \quad \mathbf{j}_3(\mathbf{R}, t)$$

(calculated 'per square unit' of the coil cross section) circulate through the coils, where $\mathbf{R}$ is the radius vector of the adjacent loop point, having the origin at the point O (the length of each vector represents the corresponding current density). The first and the second coil systems are located in the end face section Σ_1 of the semi-torus $G_1^{(i)}$ in grooves with rectangular axes parallel to Ox and Oz axes respectively, the semi-torus $G_2^{(i)}$ having similar counter-grooves. We assume characteristic size of the grooves cross section substantially greater than the air gap s.

The second coil system is a solenoid embracing the semi-torus $G_1^{(i)}$ at an interval of its length.

Denote the grooves' side surfaces parallel to Ox axis and the semi-torus' and grooves' side surfaces parallel to Oz axis by S_{jk} ($j = 1, 2$; $k = 1, 2, 3$ respectively); and the total of these surfaces denote by

$$S = \sum_{j=1}^{2} \sum_{k=1}^{3} S_{jk}.$$

Current directions assumed positive are indicated in Fig. 1.1, they correspond to positive electromagnetic forces of elements $G_1^{(i)}$ and $G_2^{(i)}$ interaction, which are usually called *ponderomotive forces* (only two coil systems are shown in Fig. 1.1 in order not to clutter up the figure).

Let σ denote electrical conductivity of ferromagnetic materials, of which the torus $G^{(i)}$ is made, μ denote relative magnetic permeability, and μ_0 denote the absolute magnetic permeability of vacuum.

The magnetic constant μ_0 is related to the electrical constant ε_0 and vacuum velocity of light c by the known expression $\varepsilon_0 \mu_0 = c^{-2}$, in SI-system $\mu_0 = 4\pi \cdot 10^{-7}$ H/m.

The high electrical conductivity ferromagnet described above, which consists of two halves (core with an air gap), can be considered a model of an electromagnetic levitation system actuator, as well as a model of AES and SC attitude control and stabilization system elements. With the presence of the active part $G_1^{(i)}$ location (with respect to the passive part $G_2^{(i)}$) automatic regulation system, the semi-torus $G_1^{(i)}$ is one of examples of a solid body with attitude hold.

Variation of currents in coils and variation of relative location of HECF core parts $G_1^{(i)}$ and $G_2^{(i)}$ causes variation of magnetic field and hence appearance of eddy currents in the HECF core, which, in turn, engender a new unsteady magnetic field. It is desirable to describe that complex process in a unified mathematical model suitable for investigation of the closed-loop system 'object – regulator' dynamics and for the choice of structure and parameters of the stabilization system. This is the problem this chapter is devoted to.

The considered problem is rather a complicated one, and an efficient solution is possible with some additional restrictions only, these are formulated below.

Suppose ω_M is the characteristic frequency of variations of magnetic field induced by external currents, ν_M is magnetic viscosity

$$\nu_M = \frac{1}{\mu_0 \mu \sigma} \tag{1.2}$$

and Ω_M is some characteristic frequency, which is defined as follows:

$$\Omega_M \sim l\,\omega_M \left(\frac{\omega_M}{\nu_M}\right)^{1/2}. \tag{1.3}$$

Let us consider dimensionless parameters

$$\mathrm{Sh}_M = \frac{\omega_M}{\Omega_M}, \quad \mathrm{Re}_M = \frac{\omega_M l^2}{\nu_M}, \quad \tilde{s} = \frac{|u|}{b}. \tag{1.4}$$

The first of them is an analogue of Strouhal number, and the second is an analogue of Reynolds number (so called *magnetic Reynolds number*). Let us consider the class of unsteady regimes characterized by the following conditions

$$\mathrm{Sh}_M \ll 1, \quad \mathrm{Re}_M \gg 1, \quad \mathrm{Sh}_M \sqrt{\mathrm{Re}_M} \sim 1, \quad \mu \gg 1, \quad \tilde{s} \ll 1. \tag{1.5}$$

Infinitesimal Sh_M makes possible 'inertialess' reconstruction of magnetic fields in domains $G_1^{(i)}$ and $G_2^{(i)}$ corresponding to variation of external field (time boundary layer), while infinitesimal Re_M^{-1} ensures small depth of space boundary layer having the order of $\delta = l\,\mathrm{Re}_M^{-1/2}$. The third condition in (1.5), which results from (1.3), assures simultaneous existence of time and space boundary layers. Finally, the last two conditions in (1.5) ensure localization of the external field predominantly in the air gap between the end face surfaces Σ_1 and Σ_2.

1.2. Asymptotic Solution of Unsteady Boundary-Value Problems for Magnetic and Electrical Fields

Let us write down Maxwell equations in quasi-steady approximation and the boundary conditions, that look in the considered case (see N.N.Fedorov [20]) as follows (SI-system).

For domains $G_j^{(i)}$:

$$\mathrm{curl}\,\mathbf{H}_{jk}^{(i)} = \sigma\,\mathbf{E}_{jk}^{(i)}, \quad \mathrm{div}\,\mathbf{B}_{jk}^{(i)} = 0,$$

$$\mathrm{curl}\,\mathbf{E}_{jk}^{(i)} = -\frac{\partial \mathbf{B}_{jk}^{(i)}}{\partial t}, \quad \mathrm{div}\,\mathbf{E}_{jk}^{(i)} = 0,$$

$$\mathbf{B}_{jk}^{(i)} = \mu_0\,\mu\,\mathbf{H}_{jk}^{(i)}; \quad j = 1, 2, \quad k = 1, 2, 3. \tag{1.6}$$

For domain $G^{(e)}$:

$$\mathrm{curl}\,\mathbf{H}_k^{(e)} = \mathbf{j}_k(\mathbf{R}, t), \quad \mathrm{div}\,\mathbf{B}_k^{(e)} = 0,$$

$$\mathrm{curl}\,\mathbf{E}_k^{(e)} = -\frac{\partial \mathbf{B}_k^{(e)}}{\partial t}, \quad \mathrm{div}\,\mathbf{E}_k^{(e)} = 0,$$

$$\mathbf{B}_k^{(e)} = \mu_0\,\mathbf{H}_k^{(e)}; \quad k = 1, 2, 3. \tag{1.7}$$

For surfaces S_{jl} and Σ_j:

$$\left(\mathbf{B}_k^{(e)} - \mathbf{B}_{jk}^{(i)}, \vec{\nu}\right) = 0, \quad \left(\mathbf{H}_k^{(e)} - \mathbf{H}_{jk}^{(i)}\right) \times \vec{\nu} = 0,$$

$$\left(\mathbf{D}_k^{(e)} - \mathbf{D}_{jk}^{(i)}, \vec{\nu}\right) = 0, \quad \left(\mathbf{E}_k^{(e)} - \mathbf{E}_{jk}^{(i)}\right) \times \vec{\nu} = 0, \tag{1.8}$$

$$\mathbf{D}_k^{(e)} = \varepsilon_0\,\varepsilon\,\mathbf{E}_k^{(e)}; \quad j = 1, 2, \quad l, k = 1, 2, 3.$$

where **B** and **H** are magnetic field induction and strength, **D** and **E** are electric field induction and strength, $\vec{\nu}$ is the unit vector of outer normal to the surfaces Σ_j and S_{jk}. Equations (1.6) - (1.8) are to be completed by the conditions of all fields attenuation at $|\mathbf{R}| \to \infty$ and their absence at $t \to -\infty$.

Using parameters of (1.4), dimensionless variables $\tilde{t}$, $\tilde{x}_n$, , normalized fields $\tilde{\mathbf{E}}^{(i)}_{jk}$, and currents $\tilde{\mathbf{j}}_k$

$$\tilde{t} = \Omega_M t, \quad \tilde{x}_n = \frac{x_n}{l}, \quad \tilde{\mathbf{E}}^{(i)}_{jk} = \frac{\mathbf{E}^{(i)}_{jk}}{l\,\Omega_M}, \quad \tilde{\mathbf{j}}_k = \mathbf{j}_k l \tag{1.9}$$

($n = 1, 2, 3$; $k = 1, 2, 3$; $j=1, 2$) and omitting equations and boundary conditions for external electric field, we can write down the boundary-value problem (1.6) - (1.8) in the following normalized form:

$$\begin{aligned}
&\widetilde{\mathrm{curl}}\,\mathbf{B}_{jk} = \frac{\mathrm{Re}_M}{\mathrm{Sh}_M}\tilde{\mathbf{E}}^{(i)}_{jk}, && \mathrm{div}\,\mathbf{B}^{(i)}_{jk} = 0,\\
&\widetilde{\mathrm{curl}}\,\tilde{\mathbf{E}}^{(i)}_{jk} = -\,\mathrm{Sh}_M\frac{\partial\,\mathbf{B}^{(i)}_{jk}}{\partial\,\tilde{t}}, && \mathrm{div}\,\tilde{\mathbf{E}}^{(i)}_{jk} = 0,\\
&\widetilde{\mathrm{curl}}\,\mathbf{H}^{(e)}_k = \tilde{\mathbf{j}}_k, && \mathrm{div}\,\mathbf{B}^{(e)}_k = 0,\\
&\left(\mathbf{B}^{(e)}_k - \mathbf{B}^{(i)}_{jk},\ \vec{\nu}\right)\Big|_{S_{jl},\,\Sigma_j} = 0, && \left(\mathbf{H}^{(e)}_k - \mathbf{H}^{(i)}_{jk}\right)\times\vec{\nu}\,\Big|_{S_{jl},\,\Sigma_j} = 0,\\
&\mathbf{B}^{(i)}_{jk} = \mu_0\mu\,\mathbf{H}^{(i)}_{jk}; \quad \mathbf{B}^{(e)}_k = \mu_0\,\mathbf{H}^{(e)}_k; && j = 1, 2, \quad l, k = 1, 2, 3.
\end{aligned} \tag{1.10}$$

The 'tilde' sign at vector operators means their execution for dimensionless variables $\tilde{x}_n$. Further on we omit this sign and subscripts j, k, using them only where it is necessary.

We seek the boundary-value Problem (1.10) solution outside the time boundary layer in the form of regular series for the external domain and in the form of sums of regular series and boundary layer functions, that can also be expanded into series, for internal domains. According to A.B.Vasilieva and V.F.Butuzov, M.N.Vishik and L.A.Lusternik [99, 102], terms of the boundary layer functions' expansion series decrease exponentially, and outside the initial interval of the order of $\mathrm{Sh}_M \ln \mathrm{Sh}_M^{-1}$ the formal solutions of the problem can be presented in the form of power series in terms of the small parameter Sh_M.

For $\mathrm{Sh}_M \to 0$ we should obtain the magnetostatics problem solution $\mathbf{B}^{(i)} = \mathbf{B}^{(ie)}_0$, $\mathbf{B}^{(e)} = \mathbf{B}^{(ei)}_0$, for which $\mathbf{E}^{(i)} \equiv 0$, $\mathbf{E}^{(e)} \equiv 0$. Besides that, for $\mathrm{Sh}_M \to 0$ solutions $\mathbf{B}^{(i)}_0$ should exist in domains $G^{(i)}_j$, which become zero at the surfaces S_{jk}, tend for $\tilde{s} \to 0$ to homogeneous solutions of internal boundary-value problems for the domain $G^{(i)}$ for $k = 1, 2, 3$, engendered by eddy currents, and corresponding solutions $\mathbf{B}^{(e)}_0$ for the domain $G^{(e)}$.

Thus vector fields $\mathbf{E}^{(i)}$, $\mathbf{B}^{(i)}$ and $\mathbf{B}^{(e)}$ power series expansions in terms of the small parameter Sh_M are to be sought in the form of

$$\mathbf{E}^{(i)} = \mathrm{Sh}_M \mathbf{E}_1^{(i)} + \mathrm{Sh}_M^2 \mathbf{E}_2^{(i)} + ...,$$
$$\mathbf{B}^{(i)} = \mathbf{B}_0^{(ie)} + \mathbf{B}_0^{(i)} + \mathrm{Sh}_M \mathbf{B}_1^{(i)} + \mathrm{Sh}_M^2 \mathbf{B}_2^{(i)} + ..., \tag{1.11}$$
$$\mathbf{B}^{(e)} = \mathbf{B}_0^{(ei)} + \mathbf{B}_0^{(e)} + \mathrm{Sh}_M \mathbf{B}_1^{(e)} + \mathrm{Sh}_M^2 \mathbf{B}_2^{(e)} + ...,$$

Substituting Expansions (1.11) into equations and boundary conditions of (1.10) we obtain a recurrent sequence of boundary-value problems. Let us consider the first two of them.

Zero-order approximation in terms of Sh_M in domain $G^{(i)} + G^{(e)}$.

$$\mathrm{curl}\, \mathbf{B}_0^{(ie)} = 0, \quad \mathrm{div}\, \mathbf{B}_0^{(ie)} = 0, \quad \mathbf{E}_0^{(ie)} \equiv 0 \quad (\text{domain } G^{(i)} = G_1^{(i)} + G_2^{(i)})$$

$$\mathrm{curl}\, \mathbf{B}_0^{(ei)} = \mu_0 \mathbf{j}, \quad \mathrm{div}\, \mathbf{B}_0^{(ei)} = 0, \quad \mathbf{E}_0^{(ei)} \equiv 0 \qquad (\text{domain } G^{(e)})$$

$$\left(\mathbf{B}_0^{(ei)} - \mathbf{B}_0^{(ie)}, \vec{\nu} \right) \Big|_{S+\Sigma} = 0, \quad \left(\mathbf{H}_0^{(ei)} - \mathbf{H}_0^{(ie)} \right) \times \vec{\nu} \Big|_{S+\Sigma} = 0, \tag{1.12}$$
$$\mathbf{B}_0^{(ei)} = \mu_0 \mathbf{H}_0^{(ei)}, \qquad \mathbf{B}_0^{(ie)} = \mu_0 \mu\, \mathbf{H}_0^{(ie)}.$$

We are to add to (1.12) conditions of vanishing for the fields $\mathbf{B}_0^{(ei)}$ and $\mathbf{H}_0^{(ei)}$ for $|\mathbf{R}| \to \infty$ and $t \to -\infty$. ($S + \Sigma$ is the total boundary of domains $G_j^{(i)}$). We are coming thus to a classic magnetostatics problem. When solving it with one of known methods (see, e.g. G.A.Greenberg [29]), it is convenient to consider various current $\mathbf{j}_k$ orientations corresponding to $k = 1, 2, 3$.

Let us write the general solution of the boundary-value Problem (1.12) for domains $G_j^{(i)}$ $\left(\mathbf{B}_{0j}^{(ie)}\right)$ and for domain $G^{(e)}$ $\left(\mathbf{B}_0^{(ei)}\right)$ in the form of

$$\mathbf{B}_{0j}^{(ie)} = \mu_0 \mu \sum_{k=1}^{3} \overset{\circ}{n}_k \vec{\beta}_{jk}^{(ie)} (\mathbf{R}, \mathbf{u})\, I_k(t),$$
$$\mathbf{B}_0^{(ei)} = \mu_0 \mu \sum_{k=1}^{3} \overset{\circ}{n}_k \vec{\beta}_{jk}^{(ei)} (\mathbf{R}, \mathbf{u})\, I_k(t), \tag{1.13}$$
$$\overset{\circ}{n}_k = \frac{\overset{\circ}{N}_k}{2\pi}, \quad j = 1, 2.$$

Here in the first expression $\mathbf{R}$ is the radius vector of domain $G_j^{(i)}$ point with the origin in point O_j and in the second expression it is the radius vector of domain $G^{(e)}$ point. $\vec{\beta}_{jk}^{(ie)}$ and $\vec{\beta}_k^{(ei)}$ are the magnetostatics boundary-value problems solutions for domains $G_j^{(i)}$ and $G^{(e)}$, corresponding to the k th coil. Their normalization is clear from (1.13). $I_k = f_k j_k$ is the current in the k th coil, f_k is the wire cross section area; $\overset{\circ}{N}_k$ is the k th coil's number of turns.

First-order approximation in terms of Sh_M in domain $G^{(i)} + G^{(e)}$.

$$\mathrm{curl}\, \mathbf{B}_0^{(i)} = \mathrm{Re}_M \mathbf{E}_1^{(i)}, \quad \mathrm{div}\, \mathbf{B}_0^{(i)} = 0, \quad \mathrm{curl}\, \mathbf{E}_1^{(i)} = -\frac{\partial \mathbf{B}_0^{(i)}}{\partial t}$$

$$(\text{domain } G^{(i)} = G_1^{(i)} + G_2^{(i)}),$$

$$\operatorname{curl} \mathbf{B}_0^{(e)} = 0, \qquad \operatorname{div} \mathbf{B}_0^{(e)} = 0 \qquad (\text{domain } G^{(e)}), \tag{1.14}$$

$$\left(\mathbf{B}_0^{(e)} - \mathbf{B}_0^{(i)}, \vec{\nu} \right) \Big|_{S+\Sigma} = 0, \quad \left(\mathbf{H}_0^{(e)} - \mathbf{H}_0^{(i)} \right) \times \vec{\nu} \Big|_{S+\Sigma} = 0.$$

Boundary-value problems of (1.14) corresponding to $k = 1, 2, 3$ require a special consideration. Their solutions $\mathbf{B}_0^{(i)}$ and $\mathbf{E}_1^{(i)}$ can be found with the boundary layer method.

Let us first consider the case $\mathbf{u} = 0$, when solutions can be presented, following A.B.Vasilieva and V.F.Butuzov, M.N.Vishik and L.A.Lusternik [99, 102], in the form of expansions

$$\mathbf{B}_0^{(i)} = \mathbf{B}_{00}^{(i)} + \frac{1}{\sqrt{\mathrm{Re}_M}} \mathbf{B}_{01}^{(i)} + \frac{1}{\mathrm{Re}_M} \mathbf{B}_{02}^{(i)} + \ldots + \widehat{\mathbf{B}}_0^{(i)},$$

$$\mathbf{E}_1^{(i)} = \mathbf{E}_{10}^{(i)} + \frac{1}{\sqrt{\mathrm{Re}_M}} \mathbf{E}_{11}^{(i)} + \frac{1}{\mathrm{Re}_M} \mathbf{E}_{12}^{(i)} + \ldots + \widehat{\mathbf{E}}_1^{(i)}. \tag{1.15}$$

Functions $\widehat{\mathbf{B}}_0^{(i)}$ and $\widehat{\mathbf{E}}_1^{(i)}$ are functions of boundary layer type, they decrease rapidly with distance inwards domain $G_j^{(i)}$ from its boundary S_{jk}. They satisfy Equations (1.12), and they also can be expanded into a power series in terms of $\mathrm{Re}_M^{-1/2}$.

Let us confine to the approximation

$$\mathbf{B}_0^{(i)} = \mathbf{B}_{00}^{(i)} + \widehat{\mathbf{B}}_0^{(i)}, \qquad \mathbf{B}_0^{(e)} = \mathbf{B}_{00}^{(e)}, \qquad \mathbf{E}_1^{(i)} = \mathbf{E}_{10}^{(i)} + \widehat{\mathbf{E}}_1^{(i)}, \tag{1.16}$$

where fields $\mathbf{E}_{10}^{(i)}$, $\mathbf{B}_{00}^{(i)}$ correspond to $\mathrm{Re}_M = \infty$, $\mathbf{u} = 0$. For $\mathbf{B}_{00}^{(i)}$ we obtain the following boundary-value problem.

Zero-order approximation in terms of $\mathrm{Re}_M^{-1/2}$ in domain $G^{(i)}$ for $\mathbf{u} = 0$.

$$\operatorname{curl} \mathbf{B}_{00}^{(i)} = 0, \qquad \operatorname{div} \mathbf{B}_{00}^{(i)} = 0, \qquad \left(\mathbf{B}_{00}^{(i)}, \vec{\nu} \right) \Big|_S = 0. \tag{1.17}$$

Function $\mathbf{B}_{00}^{(i)}$ normalization is arbitrary, so far. For $\mathbf{B}_{00}^{(e)}$ and $\mathbf{E}_{10}^{(i)}$ we obtain homogeneous boundary-value problems that we do not write down here - they have trivial solutions $\mathbf{B}_{00}^{(e)} \equiv 0$, $\mathbf{E}_{10}^{(i)} \equiv 0$.

Boundary-value Problem (1.17) has solutions that are not identically zero, for domain $G^{(i)}$ is multi-connected for $\mathbf{u} \equiv 0$. The problem of extraction of the sought single-valued solution for each value of k is not trivial, however. It is clear from physical considerations that boundary-value problems single-valued solutions corresponding to $k = 1, 2, 3$ are proximate, up to normalization, to solutions of similar boundary-value Problems (1.12), which correspond to $\mu \to \infty$, with the same values of subscript k. This implies additional conditions, that fields $\mathbf{B}_{00}^{(i)}$ are to comply with. These conditions make it possible to extract the necessary solution for each k value. Particularly, vector $\mathbf{B}_{00}^{(i)}$ circulation around specially selected paths should be either nonzero or identically zero:

$$\oint_{\Gamma_n} \left(\mathbf{B}_{00}^{(i)}, ds \right) \begin{cases} \neq 0 & \text{for } n = k, \\ = 0 & \text{for } n \neq k, \end{cases} \qquad n, k = 1, 2, 3, \tag{1.18}$$

where Γ_n is an arbitrary closed path which belongs to domain $G^{(i)} + G^{(e)}$ and embraces cylindric surface $S_{11} + S_{21}$ for $n = 1$ and cylindric surface $S_{13} + S_{23}$ for $n = 3$. The path runs in such a way, that the torus $G^{(i)}$ centre is within it for $n = 2$. ds is the arc element differential tangent to the path (here s is the arc length, certainly, not the air gap).

Conditions (1.18) enable extraction of a solution for each k value, which has a singularity outside the torus, which is similar to that of the solution's analogue engendered by external current $\mathbf{j}_k$ (magnetostatics Problem (1.12)).

Zero-order approximation in terms of $\mathrm{Re}_M^{-1/2}$ in domains $G_j^{(i)}$ and $G^{(e)}$ for $\mathbf{u} \to 0$. Let us cut the HECF core $G^{(i)}$ now into semi-torus $G_j^{(i)}$ and assume that the cut coasts (end face surfaces) Σ_j are proximate for $\mathbf{u} = 0$ to equipotential surfaces of fields $\mathbf{B}_{00}^{(i)}$ for $k = 1, 2, 3$. Let us restore the gap $\mathbf{u} \neq 0$ as described above and, supposed the last two conditions in (1.5) are obeyed, identify fields inside each semi-torus with field $\mathbf{B}_{00}^{(i)}$ within the corresponding part of domain $G^{(i)}$, assumed for surfaces Σ_j

$$\left(\mathbf{B}_{00}^{(i)} \times \vec{\nu} \right) \Big|_{\Sigma_j} \approx 0.$$

We seek an asymptotic solution of the Problem (1.14), corresponding to $\mathrm{Re}_M = \infty$, similar to (1.16) in the form of

$$\mathbf{B}_{0j}^{(i)} = \mathbf{B}_{00}^{(i)} + \mathbf{B}_{00j}^{(ie)} + \hat{\mathbf{B}}_{0}^{(i)}, \qquad \mathbf{B}_{0}^{(e)} = \mathbf{B}_{00}^{(ei)}, \tag{1.19}$$

where the field of (1.16) is the engendering one, and $\mathbf{B}_{00j}^{(ie)}$ and $\mathbf{B}_{00}^{(ei)}$ are the fields of 'secondary sources' (following G.A.Greenberg [29]). For fields $\mathbf{B}_{00j}^{(ie)}$ and $\mathbf{B}_{00}^{(ei)}$ we obtain the following boundary-value problem ($j = 1, 2,\ k = 1, 2, 3$):

$$\operatorname{curl} \mathbf{B}_{00j}^{(ie)} = 0, \qquad \operatorname{div} \mathbf{B}_{00j}^{(ie)} = 0, \qquad \operatorname{curl} \mathbf{B}_{00}^{(ei)} = 0, \qquad \operatorname{div} \mathbf{B}_{00}^{(ei)} = 0,$$

$$\left(\mathbf{B}_{00}^{(ei)} - \mathbf{B}_{00j}^{(ie)}, \vec{\nu} \right) \Big|_{S_{jk}} = 0, \qquad \left(\mathbf{H}_{00}^{(ei)} - \mathbf{H}_{00j}^{(ie)} \right) \times \vec{\nu} \Big|_{S_{jk}} = 0, \tag{1.20}$$

$$\left(\mathbf{B}_{00}^{(ei)} - \mathbf{B}_{00j}^{(ie)}, \vec{\nu} \right) \Big|_{\Sigma_i} = \left(\mathbf{B}_{00}^{(i)}, \vec{\nu} \right) \Big|_{\Sigma_j}, \quad \mathbf{B}_{00}^{(ei)} = \mu_0 \mathbf{H}_{00}^{(ei)}, \qquad \mathbf{B}_{00j}^{(ie)} = \mu_0 \mu \mathbf{H}_{00j}^{(ie)}.$$

Just as for Problem (1.12), it is necessary to add to boundary Conditions (1.20) the condition of field $\mathbf{B}_{00}^{(ei)}$ vanishing at infinity.

Let us introduce the following representation for functions $\mathbf{B}_{00}^{(i)}$, $\mathbf{B}_{00j}^{(ie)}$ and $\mathbf{B}_{00}^{(ei)}$, similar to (1.13):

$$\mathbf{B}_{00}^{(i)} = \mu_0 \mu \sum_{k=1}^{3} \overset{\circ}{\vec{n}_k} \beta_k^{(i)} (\mathbf{R})\, \alpha_k(\mathbf{u})\, J_k(t),$$

$$\mathbf{B}_{00j}^{(ie)} = \mu_0 \mu \sum_{k=1}^{3} \overset{\circ}{\vec{n}_k} \beta_{jk}^{(i)} (\mathbf{R}, \mathbf{u})\, \alpha_k(\mathbf{u})\, J_k(t),$$

$$\mathbf{B}_{00}^{(ei)} = \mu_0 \mu \sum_{k=1}^{3} \overset{\circ}{\vec{n}_k} \beta_k^{(e)} (\mathbf{R}, \mathbf{u})\, \alpha_k(\mathbf{u})\, J_k(t), \tag{1.21}$$

$$\alpha_k(\mathbf{u}) = \frac{\left(\vec{\beta}_{1k}^{(ie)}, \vec{\nu}\right)\Big|_{\Sigma O}}{\left(\vec{\beta}_{1k}^{(i)} + \vec{\beta}_{k}^{(i)}, \vec{\nu}\right)\Big|_{\Sigma O}}, \quad \alpha_k(0) = 1, \quad \vec{\beta}_k^{(e)}(\mathbf{R}, 0) = 0,$$

where $J_k(t)$ are functions of time having the dimension of current, that describe variation of fields induced by eddy currents. They correspond to external currents $I_k(t)$ with the same subscript k values. $\Big|_{\Sigma O}$ symbolizes that the function value relates to the centre of domain Σ_O, which is the cross section of the electromagnet core $G_1^{(i)}$ by the plane Oxy.

Functions $\vec{\beta}_k^{(i)}(\mathbf{R})$, $\vec{\beta}_{jk}^{(i)}(\mathbf{R}, \mathbf{u})$, $\vec{\beta}_k^{(e)}(\mathbf{R}, \mathbf{u})$ are the following boundary-value problems solutions (see (1.17) and (1.20)):

for $\mathbf{u} = 0$

$$\operatorname{curl} \vec{\beta}_k^{(i)} = 0, \qquad \operatorname{div} \vec{\beta}_k^{(i)} = 0 \qquad (\text{domain } G^{(i)}),$$

$$\left(\vec{\beta}_k^{(i)}, \vec{\nu}\right)\Big|_S = 0, \quad \left(\vec{\beta}_k^{(i)}, \vec{\nu}\right)\Big|_\Sigma = \left(\vec{\beta}_{1k}^{(ie)}, \vec{\nu}\right)\Big|_{\Sigma O} = 0, \quad \mathbf{u} = 0. \tag{1.22a}$$

for $\mathbf{u} \neq 0$, $s \neq 0$

$$\operatorname{curl} \vec{\beta}_{jk}^{(i)} = 0, \qquad \operatorname{div} \vec{\beta}_{jk}^{(i)} = 0 \qquad (\text{domain } G_j^{(i)}),$$

$$\operatorname{curl} \vec{\beta}_k^{(e)} = 0, \qquad \operatorname{div} \vec{\beta}_k^{(e)} = 0 \qquad (\text{domain } G^{(e)}), \tag{1.22b}$$

$$\left(\vec{\beta}_k^{(e)} - \vec{\beta}_{jk}^{(i)}, \vec{\nu}\right)\Big|_{S_{jk}} = 0, \quad \left(\vec{\beta}_k^{(e)} - \frac{1}{\mu}\vec{\beta}_{jk}^{(i)}\right) \times \vec{\nu}\Big|_{S_{jk}} = 0,$$

$$\left(\vec{\beta}_k^{(e)} - \vec{\beta}_{jk}^{(i)}, \vec{\nu}\right)\Big|_{\Sigma_j} = \left(\vec{\beta}_k^{(i)}, \vec{\nu}\right)\Big|_{\Sigma_j},$$

$$\left(\vec{\beta}_k^{(e)} - \frac{1}{\mu}\vec{\beta}_{jk}^{(i)}\right) \times \vec{\nu}\Big|_{\Sigma_j} = \left(\vec{\beta}_k^{(i)} \times \vec{\nu}\right)\Big|_{\Sigma_j}.$$

To obtain a solution of the first-order approximation in terms of Re_M it is necessary to eliminate discrepancy of tangential fields $\mathbf{B}_{00}^{(i)}$ components at surfaces S_{jk} ($j = 1, 2,\ k = 1, 2, 3$) for each value of subscript k. Eliminating vector $\mathbf{E}_1^{(i)} = \widehat{\mathbf{E}}_1^{(i)}$ from the first group of Equations (1.14) and adding boundary conditions resulting from (1.13), (1.17), and (1.19), that require inversion of the total field $\mathbf{B}_{00}^{(i)} + \widehat{\mathbf{B}}_0^{(i)}$ tangential component at surface S, we obtain the following new boundary-value problems for adjustment fields $\widehat{\mathbf{B}}_0^{(i)}$ of boundary layer type.

First-order approximation in terms of $\mathrm{Re}_M^{-1/2}$ in domains $G_j^{(i)}$.

$$\frac{\partial \widehat{\mathbf{B}}_0^{(i)}}{\partial t} = \frac{1}{\mathrm{Re}_M}\Delta \mathbf{B}_0^{(i)}, \quad \operatorname{div} \mathbf{B}_0^{(i)} = 0, \quad \left(\widehat{\mathbf{B}}_0^{(i)} \times \vec{\nu}\right)\Big|_S = -\left(\mathbf{B}_{00}^{(i)} \times \vec{\nu}\right)\Big|_S, \tag{1.23}$$

where Δ is Laplace operator. Let us introduce curvilinear coordinate systems q_n ($n = 1, 2, 3$), where $q_1 = \text{const}$ is the equation of surfaces S_{jm} ($j = 1,2;\ m = 1, 2, 3$)

and q_1 increases along the inner normal to surface S_{jm}. Then we perform the usual procedure of the boundary layer method and obtain the following solution of boundary-value Problem (1.23):

$$\hat{\mathbf{B}}_0^{(i)} \times \vec{\nu} = -\frac{q_1}{2}\left(\frac{\mathrm{Re}_M}{\pi}\right)^{1/2}\int_{-\infty}^{t}\frac{\mathbf{B}_{00}^{(i)} \times \vec{\nu}}{(t-\tau)\sqrt{t-\tau}}\exp\left[-\frac{q_1^2\,\mathrm{Re}_M}{4\,(t-\tau)}\right]d\tau, \tag{1.24}$$

$$\left(\hat{\mathbf{B}}_0^{(i)}, \vec{\nu}\right) = -\frac{1}{4\sqrt{\pi}\,\mathrm{Re}_M}\int_{-\infty}^{t}\frac{\operatorname{div}\left(\mathbf{B}_{00}^{(i)} \times \vec{\nu}\right)}{\sqrt{t-\tau}}\exp\left[-\frac{q_1^2\,\mathrm{Re}_M}{4\,(t-\tau)}\right]d\tau,$$

in which the role of $\mathbf{B}_{00}^{(i)}$ is played by the Problem (1.17) solutions corresponding to $k = 1, 2, 3$.

The second of Expressions (1.24) demonstrates that the newly appeared discrepancy of the boundary condition at surface S for the normal component of field $\mathbf{B}_0^{(i)} = \mathbf{B}_{00}^{(i)} + \hat{\mathbf{B}}_0^{(i)}$ is a higher order infinitesimal that is characteristic of the boundary layer asymptotics (see A.B.Vasilieva and V.F.Butuzov, M.N.Vishik and L.A.Lusternik [99, 102]).

Electric field in domain $G^{(i)}$ near boundary S is determined from the following expression resulting from (1.14):

$$\hat{\mathbf{E}}_1^{(i)} \times \vec{\nu} = -\frac{1}{\mathrm{Re}_M}\left[\vec{\nu}\,\frac{\partial}{\partial q_1} \times \left(\hat{\mathbf{B}}_0^{(i)} \times \vec{\nu}\right)\right] \tag{1.25}$$

(it is taken into consideration that the curvilinear coordinate system q_1, q_2, q_3 unit vector $\mathbf{i}_1$ is directed along the inner normal to elements S_{jm} of boundary S , i.e. opposite to $\vec{\nu}$). Let us substitute the first of Expressions (1.24) into (1.25) and carry out the differential operations with respect to q_1 and integration by parts with respect to τ, after that assume $q_1 = 0$. We obtain then at surface S , restoring sign ~ at the normalized variables

$$\left(\hat{\tilde{\mathbf{E}}}_1^{(i)} \times \vec{\nu}\right)\Big|_S = \frac{1}{\sqrt{\pi}\,\mathrm{Re}_M}\int_{-\infty}^{t}\frac{\partial}{\partial\tilde{\tau}}\left[\vec{\nu} \times \left(\hat{\mathbf{B}}_{00}^{(i)} \times \vec{\nu}\right)\right]\frac{d\tilde{\tau}}{\sqrt{\tilde{t}-\tilde{\tau}}}. \tag{1.26}$$

Now we can return to non-normalized fields $\mathbf{E}^{(i)}$ and $\mathbf{B}_{00}^{(i)}$ and present them in the considered approximation in the form:

$$\mathbf{E}^{(i)} = l\,\Omega_M\,\mathrm{Sh}_M\,\hat{\mathbf{E}}_1^{(i)}, \qquad \mathbf{B}_{00}^{(i)}\Big|_S = \mu_0\,\mu\left(\mathbf{H}^{(i)} - \mathbf{H}^{(e)}\right). \tag{1.27}$$

It follows that

$$\left(\vec{\nu} \times \mathbf{E}^{(i)}\right)\Big|_S = \mu_0\,\mu\left(\frac{\nu_M}{\pi}\right)^{1/2}\int_{-\infty}^{t}\frac{\partial}{\partial\tau}\left[\vec{\nu} \times \left(\vec{\nu} \times \left(\mathbf{H}^{(i)} - \mathbf{H}^{(e)}\right)\right)\right]\frac{d\tau}{\sqrt{t-\tau}}. \tag{1.28}$$

Expression (1.28) is a fundamental one and plays an important role in constructing mathematical models of unsteady eddy currents. Suppose $\mathbf{H}^{(i)} - \mathbf{H}^{(e)} = \mathbf{H}e^{i\omega t}$. Using a familiar quadrature

$$\int_{-\infty}^{t} \frac{de^{i\omega\tau}}{d\tau} \frac{d\tau}{\sqrt{t-\tau}} = \left(\frac{\pi\omega}{2}\right)^{1/2} (1+i)\, e^{i\omega t},$$

we obtain in this case from (1.28)

$$\left(\vec{\nu} \times \mathbf{E}^{(i)}\right)\Big|_S = -\mu_0 \mu \left(\frac{\omega\nu_M}{2}\right)^{1/2} (1+i) \left[\vec{\nu} \times \left(\vec{\nu} \times \left(\mathbf{H}^{(i)} - \mathbf{H}^{(e)}\right)\right)\right]\Big|_S . \quad (1.29)$$

In this form the expression coincides with Leontovich's condition widely used in problems associated with energy dissipation in metal with particularly pronounced skin effect (see N.N.Fedorov [20]).

Thus, (1.28) is a generalization of Leontovich's condition to arbitrary unsteady fields. It should be noted that Expression (1.28) appears as a natural result of the unsteady boundary-value Problem (1.10) solution representation for the presence of space and time boundary layers in the form of Expansions (1.11) and (1.15) without any additional assumptions.

Taking into consideration that $\left[\vec{\nu} \times \left(\vec{\nu} \times \mathbf{B}_{00}^{(i)}\right)\right]\Big|_S = -\mathbf{B}_{00}^{(i)}\Big|_S$, we can reduce the right side of (1.28) to the following final form:

$$\left(\vec{\nu} \times \mathbf{E}^{(i)}\right)\Big|_S = \left(\frac{\nu_M}{\pi}\right)^{1/2} \int_{-\infty}^{t} \frac{\partial \mathbf{B}_{00}^{(i)}}{\partial\tau} \frac{d\tau}{\sqrt{t-\tau}} . \quad (1.30)$$

The obtained approximate expression includes discrepancies of tangential and normal components at surface S, that are infinitesimal of the order of the following (neglected) terms of asymptotic Expansions (1.15) or of higher order.

1.3. Eddy Currents Mathematical Model*

Under mathematical model of eddy currents in cores (of HECF elements) we shall mean a set of integrodifferential equations describing variation with time of both 'external' and 'internal' (eddy) currents and, hence, of magnetic fields induced by them.

Let us consider first a model problem with absence of air gap between the two parts of the core $G_1^{(i)}$ and $G_2^{(i)}$ ($\mathbf{u} \equiv 0$) and absence of external currents ($\mathbf{j}_k \equiv 0$ for $t \geq 0$, they can exist for $t < 0$). In the considered first-order approximation

$$\mathbf{B}^{(i)} = \mathbf{B}_{00}^{(i)}, \qquad \mathbf{B}^{(e)} = \mathbf{B}_{00}^{(e)} \equiv 0. \quad (1.31)$$

We relate fields $\hat{\mathbf{B}}_0^{(i)}$ and $\hat{\mathbf{E}}_1^{(i)}$ to infinitely thin layer within $G^{(i)}$ at the surface S.

* Essentially, this article is about construction of eddy currents mathematical model with unsteady skin effect.

To obtain the adjacent mathematical model let us take advantage of a variational principle similar to Hamilton-Ostrogradsky principle, which instead of 'action', minimizes variations per time unit of the system's total energy $\dot{W}_M = \frac{dW_M}{dt}$:

$$\int_{t_1}^{t_2} \delta \dot{W}_M \, dt = 0, \tag{1.32}$$

where t_1, t_2 are arbitrary time moments, $\delta \dot{W}_M$ corresponding to system trajectory variation in configuration space.

The variational principle of that form, equivalent to that used by A.I.Lurie [59], allows consideration of dissipative systems. In this case

$$\delta \dot{W}_M = \frac{1}{\mu_0 \mu} \int_{G^{(i)}} \left(\frac{\partial \mathbf{B}_{00}^{(i)}}{\partial t}, \delta \mathbf{B}_{00}^{(i)} \right) dQ + \oint_S \Pi_\nu \, dS. \tag{1.33}$$

Here Π_ν is the projection to outer normal to S of Poynting vector $\vec{\Pi}$, which determines the electromagnetic field energy flux through unity area (see N.N.Fedorov [20]):

$$\Pi_\nu = \left(\vec{\Pi}, \vec{\nu} \right) = \left(\mathbf{E}^{(i)} \times \mathbf{H}^{(i)}, \vec{\nu} \right) = - \left(\mathbf{E}^{(i)} \times \vec{\nu}, \mathbf{H}^{(i)} \right). \tag{1.34}$$

Substituting $\mathbf{E}^{(i)} \times \vec{\nu}$ from (1.30) into (1.34) we obtain the following expression for vector $\vec{\Pi}$ flux from $G^{(i)}$ domain to infinitely thin layer at the inner side of surface S :

$$\oint_S \delta \Pi_\nu \, dS = - \oint_S \left(\mathbf{E}^{(i)} \times \vec{\nu}, \delta \mathbf{H}^{(i)} \right) dS =$$

$$= \frac{1}{\mu_0 \mu} \left(\frac{\nu_M}{\pi} \right)^{1/2} \oint_S \left(\int_{-\infty}^{t} \frac{\partial \mathbf{B}_{00}^{(i)}}{\partial \tau} \frac{d\tau}{\sqrt{t-\tau}}, \delta \mathbf{B}_{00}^{(i)} \right) dS, \tag{1.35}$$

where $\mathbf{B}_{00}^{(i)}$ is the field from (1.21) for $\mathbf{u} = 0$.

Let us denote coordinates $J_k(t)$ variations by $\delta J_k(t)$, and suppose the following boundary conditions are obeyed:

$$\delta J_k(t_1) = \delta J_k(t_2) = 0. \tag{1.36}$$

Assuming $t_1 = -\infty$ we obtain from (1.32) - (1.35) the following variational equation

$$\sum_{n=1}^{3} \int_{-\infty}^{t_2} \left(\sum_{m=1}^{3} \overset{\circ}{M}_{nm} \frac{dJ_m}{dt} + \overset{\circ}{\gamma}_{nm} \int_{-\infty}^{t} \frac{dJ_m}{d\tau} \frac{d\tau}{\sqrt{t-\tau}} \right) \delta J_n(t) \, dt = 0, \tag{1.37}$$

where

$$\overset{\circ}{M}_{nm} = \overset{\circ}{M}_{mn} = \rho_{mn} \int_{G^{(i)}} \left(\vec{\beta}_n^{(i)}, \vec{\beta}_m^{(i)} \right) dQ,$$

$$\overset{\circ}{\gamma}_{nm}=\overset{\circ}{\gamma}_{mn}=\rho_{mn}\left(\frac{\nu_M}{\pi}\right)^{1/2}\oint_S\left(\vec{\beta}_n^{(i)},\ \vec{\beta}_m^{(i)}\right)dS, \tag{1.38}$$

$$\rho_{nm}=\rho_{mn}=\mu_0\,\mu\,\overset{\circ}{n}_m\overset{\circ}{n}_n=\frac{\overset{\circ}{N}_m\overset{\circ}{N}_n}{4\pi^2}\,\mu\,\mu_0,$$

$G^{(i)}$ is the domain occupied by the ferromagnet (torus without a cut); S is its total surface. Since $\delta J_n(t)$ are arbitrary functions satisfying boundary Conditions (1.36), the left-hand part of (1.37) is identically zero, if the coefficients at $\delta J_0(t)$ in the integrand expression are equal to zero. This yields the sought equations connecting generalized currents J_m and generalized voltages $V_n^{(i)}$ (analogues of generalized velocities and generalized forces):

$$\sum_{m=1}^{3}\overset{\circ}{M}_{nm}\frac{dJ_m}{dt}=V_n^{(i)},\quad V_n^{(i)}=-\sum_{m=1}^{3}\overset{\circ}{\gamma}_{nm}\int_{-\infty}^{t}\frac{dJ_m}{d\tau}\frac{d\tau}{\sqrt{t-\tau}} \tag{1.39}$$

($n = 1, 2, 3$). Integrodifferential equations of the type of (1.39) occur in a number of problems for systems with aftereffect (see N.Kh.Arutiunian and V.B.Kolmanovsky, V.B.Kolmanovsky and V.R.Nosov, A.I.Lurie [5, 41, 58]). In the considered case they are reduced to a set of integral equations of Volterra type with respect to unknown functions $\dot{J}_m(t)=\dfrac{dJ_m}{dt}$.

Equations (1.39) can gain a more compact form, if we introduce new vectors $\mathbf{J}$, $\mathbf{V}^{(i)}$, and symmetric tensors M°, T° of rank 2:

$$\mathbf{J}=\{J_1, J_2, J_3\},\quad \mathbf{V}^{(i)}=\left\{V_1^{(i)}, V_2^{(i)}, V_3^{(i)}\right\},\quad \mathsf{M}^\circ=\{\overset{\circ}{M}_{mn}\},\quad \mathsf{T}^\circ=\{\overset{\circ}{\gamma}_{mn}\} \tag{1.40}$$

($m, n = 1, 2, 3$). Then (1.39) can be written as two vector equations

$$(\mathsf{M}^\circ, \dot{\mathbf{J}})=\mathbf{V}^{(i)},\quad \mathbf{V}^{(i)}=-\left(\mathsf{T}^\circ, \int_{-\infty}^{t}\frac{\dot{\mathbf{J}}(\tau)\,d\tau}{\sqrt{t-\tau}}\right), \tag{1.41}$$

equivalent to one equation

$$(\mathsf{M}^\circ, \dot{\mathbf{J}})+\left(\mathsf{T}^\circ, \int_{-\infty}^{t}\frac{\dot{\mathbf{J}}(\tau)\,d\tau}{\sqrt{t-\tau}}\right)=0, \tag{1.42}$$

where $(\mathsf{M}^\circ, \dot{\mathbf{J}})$ and other similar expressions denote right scalar product of a tensor by a vector.

Equation (1.42) can be considered a mathematical model of eddy currents in HECF elements, valid for small magnetic Strouhal numbers and great magnetic Reynolds numbers.

1.4. General Equations of System Dynamics with External and Eddy Currents and Mechanical Degrees of Freedom

Let us consider the general case $\mathbf{u} \neq 0$, $\mathbf{j}_k \neq 0$ with $t \geq 0$ in terms of the first order approximation equations. We shall assume, that each current I_k with density j_k exists in the coil having N_k° loops and the ohmic resistance R_k. To synthesize an appropriate general mathematical model we shall make use of Lagrange-Maxwell equations.

Let us introduce three generalized coordinate systems corresponding to the three mechanical degrees of freedom ($q_1 \equiv \xi$; $q_2 \equiv \eta$; $q_3 \equiv \zeta$), three 'external' electrical degrees of freedom (I_1 , I_2 , I_3), and three 'internal' electrical degrees of freedom (J_1 , J_2 , J_3), as well as generalized forces (Q_1 , Q_2 , Q_3) and generalized voltages ($V_1^{(e)}$, $V_2^{(e)}$, $V_3^{(e)}$).

Total energy of the system, in the considered approximation, consists of kinetic energy T of the mechanical part of the system with mass m, magnetic field energy T_M, and elastic strain potential energy U (we are not considering gravitational field potential energy so far, for it may be absent in the accepted coordinate system).

Further it is necessary to take into account the mechanical dissipative function, corresponding to the damping elements of elastic constraints, and the electrical dissipative function, corresponding to the resistance R_k of the windings with external currents I_k. We obtain the following expressions for the mentioned functions.

Kinetic and potential energy

$$T = \frac{m}{2} \sum_{k=1}^{3} \dot{q}_k^2, \quad U = \frac{1}{2} \sum_{k=1}^{3} \sum_{m=1}^{3} c_{mk}\, q_m\, q_k , \tag{1.43}$$

where m is the mass of the mobile part of the system, $c_{mk} = c_{km}$ are the generalized rigidities of elastic constraints (elements of the rigidity matrix).

Magnetic field energy

$$T_M = \frac{1}{2\mu_0\mu} \sum_{j=1}^{2} \int_{G_j^{(i)}} \left(\overset{(i)}{\mathbf{B}}_j, \overset{(i)}{\mathbf{B}}_j \right) dQ + \frac{1}{2\mu_0} \int_{G^{(e)}} \left(\overset{(e)}{\mathbf{B}}, \overset{(e)}{\mathbf{B}} \right) dQ, \tag{1.44}$$

$$\overset{(i)}{\mathbf{B}}_j = \overset{(i)}{\mathbf{B}}_{00} + \overset{(ie)}{\mathbf{B}}_{0j} + \overset{(ie)}{\mathbf{B}}_{00j} , \quad \overset{(e)}{\mathbf{B}} = \overset{(ei)}{\mathbf{B}}_0 + \overset{(ei)}{\mathbf{B}}_{00} .$$

Expression (1.44) together with (1.13) and (1.21) yields

$$T_M = \frac{1}{2} \sum_{m=1}^{3} \sum_{n=1}^{3} \left(L_{mn}\, I_m\, I_n + K_{mn}\, I_m\, J_n + K_{nm}\, J_m\, I_n + M_{mn}\, J_m\, J_n \right), \tag{1.45}$$

where

$$L_{mn} = L_{nm} = \rho_{mn} \left[\sum_{j=1}^{2} \int_{G_j^{(i)}} \left(\vec{\beta}_{jm}^{(ie)}, \vec{\beta}_{jn}^{(ie)} \right) dQ + \frac{1}{\mu} \int_{G^{(e)}} \left(\vec{\beta}_m^{(ei)}, \vec{\beta}_n^{(ei)} \right) dQ \right],$$

$$K_{mn} = \rho_{mn}\left[\sum_{j=1}^{2} \int_{G_j^{(i)}} \left(\alpha_n \vec{\beta}_n^{(i)} + \vec{\beta}_{jn}^{(i)}, \vec{\beta}_{jm}^{(ie)}\right) dQ + \frac{1}{\mu} \int_{G^{(e)}} \alpha_n \left(\vec{\beta}_m^{(ei)}, \vec{\beta}_n^{(e)}\right) dQ\right],$$

$$M_{mn} = M_{nm} = \rho_{mn}\left[\sum_{j=1}^{2} \int_{G_j^{(i)}} \left(\alpha_m \vec{\beta}_m^{(i)} + \vec{\beta}_{jm}^{(i)}, \alpha_n \vec{\beta}_n^{(i)} + \vec{\beta}_{jn}^{(i)}\right) dQ + \right.$$

$$\left. + \frac{1}{\mu} \cdot \int_{G^{(e)}} \alpha_n \alpha_m \left(\vec{\beta}_m^{(e)}, \vec{\beta}_n^{(e)}\right) dQ\right], \qquad (1.46)$$

$$n, m = 1, 2, 3.$$

Total mechanical and electrical dissipative function

$$\chi = -\frac{1}{2}\sum_{m=1}^{3}\left(\sum_{k=1}^{3} b_{mk}\, \dot{q}_m \dot{q}_k + R_m I_m^2\right), \qquad (1.47)$$

where b_{mk} are the mechanical dissipative function coefficients, R_m is the ohmic resistance of the windings.

Generalized voltages. In the general case of $\mathbf{u} \neq 0$ the field (1.21) in the expression of the Poynting's vector flux (1.35) depends on $\mathbf{u}$, and hence it depends on time. Therefore, changing in (1.35) the order of integration over surface S and with respect to time, we obtain, analogous to (1.39), the following expression for the generalized voltage $V_n^{(i)}$ $(n = 1, 2, 3)$:

$$V_n^{(i)} = -\sum_{m=1}^{3} \gamma^{\circ}{}_{mn}\, \alpha_n \int_{-\infty}^{t} \frac{d\,(\alpha_m J_m)}{d\tau} \frac{d\tau}{\sqrt{t-\tau}}, \qquad (1.48)$$

where (see (1.38))

$$\gamma^{\circ}{}_{mn} = \gamma^{\circ}{}_{nm} = \rho_{mn}\left(\frac{\nu_M}{\pi}\right)^{1/2} \sum_{j=1}^{2}\sum_{k=1}^{3} \oint_{S_{jk}} \left(\vec{\beta}_n^{(i)}, \vec{\beta}_m^{(i)}\right) dS, \qquad (1.49)$$

$$m, n = 1, 2, 3.$$

Lagrange-Maxwell equations. Applying Hamilton-Ostrogradsky variational principle we obtain the following Lagrange-Maxwell equations for the considered system:

$$\frac{d}{dt}\frac{\partial W}{\partial \dot{q}_k} - \frac{\partial W}{\partial q_k} + \frac{\partial U}{\partial q_k} - \frac{\partial \chi}{\partial \dot{q}_k} = Q_k, \quad \frac{d}{dt}\frac{\partial W}{\partial I_m} - \frac{\partial \chi}{\partial I_m} = V_m^{(e)},$$

$$\frac{d}{dt}\frac{\partial W}{\partial J_n} = V_n^{(i)}, \quad k, m, n = 1, 2, 3, \qquad (1.50)$$

where $W = T + T_M$, U, χ, $V_n^{(i)}$ are determined from Expressions (1.43), (1.45), (1.47), and (1.48). The values

$$\Phi_m = \frac{\partial W}{\partial I_m}, \quad m = 1, 2, 3 \tag{1.51}$$

are the magnetic flux (magnetic induction flux, magnetic-flux linkage) similar to generalized momenta;

$$F_k = \frac{\partial W}{\partial q_k}, \quad k = 1, 2, 3, \tag{1.52}$$

are the forces of mechanical action on the system of electromagnetic fields (ponderomotive forces). Substituting appropriate expressions in (1.50) and omitting the superscript in $V_m^{(e)}$, we can transform Equations (1.50) to the following:

$$m\frac{d^2 q_k}{dt^2} + \sum_{m=1}^{3}\left(b_{mk}\frac{dq_m}{dt} + c_{mk}\, q_m\right) -$$

$$-\frac{1}{2}\sum_{m=1}^{3}\sum_{n=1}^{3}\left(\frac{\partial L_{mn}}{\partial q_k} I_m I_n + 2\frac{\partial K_{mn}}{\partial q_k} I_m J_n + \frac{\partial M_{mn}}{\partial q_k} J_m J_n\right) = Q_k, \tag{1.53}$$

$$\sum_{n=1}^{3}\frac{d}{dt}\left(L_{mn} I_n + K_{mn} J_n\right) + R_m I_m = V_m,$$

$$\sum_{m=1}^{3}\frac{d}{dt}\left(M_{nm} J_m + K_{mn} I_m\right) + \sum_{m=1}^{3}\gamma^{\circ}_{nm}\,\alpha_n \int_{-\infty}^{t}\frac{d\,(\alpha_m J_m)}{d\tau}\frac{d\tau}{\sqrt{t-\tau}} = 0,$$

$$k, m, n = 1, 2, 3.$$

These are the general equations of the considered electromechanical system dynamics.

1.5. Mathematical Models of Stabilized Ferromagnetic Elements as Object of Control

The general mathematical Model (1.53) considering eddy currents in HECF elements is nonlinear, that considerably hampers its for synthesis of automatic control systems and analysis of the closed-loop system 'object – regulator'. But for a number of practical problems we can use an essentially simpler linear model related to small deviations of all generalized coordinates from some stationary or 'slowly varying' programmed values.

Let us introduce a new intermediate coordinate system $O^\circ x^\circ y^\circ z^\circ$ coinciding with $Oxyz$ for $\mathbf{u} = \mathbf{u}^\circ = -\mathbf{i}_2 s^\circ(t)$, where $s^\circ(t)$ is a slowly varying function of time (constant in particular case) corresponding to the programmed motion of the system, which we shall assume unperturbed motion; $\mathbf{i}_2$ is the CY axis unit vector. By *perturbed motion* we mean the motion of the coordinate system $Oxyz$ with respect to $O^\circ x^\circ y^\circ z^\circ$, which is characterized by small variations of all the generalized coordinates relative to their unperturbed values. Thus,

$$q_k = \overset{\circ}{q}_k + \delta q_k(t)\,, \quad Q_k = \overset{\circ}{Q}_k + \delta Q_k(t)\,, \tag{1.54}$$
$$I_m = \overset{\circ}{I}_m + \delta I_m(t)\,, \quad J_n = \overset{\circ}{J}_n + \delta J_n(t)\,, \quad V_m = \overset{\circ}{V}_m + \delta V_m(t)\,,$$

where all the quantities labeled with superscript $^\circ$ are slowly varying functions of time or constants, and normally

$$|\overset{\circ}{q}_k| < |\overset{\circ}{q}_2|, \quad |\overset{\circ}{Q}_k| < |\overset{\circ}{Q}_2|, \quad |\overset{\circ}{I}_m| < |\overset{\circ}{I}_2|, \quad |\overset{\circ}{V}_m| < |\overset{\circ}{V}_2|, \quad k, m = 1, 3,$$
$$\overset{\circ}{J}_n \equiv 0, \tag{1.55}$$
$$\dot{\overset{\circ}{q}}_k \approx 0, \quad \ddot{\overset{\circ}{q}}_k \approx 0, \quad \dot{\overset{\circ}{I}}_m \approx 0, \quad k, m, n = 1, 2, 3.$$

Variations of generalized coordinates and generalized forces satisfy the following conditions

$$|\delta q_k| \ll |\overset{\circ}{q}_k|, \quad |\delta Q_k| \ll |\overset{\circ}{Q}_k|, \quad |\delta I_m| \ll |\overset{\circ}{I}_m|,$$
$$|\delta J_n| \ll |\overset{\circ}{I}_n|, \quad |\delta V_m| \ll |\overset{\circ}{V}_m|, \quad k, m, n = 1, 2, 3. \tag{1.56}$$

Supposed

$$q_k = \overset{\circ}{q}_k, \quad Q_k = \overset{\circ}{Q}_k, \quad I_m = \overset{\circ}{I}_m, \quad V_m = \overset{\circ}{V}_m, \quad J_n = 0, \quad k, m, n = 1, 2, 3, \tag{1.57}$$

we obtain equations of unperturbed motion (steady state, in particular), that connect slowly varying (steady, in particular) values of generalized forces $\overset{\circ}{Q}_k$ and voltages $\overset{\circ}{V}_m$ with displacements $\overset{\circ}{q}_k$ and currents $\overset{\circ}{I}_m$:

$$\sum_{m=1}^{3} c_{mk}\overset{\circ}{q}_k - \frac{1}{2}\sum_{m=1}^{3}\sum_{n=1}^{3}\left(\frac{\partial L_{mn}}{\partial q_k}\right)^{\circ} \overset{\circ}{I}_m \overset{\circ}{I}_n = \overset{\circ}{Q}_k, \tag{1.58}$$

$$R_m \overset{\circ}{I}_m = \overset{\circ}{V}_m, \quad k, m = 1, 2, 3,$$

where

$$\overset{\circ}{q}_2 = -\overset{\circ}{s}, \quad \left(\frac{\partial L_{mn}}{\partial q_1}\right)^{\circ} = \left(\frac{\partial L_{mn}}{\partial \xi}\right)^{\circ}, \quad \left(\frac{\partial L_{mn}}{\partial q_2}\right)^{\circ} = \left(\frac{\partial L_{mn}}{\partial \eta}\right)^{\circ}, \quad \left(\frac{\partial L_{mn}}{\partial q_3}\right)^{\circ} = \left(\frac{\partial L_{mn}}{\partial \zeta}\right)^{\circ} \tag{1.59}$$

Sign $^\circ$ here and further means that the adjacent quantities are calculated for unperturbed values of all the generalized coordinates.

Further calculations are performed according to the standard scheme; Expressions (1.54) are substitueted into Equation (1.53), their left-side and right-side terms are expanded into Taylor series in the vicinity of unperturbed values of generalized coordinates and forces (1.57), time differentiation is carried out considering dependence of coefficients L_{mn}, M_{mn} and others of generalized coordinates and linearization is performed considering (1.58), i. e. all the terms of order 2 and higher infinitesimal, relative to all variations, are neglected. Resulting is the following set of linearized equations of perturbed motion:

$$m\,\delta\ddot{q}_k + \sum_{m=1}^{3}\left[b_{mk}\,\delta\dot{q}_m + (c_{mk} - d_{mk})\,\delta q_m\right] - \sum_{m=1}^{3} e_{km}\,\delta I_m - \sum_{n=1}^{3} g_{kn}\,\delta J_n = \delta Q_k,$$

$$\sum_{n=1}^{3} (L_{mn}\,\delta\dot{I}_n + K_{mn}\,\delta\dot{J}_n) + \sum_{k=1}^{3} e_{km}\,\delta\dot{q}_k + R_m\,\delta I_m = \delta V_m\,, \tag{1.60}$$

$$\sum_{m=1}^{3} (M_{nm}\,\delta\dot{J}_m + K_{mn}\,\delta\dot{I}_m) + \sum_{k=1}^{3} g_{kn}\,\delta\dot{q}_k + \sum_{m=1}^{3} \gamma_{nm} \int_{-\infty}^{t} \frac{\delta\dot{J}_m(\tau)\,d\tau}{\sqrt{t-\tau}} = 0,$$

$$k,\ m,\ n = 1, 2, 3,$$

where L_{mn}, M_{mn}, K_{mn} are parameters of (1.46) for $s = s^{\circ}$,

$$d_{mk} = d_{km} = \frac{1}{2}\sum_{i=1}^{3}\sum_{j=1}^{3}\left(\frac{\partial^2 L_{ij}}{\partial q_m\,\partial q_k}\right)^{\circ} \overset{\circ}{I_i}\,\overset{\circ}{I_j},$$

$$e_{km} = \sum_{n=1}^{3}\left(\frac{\partial L_{nm}}{\partial q_k}\right)^{\circ} \overset{\circ}{I_n}, \quad g_{kn} = \sum_{m=1}^{3}\left(\frac{\partial K_{mn}}{\partial q_k}\right)^{\circ} \overset{\circ}{I_m}, \tag{1.61}$$

$$R_m = R_{mm}\,, \quad \gamma_{nm} = \gamma_{mn} = \overset{\circ}{\alpha}_m\,\overset{\circ}{\alpha}_n\,\overset{\circ}{\gamma}_{mn}\,.$$

Equations (1.60) can be written in a more compact form of vector and tensor symbols are used. Let us omit the variation sign δ at all generalized coordinates and forces and the superscript $^{\circ}$ at coefficients and introduce the following vectors and tensors of rank 2

$$\mathbf{u} = \{q_1, q_2, q_3\}, \quad \mathbf{Q} = \{Q_1, Q_2, Q_3\}, \quad \mathbf{I} = \{I_1, I_2, I_3\},$$
$$\mathbf{J} = \{J_1, J_2, J_3\}, \quad \mathbf{V} = \{V_1, V_2, V_3\}. \tag{1.62a}$$

$$\mathsf{L} = \{L_{mn}\}, \quad \mathsf{M} = \{M_{mn}\}, \quad \mathsf{K} = \{K_{mn}\},$$
$$\mathsf{B} = \{b_{mk}\}, \quad \mathsf{C} = \{c_{mk}\}, \quad \mathsf{D} = \{d_{mk}\}, \tag{1.62b}$$
$$\mathsf{G} = \{g_{kn}\}, \quad \mathsf{E} = \{e_{km}\}, \quad \mathsf{R} = \{R_{mm}\}, \quad \mathsf{T} = \{\gamma_{mn}\},$$
$$m,\ n,\ k = 1, 2, 3.$$

The components of introduced vectors with new denominations represent variations of the adjacent generalized coordinates. The set of 9 scalar equations of (1.60) can now be written in the form of 3 vector equations:

$$m\,\ddot{\mathbf{u}} + (\mathsf{B}, \dot{\mathbf{u}}) + (\mathsf{C} - \mathsf{D}, \mathbf{u}) - (\mathsf{E}, \mathbf{I}) - (\mathsf{G}, \mathbf{J}) = \mathbf{Q},$$
$$(\mathsf{L}, \dot{\mathbf{I}}) + (\mathsf{K}, \dot{\mathbf{J}}) + (\overline{\mathsf{E}}, \dot{\mathbf{u}}) + (\mathsf{R}, \mathbf{I}) = \mathbf{V}, \tag{1.63}$$
$$(\mathsf{M}, \dot{\mathbf{J}}) + (\overline{\mathsf{K}}, \dot{\mathbf{I}}) + (\overline{\mathsf{G}}, \dot{\mathbf{u}}) + \left(\mathsf{T}, \int_{-\infty}^{t} \frac{\dot{\mathbf{J}}(\tau)\,d\tau}{\sqrt{t-\tau}}\right) = 0.$$

Expressions in parentheses symbolize right products of tensors by a vector or scalar product of tensors. The bar over tensor E or G indicates a conjugated tensor (i.e. the tensor, the matrix of which is transposed relative to the matrix of the source tensor).

Equations (1.60) and (1.63) have a wonderful symmetry and antisymmetry that reflect profound internal features of the considered electromechanical system. The mentioned symmetry in caused by symmetry of mutual induction tensors L, M, mechanical rigidity tensor C, and electromagnetic rigidity tensor D (Maxwell's tensor), as well as tensors B and T. Tensor B elements are the coefficients of the mechanical dissipative function. Antisymmetry is caused by opposite signs of products of tensors E and G by vectors **I** and **J** in the first equation of (1.63) and products of conjugated tensors $\overline{\mathsf{E}}$ and $\overline{\mathsf{G}}$ by vector $\dot{\mathbf{u}}$ in the second and third equations of (1.63). It reflects the gyroscopic character of ponderomotive forces and of electromotive forces associated with them. The mentioned features of Equations (1.63) and similar features of Equations (1.60) provide for compliance with the law of conservation of energy with absence of dissipative forces, i. e. for $\mathsf{B} \equiv 0$, $\mathsf{T} \equiv 0$.

Let us consider some particular cases now.

1.5.1. The General Model of Eddy Currents.

a) $G_j^{(i)}$ domains and system of elastoviscous ties with two vertical symmetry planes. It is clear from the considerations of symmetry, that in this case all non-diagonal elements of tensors (1.62b) are zero, and the set of Equations (1.63) splits to three independent systems. Changing double subscripts of diagonal elements of all tensors to single ones, we can write those sets of equations in the form of

$$m\ddot{q}_n + b_n\dot{q}_n + (c_n - d_n)\, q_n - e_n I_n - g_n J_n = Q_n\,,$$
$$L_n\dot{I}_n + K_n\dot{J}_n + e_n\dot{q}_n + R_n I_n = V_n\,, \tag{1.64}$$
$$M_n\dot{J}_n + K_n\dot{I}_n + g_n\dot{q}_n + \gamma_n \int_{-\infty}^{t} \frac{\dot{J}_n(\tau)\, d\tau}{\sqrt{t-\tau}} = 0, \quad n = 1, 2, 3.$$

Subscript value $n = 1$ corresponds to the motion along CX axis; $n = 2$ corresponds to motion along CY axis, $n = 3$ corresponds to motion along CZ axis.

b) Constant relative displacement $u^{\circ} = -s^{\circ}$ of elements $G_1^{(i)}$ and $G_2^{(i)}$. Supposed in Equations (1.63) $\dot{\mathbf{u}} = 0$, we obtain

$$(\mathsf{C} - \mathsf{D}, \mathbf{u}^{\circ}) - (\mathsf{E}, \mathbf{I}) - (\mathsf{G}, \mathbf{J}) = \mathbf{Q}, \quad (\mathsf{L}, \dot{\mathbf{I}}) + (\mathsf{K}, \dot{\mathbf{J}}) + (\mathsf{R}, \mathbf{I}) = \mathbf{V},$$
$$(\mathsf{M}, \dot{\mathbf{J}}) + (\overline{\mathsf{K}}, \dot{\mathbf{I}}) + \left(\mathsf{T}, \int_{-\infty}^{t} \frac{\dot{\mathbf{J}}(\tau)\, d\tau}{\sqrt{t-\tau}}\right) = 0. \tag{1.65}$$

This model is the closed to its hydrodynamic analogue (see Article 4.1 below).

c) LECF element with $\sigma = 0$. It can be a model of laminated or ferrite core ($\sigma \to 0$). This corresponds to practical absence of eddy currents, that is beyond applicability limits of the initial assumptions formulated at the beginning of this chapter. The proper model can, however, be obtained from (1.63) letting $\mathbf{J} \equiv 0$:

$$m\,\ddot{\mathbf{u}} + (\mathsf{B}, \dot{\mathbf{u}}) + (\mathsf{C} - \mathsf{D}, \mathbf{u}) - (\mathsf{E}, \mathbf{I}) = \mathbf{Q},$$
$$(\mathsf{L}, \dot{\mathbf{I}}) + (\overline{\mathsf{E}}, \dot{\mathbf{u}}) + (\mathsf{R}, \mathbf{I}) = \mathbf{V}, \tag{1.66}$$

This case corresponds to 'solidified liquid' in the hydrodynamic model.

1.5.2. Asymptotic Models of Eddy Currents.

a) HECF element with $\mu \to \infty$, $\sigma \to \infty$ for $u \equiv 0$. In this case we can use the model of (1.63) letting $|\mathbf{u}| \equiv 0$, $\mathsf{K} = \overline{\mathsf{K}} = \mathsf{M}$. Substituting $(\mathsf{M}, \dot{\mathbf{J}} + \dot{\mathbf{I}})$ from the third equation into the second one setting then $\mathbf{u}^{\circ} \equiv 0$, $\mathbf{I} + \mathbf{J} \equiv 0$, we obtain

$$(\mathsf{L}^{\circ} - \mathsf{M}^{\circ}, \dot{\mathbf{I}}) + \left(\mathsf{T}^{\circ}, \int_{-\infty}^{t} \frac{\dot{\mathbf{I}}(\tau)\, d\tau}{\sqrt{t-\tau}}\right) + (\mathsf{R}, \mathbf{I}) = \mathbf{V}, \tag{1.67}$$

where

$$\mathsf{M}^{\circ} = \mathsf{M}\,|_{\mathbf{u}^{\circ}=0}\,, \quad \mathsf{L}^{\circ} = \mathsf{L}\,|_{\mathbf{u}^{\circ}=0}\,, \quad \mathsf{T}^{\circ} = \mathsf{T}\,|_{\mathbf{u}^{\circ}=0}\,. \tag{1.68}$$

Here total magnetic field in the ferromagnet is absent everywhere except a very thin boundary layer (sharply pronounced skin effect).

b) Superconductor ($\sigma = \infty$). For this case it is possible to obtain the mathematical model from (1.63) and (1.49) setting $\sigma = \infty$, that results in $\mathsf{T} \equiv 0$:

$$\begin{aligned} &m\,\ddot{\mathbf{u}} + (\mathsf{B}, \dot{\mathbf{u}}) + (\mathsf{C} - \mathsf{D}, \mathbf{u}) - (\mathsf{E}, \mathbf{I}) - (\mathsf{G}, \mathbf{J}) = \mathbf{Q}\,, \\ &(\mathsf{L}, \dot{\mathbf{I}}) + (\mathsf{K}, \dot{\mathbf{J}}) + (\overline{\mathsf{E}}, \dot{\mathbf{u}}) + (\mathsf{R}, \mathbf{I}) = \mathbf{V}\,, \\ &(\mathsf{M}, \dot{\mathbf{J}}) + (\overline{\mathsf{K}}, \dot{\mathbf{I}}) + (\overline{\mathsf{G}}, \dot{\mathbf{u}}) = 0. \end{aligned} \tag{1.69}$$

This model is an analogue of the hydrodynamic Model (2.69) corresponding to ideal liquid.

1.6. Mathematical Model of the Electromagnetic Levitation System Actuator

Let us consider, for example, a test facility including an electromagnetic system affording levitation, i.e. «free suspension» (similar to the legendary Mohammed's coffin in Medina) of some rigid platform upon which the object of dynamical testing or its elements are situated.

The actuator of this system consists of controlled electromagnets interacting with ferromagnetic rail.

1.6.1. Preliminary remarks. Suppose that the electromagnets and ferromagnetic rails are of prismatic shape (considerably elongated in one direction) and have two vertical planes of symmetry. Assume that the electromagnet core's cross section, as well as the rail's one, is of U-shape or C-shape, of uniform width (the last assumption is not a matter of principle).

Consider one of the electromagnets connected with the platform by a system of elastoviscous ties, and the ferromagnetic rail similarly connected with the motionless foundation (see Fig. 1.2). In the figure those ties are shown symbolically with some springs. The electromagnet by itself is highly instable. Therefore to afford stability a special coil current control system is necessary, that. together with the electromagnet acting as an actuator, makes a closed-loop air gap automatic control system. A set of such systems, responding usually to air gap and some other observed coordinates variation, is to afford free suspension (levitation) of both magnets and the platform for small (in the sense mentioned at the beginning of the chapter) gaps between the electromagnets' poles and ferromagnetic rails.

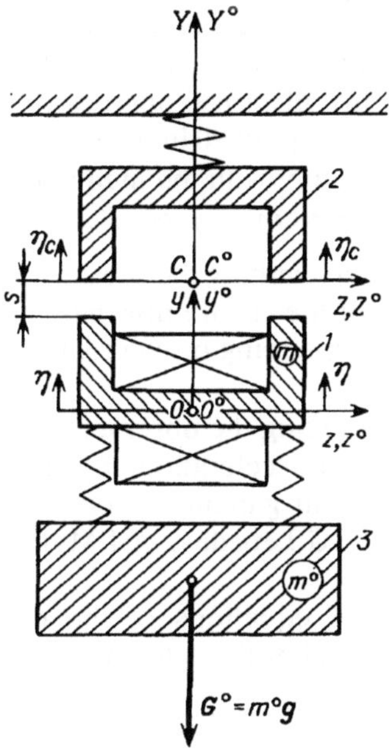

Fig. 1.2. The scheme of the levitation system electromagnet: 1 - the electromagnet; 2 - the ferromagnetic rail; 3 - the passive mass

Suppose partial frequency of the ferromagnetic rail natural oscillations with respect to the foundation is much greater, and partial frequency of the platform's natural oscillations with respect to electromagnets is much less, than the characteristic natural frequency of the closed-loop system 'electromagnet – regulator'.

Then for each electromagnet we can confine the consideration to only one mechanical degree of freedom, namely vertical displacement of electromagnet. The platform will be considered motionless and the ferromagnetic rail will be performing forced motion in vertical direction in a specified manner.

Let us introduce one more moving coordinate system $C^\circ X^\circ Y^\circ Z^\circ$ related to the ferromagnetic rail performing forced vertical motion with respect to the coordinate system $C\,X\,Y\,Z$ related to the foundation. Without this motion the coordinate systems $C^\circ X^\circ Y^\circ Z^\circ$ and $C\,X\,Y\,Z$ coincide. The axes $C\,X$ and $C^\circ X^\circ$ are directed along the longitudinal axis of the rail; the axes $O^\circ X^\circ$ and $O\,X$ are directed along the longitudinal axis of the electromagnet.

Introduce the following symbols:

l_1 , l_2 are the length of the electromagnet core and of the ferromagnetic rail;

l , b are width and thickness of the electromagnet core;

$-y_0$ is the distance between the poles' end face and the electromagnets' centre of mass;

s is the air gap;

η , $\eta_c(t)$ are vertical displacements of the electromagnet and rail centre of mass;

m°, m are the platform mass related to one electromagnet, and total mass of the electromagnet and construction elements connected to it;

$G = (m^\circ + m)\,g$ is the total weight counterbalanced by the lifting force of the electromagnet (in the state of balance);

I is current in the electromagnet coil;

J is the generalized coordinate describing time variation of eddy current;
V is voltage applied to the electromagnet's coil;
$s^\circ, I^\circ, V^\circ$ are programmed values of the air gap, current, and voltage;
N° is the number of turns.

1.6.2. General Equations. Suppose magnetic permeability and other characteristics of the core ferromagnetic material, as well as the manner of the electromagnet's motion are such, that the mathematical model holds true, that is a particular case of (1.53), which corresponds to one mechanical and two electrical degrees of freedom ($q_1 = q_3 \equiv 0,\ \ I_1 = I_3 \equiv 0,\ \ J_1 = J_3 \equiv 0$), and to air gap, small in the sense of (1.5). Suppose further, that parameters introduced above comply with the inequalities

$$l_1 \ll l_2\,, \quad l \ll l_1\,, \quad b \ll l\,, \quad s \ll b\,, \quad m \ll m^\circ, \quad \eta_c \ll s^\circ. \tag{1.70}$$

Let us apply Equations (1.53) to the considered system, omitting the subscripts that are not necessary. Consider the fact that the two coasts of the cut are now moving and so the self-induction and mutual induction coefficients depending on the real air gap are now functions of the argument $s\,(\eta,\ \eta)$.

Let us introduce further the control equation including the observed coordinates of the air gap, electromagnet coil current, and vertical acceleration (a simple set of measurements is possible, e. g. air gap and current or air gap and acceleration, as in T.I.Katsan, V.G,Lebedev, and A.I.Mytarev [35],V.G.Lebedev, A.I.Mytarev, and B.I.Rabinovich [52], and E.Gottzein [25]).

Resulting from (1.53) is the following set of equations of the controlled electromagnet dynamics (non-linear mathematical model of the actuator):

$$m\,\frac{d^2\eta}{dt^2} + \frac{1}{2}\,[L'(s)\,I^2 + 2K'(s)\,I\,J + M'(s)\,J^2\ ] + G^\circ = P\,(t)\,,$$

$$\frac{d}{dt}\,[L\,(s)\,I + K\,(s)\,J\,] + R\,I - V^\circ = \delta\,, \tag{1.71}$$

$$\frac{d}{dt}\,[M\,(s)\,J + K\,(s)\,I\,] + \gamma^\circ\alpha\,(s)\int_{-\infty}^{t}\frac{d}{dt}\,[\alpha\,J\,(\tau)\,]\,\frac{d\tau}{\sqrt{t-\tau}} = 0\,,$$

$$\delta = \Im_1(\Delta s) + \Im_2(\Delta I\,) + \Im_3(\,\ddot{\eta}\,)\,, \quad s = -\,\eta + \eta_c(t\,)\,,$$

$$\Delta s = s - s^\circ, \quad \Delta I = I - I^\circ.$$

Here $K\,(s)$, $L\,(s)$, $M\,(s)$, $\alpha\,(s)$, $\eta_c(t)$, $P\,(t)$ are known functions, $\Im_j$ (j =1, 2, 3) are some linear differential operators determined by the control law. Sign $'$ denotes derivative with respect to s.

We assume further, that the terms $m\,\dfrac{d^2\eta}{dt^2}$, $P\,(t)$, $\gamma^\circ\alpha\,(s)$ of Equations (1.71) are small perturbations in the sense of the following inequalities:

$$m\left|\frac{d^2\eta}{dt^2}\right| \ll G^\circ; \quad |P\,(t)| \ll G^\circ; \quad \frac{\gamma^\circ\alpha\,(s)}{\sqrt{T}} \ll L\,(s^\circ)\,, \tag{1.72}$$

where T is the characteristic time, for which we can assume, in particular case, the characteristic oscillations period for the closed-loop system 'object – regulator'. The physical interpretation of Inequalities (1.72) is as follows: they outline a class of perturbed motions characterized by small variations of ponderomotive forces and of the total magnetic field induction (external field and the one induced by eddy currents).

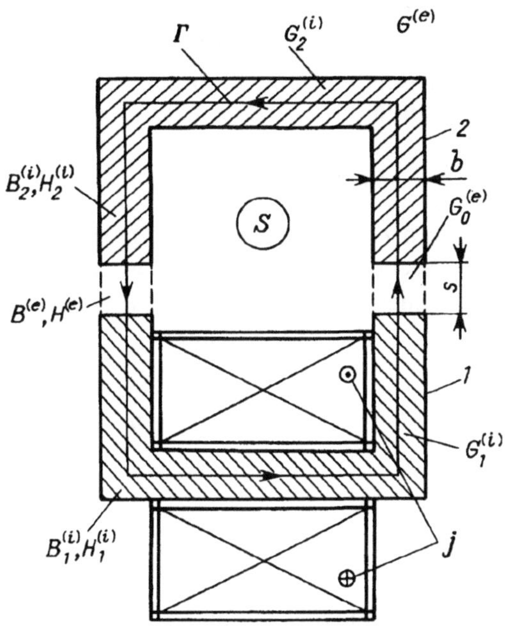

Fig. 1.3. The scheme illustrating approximate calculation of $G_0^{(e)}$ induction in the air gap: 1 - the electromagnet; 2 - the ferromagnetic rail

1.6.3. Approximate expressions of mutual induction and self-induction coefficients. Let us consider the magnetostatics Problem (1.12), supposed $\mathbf{j} \equiv \mathbf{j}_2$, $\mathbf{j}_1 \equiv 0$, $\mathbf{j}_3 \equiv 0$. Draw a contour being the medium line of the core cross section, as shown in Fig. 1.3, and introduce simpler symbols for induction and intensity fields of (1.13) averaged over the core section:

$B_j^{(i)}$, $H_j^{(i)}$ in domains $G_j^{(i)}$, $j = 1, 2$;
$B^{(e)}$, $H^{(e)}$ in the air gap domain $G_0^{(e)}$, boundary effects not considered.

Let us calculate the circulation of the magnetic field intensity vector $\mathbf{H}$ around the closed path Γ (in the magnetostatic approximation) with the help of Stokes theorem:

$$\oint_{\Gamma} (\mathbf{H}, \mathbf{s})\, ds = \int_{S} (\mathbf{j}, \vec{\nu})\, dS, \tag{1.73}$$

where $\mathbf{s}$ is the unit vector tangential to contour Γ, $\vec{\nu}$ is the unit vector normal to the surface element S bounded by the contour Γ in the plane of the core cross section. Taking into account, that $l \ll l_1$ and $b \ll l$, let us substitute the fields H and B by their averages over the core section. Then (1.73) and boundary Conditions (1.12) yield the following approximate equalities:

$$2sH^{(e)} + H_1^{(i)}s_1 + H_2^{(i)}s_2 = N^{\circ}I,$$

$$H^{(e)} = \frac{B^{(e)}}{\mu_0}, \quad H_1^{(i)} = \frac{B_1^{(i)}}{\mu_0 \mu}, \quad H_2^{(i)} = \frac{B_2^{(i)}}{\mu_0 \mu}, \quad B_1^{(i)} = B_2^{(i)} = B^{(e)}, \tag{1.74}$$

where s_1 and s_2 are lengths of contour Γ parts belonging to the domains $G_1^{(i)}$ and $G_2^{(i)}$. Further,

$$B_1^{(i)} = B_2^{(i)} = B^{(e)} = \frac{\mu_0 N^\circ h(s) I}{2 s_\mu}, \quad h(s) = \frac{s_\mu}{s + s_\mu}, \quad s_\mu = \frac{s_1 + s_2}{2\mu}. \tag{1.75a}$$

We also substitute induction fields induced by eddy currents in the domains $G_j^{(i)}$ by the core cross-section average value denoted by $B_0^{(i)}$, $B_{0j}^{(i)}$. Including (1.21) we obtain similarly an expression for the induction fields, analogous to (1.75a), and the normalizing factor α, independent of air gap in this approximation:

$$B_0^{(i)} + B_{01}^{(i)} = B_0^{(i)} + B_{02}^{(i)} = \frac{\mu_0 N^\circ h(s) J}{2 s_\mu}, \quad \alpha(s) \equiv \alpha(0) = 1. \tag{1.75b}$$

Let us use Expressions (1.46), (1.49) and (1.13), (1.21), (1.75). In order to compensate to some extent the applied core-section averaging of fields, we include constants L°, M°, K° into the expressions of functions $L(s)$, $M(s)$, $K(s)$ and use the general Expression (1.61) for γ with $\alpha(s) = \alpha^\circ \neq 1$. We obtain

$$L(s) = L^\circ + \frac{h(s)\lambda}{s_\mu}, \quad K(s) = K^\circ + \frac{h(s)\lambda}{s_\mu}, \quad M(s) = M^\circ + \frac{h(s)\lambda}{s_\mu}, \tag{1.76}$$

$$\gamma^\circ = \frac{\mu_0 l_1 N^{\circ 2}}{s_\mu}\left(\frac{\nu_M}{\pi}\right)^{1/2}, \quad \gamma(s) = \alpha^{\circ 2}\gamma^\circ, \quad \lambda = \frac{\mu_0 b l_1 N^{\circ 2}}{2}.$$

Note that for real values of the levitation system parameters $s_\mu \ll s^\circ$, $L^\circ \ll L$, $M^\circ \ll M$, $K^\circ \ll K$.

1.6.4. Linearization of the controlled electromagnet dynamics equations. We can obtain linearized equations of the considered system perturbed motion from general Equations (1.71). We shall, however, take another way and derive them anew, taking into account some specific features of the considered system and functions $K(s)$, $L(s)$, $M(s)$, and $h(s)$ (1.75a), (1.76). It appears, that if you step a little aside from the classical linearization scheme used above, you can widen the applicability limits of the linearized equations in the considered particular case.

Let us make use of Expressions (1.76), substitute them in Equations (1.71), and introduce new variables

$$\Phi = \frac{I}{s + s_\mu}, \quad \Psi = \frac{J}{s + s_\mu}. \tag{1.77}$$

Besides that, in order to emphasize infinitesimal $|P(t)|$ and γ, we introduce formally small parameter ε into Equations (1.71) denoting

$$\gamma = \varepsilon\tilde{\gamma} = \varepsilon\alpha^{\circ 2}\gamma^\circ, \quad P(t) = \varepsilon\tilde{P}(t). \tag{1.78}$$

Consequently the first three Equations (1.71) gain the form

$$m\ddot{\eta} - \frac{\lambda}{2}(\Phi + \Psi)^2 + (m^\circ + m)g = \varepsilon\tilde{P}(t),$$

$$L^\circ\dot{I} + K^\circ\dot{J} + \lambda(\dot{\Phi} + \dot{\Psi}) + RI - V^\circ = \delta, \tag{1.79}$$

$$M^\circ \dot{J} + K^\circ \dot{I} + \lambda(\dot{\Phi} + \dot{\Psi}) + \varepsilon\tilde{\gamma} \int_{-\infty}^{t} \frac{\dot{\Psi}(\tau)\,d\tau}{\sqrt{t-\tau}} = 0 .$$

In the programmed state assumed unperturbed

$$s = s^\circ,\quad \eta_c \equiv 0,\quad I = I^\circ,\quad \eta = -s^\circ,\quad P(t) \equiv 0,$$

$$\Phi = \Phi^\circ = \frac{I^\circ}{s^\circ + s_\mu},\quad \Psi = \Psi^\circ \equiv 0 . \tag{1.80}$$

This yields

$$\frac{\lambda\,\Phi^{\circ 2}}{2} = (m^\circ + m)\,g,\quad V^\circ = R\,I^\circ. \tag{1.81}$$

Equations (1.79) can now be written in the form

$$m\ddot{\eta} + \frac{\lambda}{2}\left[(\Phi+\Psi)^2 - \Phi^{\circ 2}\right] = \varepsilon\tilde{P}(t),$$

$$L^\circ \dot{I} + K^\circ \dot{J} + \lambda(\dot{\Phi} + \dot{\Psi}) + R\,I = \delta, \tag{1.82}$$

$$M^\circ \dot{J} + K^\circ \dot{I} + \lambda(\dot{\Phi} + \dot{\Psi}) + \varepsilon\tilde{\gamma} \int_{-\infty}^{t} \frac{\dot{\Psi}(\tau)\,d\tau}{\sqrt{t-\tau}} = 0 .$$

Let us introduce dimensionless variables

$$\Delta\tilde{\Phi} = \frac{\Delta\Phi}{\Phi^\circ},\quad \Delta\tilde{\Psi} = \frac{\Delta\Psi}{\Psi^\circ},\quad \Delta\tilde{I} = \frac{\Delta I}{I^\circ},\quad \Delta\tilde{J} = \frac{J}{I^\circ},\quad \Delta\tilde{s} = \frac{\Delta s}{s^\circ},$$

$$\Delta\Phi = \Phi - \Phi^\circ,\quad \Delta\Psi = \Psi - \Psi^\circ \equiv \Psi,\quad \Delta\eta = \eta + s^\circ \tag{1.83}$$

and consider a dimensionless increment of the function $\Phi + \Psi$

$$\Delta\tilde{\Phi} + \Delta\tilde{\Psi} = (\Phi + \Psi - \Phi^\circ)\frac{1}{\Phi^\circ} = \frac{1}{\tilde{\Phi}^\circ}\left[\frac{I^\circ + \Delta I + J}{s^\circ + s_\mu + \Delta s} - \frac{I^\circ}{s^\circ + s_\mu}\right] = \frac{\Delta\tilde{I} + \Delta\tilde{J} - \Delta\tilde{s}}{1 + \Delta\tilde{s}} .$$

This expression can be rewritten :

$$\Delta\tilde{\Phi} + \Delta\tilde{\Psi} = \Delta\tilde{I} + \Delta\tilde{J} - \Delta\tilde{s}(1 + \Delta\tilde{\Phi} + \Delta\tilde{\Psi}) . \tag{1.84}$$

Since

$$(\Phi + \Psi)^2 - \Phi^{\circ 2} = \Phi^\circ(\Phi + \Psi + \Phi^\circ)(\Delta\tilde{\Phi} + \Delta\tilde{\Psi}),$$

and taking into consideration infinitesimal $m\,|\ddot{\eta}|$ and $\varepsilon\,|\tilde{P}(t)|$ of (1.72) we obtain the following inequality

$$|\Delta\tilde{\Phi} + \Delta\tilde{\Psi}| \ll 1. \tag{1.85}$$

Now we can perform the linearization of Equations (1.82) and (1.84) with respect to $\Delta\tilde{\Phi} + \Delta\tilde{\Psi}$, that results in the following set of equations :

$$m\ddot{\eta} - \Phi^{o2}(\Delta\tilde{\Phi} + \Delta\tilde{\Psi}) = \varepsilon\tilde{P}(t),$$

$$L^{\circ}I^{\circ}\Delta\dot{\tilde{I}} + K^{\circ}I^{\circ}\Delta\dot{\tilde{J}} + \lambda\Phi^{\circ}(\Delta\dot{\tilde{\Phi}} + \Delta\dot{\tilde{\Psi}}) + RI^{\circ}\Delta\tilde{I} = \delta,$$

$$M^{\circ}I^{\circ}\Delta\dot{\tilde{J}} + K^{\circ}I^{\circ}\Delta\dot{\tilde{I}} + \lambda\Phi^{\circ}(\Delta\dot{\tilde{\Phi}} + \Delta\dot{\tilde{\Psi}}) + \varepsilon\,\Phi^{\circ}\tilde{\gamma}\int_{-\infty}^{t}\frac{\dot{\tilde{\Psi}}(\tau)\,d\tau}{\sqrt{t-\tau}} = 0, \qquad (1.86)$$

$$\delta = s^{\circ}\Im_1(\Delta\tilde{s}) + I^{\circ}\Im_2(\Delta\tilde{I}) + s^{\circ}\Im_3(\ddot{\tilde{\eta}}),$$

$$\Delta\tilde{\Phi} + \Delta\tilde{\Psi} = \Delta\tilde{I} + \Delta\tilde{J} - \Delta\tilde{s}, \quad \Delta\tilde{s} + \Delta\tilde{\eta} = \frac{\eta_c(t)}{s^{\circ}}.$$

Equations (1.86) are linear, though they are obtained with the assumption of infinitesimal sum $\Delta\tilde{\Phi} + \Delta\tilde{\Psi}$ Modulus (1.85) only, but not variations of gap and current as well as eddy current. In that sense the limits of applicability of Equations (1.86) are wider than of the equations of the previous article obtained with regular linearization technique.

The physical pattern of perturbed motion described by Equations (1.86) for $\varepsilon \to 0$ is as follows: for non-small oscillations of current and air gap, the gap and external current oscillate in phase, while the gap and the magnet displacement, and external current and eddy current oscillate with opposite phases, difference of amplitudes of relative of gap and current, gap and magnet displacement, external and eddy current are small quantities (compared to unity).

1.6.5. Simplified mathematical models. The mathematical Model (1.86) can be directly used for description of the electromagnet dynamics in the control loop and of the whole closed-loop system 'object – regulator' dynamics. For preliminary analysis and control law synthesis, however, the reduction of Model (1.86) to the standard form (see Article 1.5) and some further reductions may be useful.

Let us convert (1.86) to the variables Δs, ΔI, ΔJ. Substitute Expressions (1.80), (1.83), (1.84), (1.77) for Φ°, $\Delta\tilde{\Phi}$, $\Delta\tilde{\Psi}$ and perform the linearization of the last term of the third equation, supposed $\Delta\tilde{s} \ll 1$. After appropriate calculations, omitting Δ sign at increments of variables, we obtain the following set of linearized equations:

$$m\ddot{\eta} - c_I(I+J) + c_s s = P(t), \quad c_{II}\dot{I} + c_{IJ}\dot{J} - c'_s\dot{s} + RI = \delta,$$

$$c_{JJ}\dot{J} + c_{JI}\dot{I} - c'_s\dot{s} + c_{\gamma}\int_{-\infty}^{t}\frac{\dot{J}(\tau)\,d\tau}{\sqrt{t-\tau}} = 0, \qquad (1.87)$$

$$\delta = \Im_1(s) + \Im_2(I) + \Im_3(\ddot{\eta}), \quad s = -\eta + \eta_c(t).$$

where all the variables are variations of the adjacent unperturbed values. The coefficients of Equations (1.87) are specified with the following expressions :

$$c_I = \frac{\lambda I^{\circ}}{(s^{\circ} + s^*_{\mu})^2}, \quad c_s = \frac{\lambda I^{o2}}{(s^{\circ} + s_{\mu})^3}, \quad c'_s = c_I, \quad c_{II} = L^{\circ} + \frac{\lambda}{s^{\circ} + s_{\mu}},$$

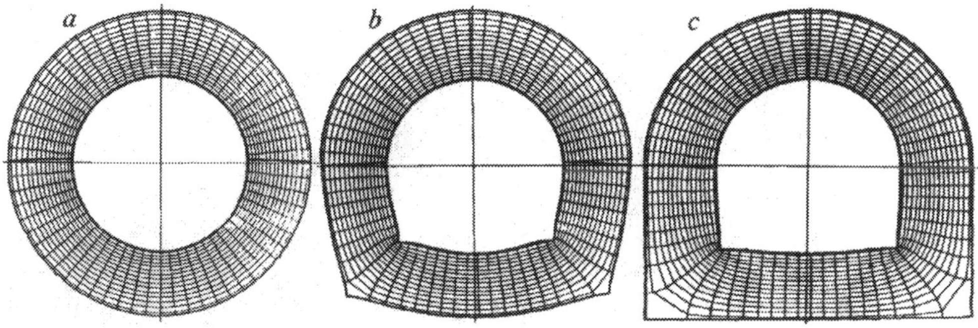

Fig. 1.4. Conformal mapping of the core domain with zero air gap onto a circular annulus: *a* - polar coordinate grid; *b* - coordinate grid at an intermediate stage of mapping, conformally equivalent to *a*; *c* - the final conformally equivalent grid

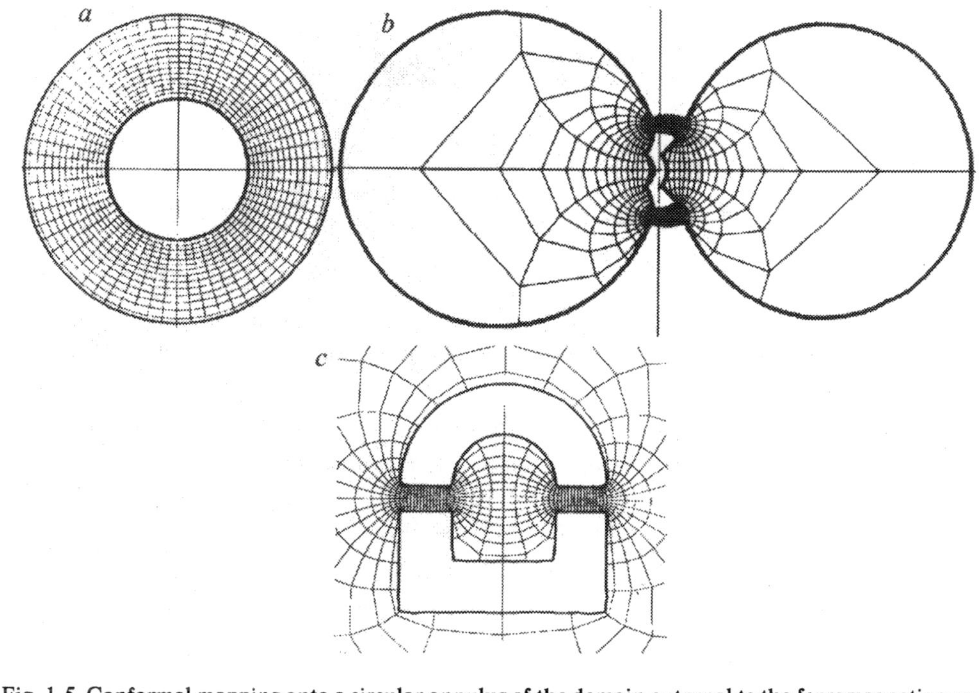

Fig. 1.5. Conformal mapping onto a circular annulus of the domain external to the ferromagnetic core for symmetric displacement of the electromagnet. Denominations are the same as in Fig. 1.4. Conversion from *b* to *c* is made with the help of linear-fractional transform.

$$c_{JJ} = M^\circ + \frac{\lambda}{s^\circ + s_\mu}, \quad c_{IJ} = c_{JI} = K^\circ + \frac{\lambda}{s^\circ + s_\mu}, \quad c_\gamma = \gamma^\circ \alpha^{\circ\,2}. \qquad (1.88)$$

The set of Equations (1.87) can be directly obtained from the set of Equations (1.63) as being a particular case of the latter. Let us perform some further

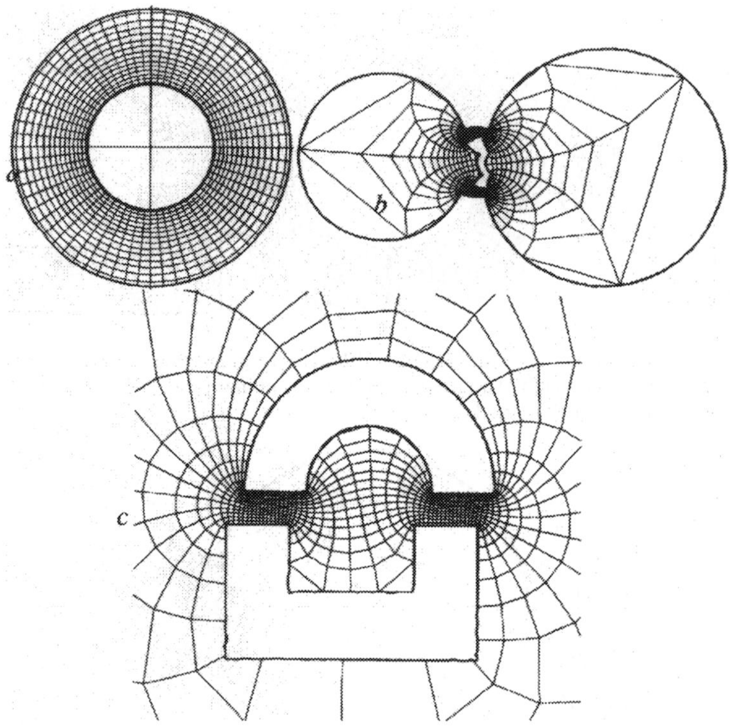

Fig. 1.6. Conformal mapping onto a circular annulus of the domain external to the ferromagnetic core for the case of additional side shift of the electromagnet. Denominations are the same as in Fig. 1.5

transformations of Equations (1.87). We introduce a new variable J^* instead of J; $J = \kappa J^*$. We choose the normalizing factor κ so that coefficients at $\dot{J}^*$ and $\dot{I}$ coincide in the third equation. Perform change of variables in (1.87) and multiply the third equation by κ. The first three equations gain the following form :

$$m\ddot{\eta} - c_I I - \kappa c_I J^* + c_s s = P(t), \quad c_{II}\dot{I} + \kappa c_{IJ}\dot{J}^* - c'_s\dot{s} + RI = \delta,$$

$$\kappa^2 c_{JJ}\dot{J}^* + \kappa c_{JI}\dot{I} - \kappa c'_s\dot{s} + \kappa^2 c_\gamma \int_{-\infty}^{t} \frac{\dot{J}^*(\tau)\, d\tau}{\sqrt{t-\tau}} = 0. \tag{1.89}$$

Suppose $\kappa c_{JI} = \kappa^2 c_{JJ}$, hence $\kappa = \dfrac{c_{JI}}{c_{JJ}}$. Now the complete set of Equations (1.89) can be written in the form of

$$m\ddot{\eta} - c_I I - c_J J^* + c_s s = P(t), \quad c'_I\dot{I} + c'_J\dot{J}^* - c_s\dot{s} + RI = \delta,$$

$$c'_J(\dot{J}^* + \dot{I}) - c_s\dot{s} + c'_\gamma \int_{-\infty}^{t} \frac{\dot{J}^*(\tau)\, d\tau}{\sqrt{t-\tau}} = 0, \tag{1.90}$$

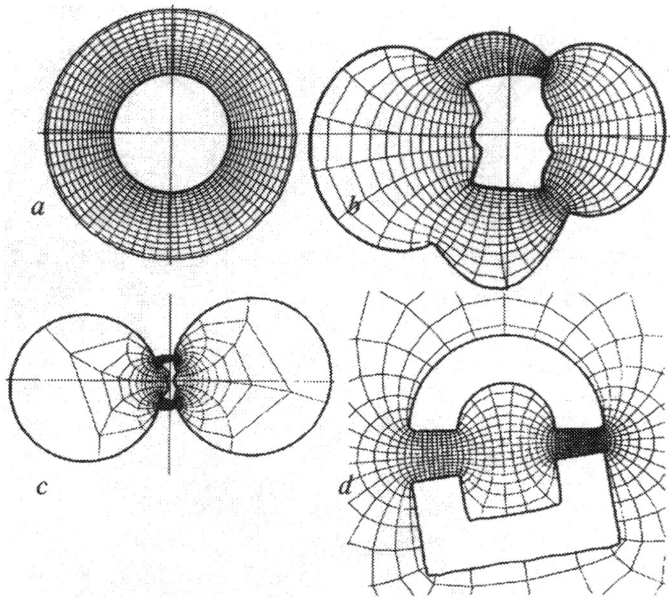

Fig. 1.7. Conformal mapping onto a circular annulus of the domain external to the ferromagnetic core for the case of additional 'cocking' of the electromagnet. Denominations are the same ase in Fig. 1.5. Conversion from b to c is made with the help of Joukowski-type transformation, and from c to d with linear-fractional transformation

$$\delta = \mathfrak{I}_1(s) + \mathfrak{I}_2(I) + \mathfrak{I}_3(\ddot{\eta}), \quad s = -\eta + \eta_c(t),$$

where

$$c_J = c_I \left(\frac{c_{JI}}{c_{JJ}}\right), \qquad c'_I = c_{II}, \qquad c'_J = \frac{c_{JI}^2}{c_{JJ}}, \qquad c'_\gamma = c_\gamma \left(\frac{c_{JI}}{c_{JJ}}\right)^2. \tag{1.91}$$

The mathematical Model (1.90) is a basic one for magnetic levitation systems investigation. Maxwell approximation in calculation of all fields corresponds to $L^\circ = M^\circ = 0$, $\alpha^\circ = 1$, $\kappa = 1$. In this case we obtain

$$m\ddot{\eta} - c_I(I + J) + c_s s = P(t), \quad c'_I(\dot{I} + \dot{J}) - c'_s \dot{s} + RI = \delta,$$

$$c'_I(\dot{J} + \dot{I}) - c'_s \dot{s} + c_\gamma \int_{-\infty}^{t} \frac{\dot{J}(\tau)\, d\tau}{\sqrt{t - \tau}} = 0, \tag{1.92}$$

$$\delta = \mathfrak{I}_1(s) + \mathfrak{I}_2(I) + \mathfrak{I}_3(\ddot{\eta}), \quad s = -\eta + \eta_c(t).$$

Coefficients c'_I, c_s, c_I, c'_s, satisfy the identity

$$c'_I c_s - c'_s c_I = 0. \tag{1.93}$$

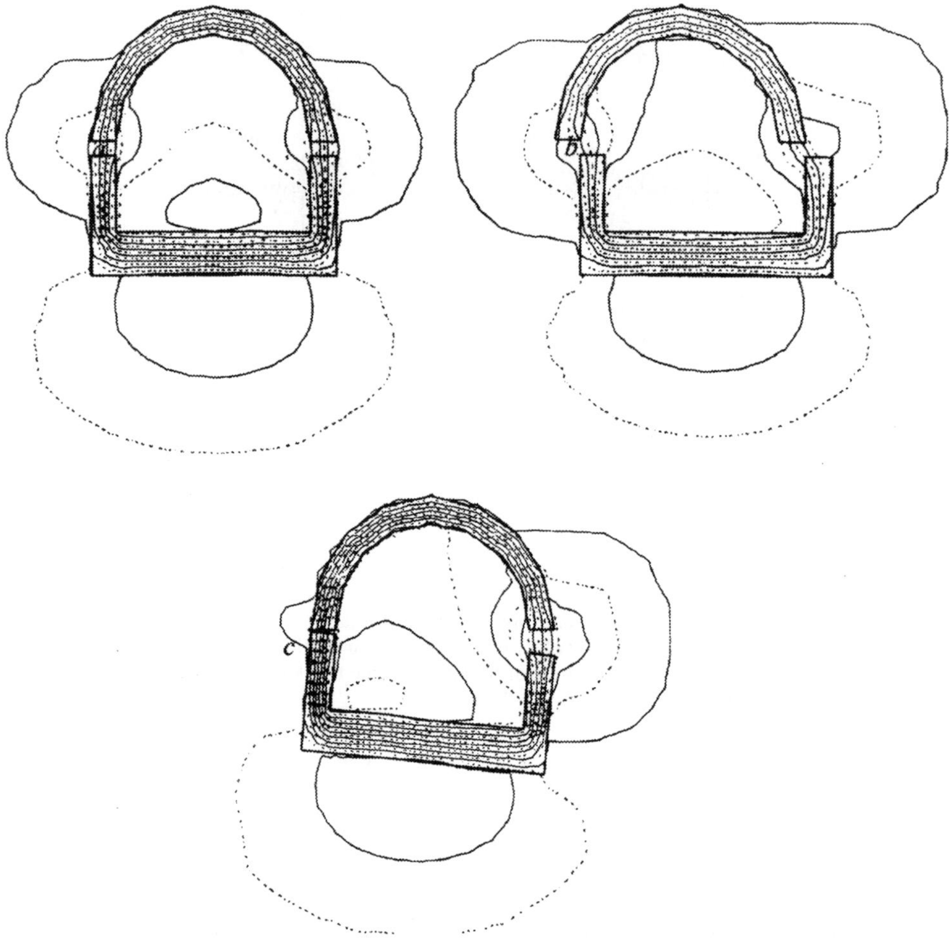

Fig. 1.8. Field lines of the magnetic field induced by external current in the electromagnet's coil; *a* - symmetric displacement of the electromagnet; *b* - additional side shift of the electromagnet; *c* - additional 'cocking' of the electromagnet

If we suppose that the ferromagnetic material of the core has not great, but, on the contrary, very low electrical conductivity (laminated iron), we can ignore eddy currents in (1.92), i.e. set $\sigma \to 0$ (see (1.66)). This very mathematical model is usually cited in the works by E.Gottzein *et al.* [25, 26, 27].

Let us consider a model core corresponding to a long electromagnet with U-shaped core and C-shaped ferromagnetic rail. Boundary-value Problems (1.12), (1.22a), and (1.22b) can be considered two-dimensional in this case. To solve them we can apply a numerical conformal mapping technique based on RT-algorithm by B.I.Rabinovich and Yu.V.Tyurin [83, 84] and finite element method (FEM). The central item of this approach is conformal mapping of a complex configuration two-dimensional irregular annulus domain onto a circular annulus.

In the boundary-value Problem (1.22a) such a domain results from the section of the closed core by a plane perpendicular to the magnet's longitudinal axis.

In the boundary-value Problem (1.22b) we deal with an infinite domain external to the cross sections of the electromagnet's core and the ferromagnetic rail. This domain is mapped onto an irregular annulus having piecewise smooth bound with the help of linear-fractional transform.

Figures 1.4. – 1.7. illustrate successive stages of conformal mapping with RT-algorithm and generation of grids conformally equivalent to the polar coordinates grid. Cases without air gap (Fig.1.4.), with air gap, with additional side shift, and, for comparison, with 'cocking' of the magnet (Figures 1.5. - 1.7.). These grids present a qualitative picture of field lines and equipotential lines of magnetic fields corresponding to boundary-value Problems (1.22a) and (1.22b) solutions.

Fig.1.8. presents a picture of field lines of the magnetic field induced by external current (boundary-value Problem (1.12)) for one of real electromagnets used for experimental investigation (described in detail in Chapter 4). The field lines are calculated with FEM (for the some configurations as in Figs. 1.5. – 1.7.)*.

Note, in conclusion, the following. Both, complete nonlinear Model (1.71), and its linear counterpart (1.86), do not provide for a number of important physical effects, particularly saturation effects and magnetic hysteresis, that can influence the final selection of structure and parameters of control laws. Therefore, one of the goals of experimental investigations (some of which will be discussed in Chapter 4) is checking of models of various degrees of completeness and their correlation with experimental results in the operating range of system parameters and frequency (in the state of steady harmonic oscillations), allowing for additional factors mentioned above.

---

* The presented results were obtained by Dr.Yu.V.Tyurin, R.I.Ashkinazi, and Dr.V.K.Chudnov (see, particularly, [97, 16]).

Chapter 2

Mathematical Models of Stabilized Objects with Compartments Containing Vortex Low-Viscous Liquid

In order to penetrate into the inner and further recesses of nature, it is necessary that both notions and axioms be derived from things by a more sure and guarded way; and that a method of intellectual operation be introduced altogether better and more certain.

Francis Bacon. *Novum Organum*, XVIII*

With complication of flying vehicles with LPRE and their control systems, still sharper becomes the problem of affording their stability in the frequency range of natural oscillations of liquid fuel components partially filling the tanks. This relates to RC as well as SC. Conventionally a way of complete or, at least, partial solution of the problem is to apply more and more developed and complicated dampers of liquid oscillations. This, in turn, presents still higher requirements to adequacy of mathematical models describing the system case liquid perturbed motion.

Up to recent times side by side with a number of simplifying assumptions and axioms, the following ones are used for solution of applied problems (see G.S.Narimanov [70] and B.I.Rabinovich [76]).

1. Conception of nonvortex motion of liquid throughout the space occupied by it, except for a thin wall layer in the vicinity of sharp edges of in-tank devices.

2. Assumption of the possibility to consider only the dissipative component of the adjacent vortex fields, ignoring kinetic energy of the liquid vortex motion, in mathematical models used for RC and SC dynamics investigation.

Further refinement of those vehicles' mathematical models, conforming to modern problems, requires revision of both mentioned hypotheses.

General equations of dynamics of a solid body with an arbitrary configuration cavity partially filled with ideal liquid were first published in the work by G.S.Narimanov [68]. They became the base for mathematical models of perturbed

* The Works of Francis Bacon, Baron of Verulam, Viscount St. Alban, and Lord High Chancellor of England. Collected and edited by James Spedding, M.A., of Trinity College, Cambridge; Robert Leslie Ellis, M.A., Late Fellow of Tr. Coll., Cambr.; Douglas Denon Heath, Barrister-at-Law; Late Fellow of Tr. Coll., Cambr. Vol. IV. Translations of philosophical works, Vol. I, London, 1870, p. 50

motion of stabilized vehicles with compartments containing liquid and played a great role in solution of a number of problems of those vehicles' dynamics. Literature devoted to these questions, that appeared since the publication of G.S.Narimanov's work [68] is practically boundless, it counts many hundred items. But as a rule, all those works do not consider liquid vortex motion kinetic energy (except for rotating objects in H.P.Greenspan's book [28]).

This chapter describes perturbed motion of vehicles with LPRE, with attitude stabilization, with tanks having damping devices in the form of radial or coaxial ring-shaped ribs which cause highly vortical motion of LV liquid. Adequate description of such vehicles' dynamics requires "that both notions and axioms be derived from things", i.e. to introduce a new axiomatics. The assumed phenomenological model of LV liquid vortex motion is a further development of models by B.I.Rabinovich and V.M.Rogovoy [81] and by B.I.Rabinovich [77] in the direction of considering higher harmonics of this motion. On this foundation B.Rabinovich in [78] constructed equations of the body – liquid system perturbed motion with consideration of vortex motion of liquid. The main mathematical models in this chapter are based on those equations.

First in Article 2.1 considered are conventional mathematical models of a solid body with a cavity partially filled with ideal liquid, perturbed motion of which in an 'absolute' (non-rotational) coordinate system is potential (see N.N.Moiseyev and V.V.Rumiantsev [66], G.S.Narimanov [70], and B.I.Rabinovich [76]).

2.1. Mathematical Models of a Solid Body with Cavity Partially Filled with Ideal Liquid.

2.1.1. Main assumptions. Coordinate systems. Generalized coordinates. Let us consider a solid, absolutely rigid body having a cavity with internal radial or coaxial ring-shaped ribs, which is partially filled with liquid. Suppose the liquid is ideal and incompressible. We shall introduce three coordinate systems $CXYZ$, $O^\circ x^\circ y^\circ z^\circ$, and $Oxyz$. The first of them is inertial, the second is a body related coordinate system, the body performing some programmed motion with nonzero acceleration. This motion, which we suppose unperturbed, is characterized by translational displacement with constant acceleration vector direction. In the coordinate system $O^\circ x^\circ y^\circ z^\circ$ some mass forces field is in effect, namely the field of gravitational forces and inertial forces of unperturbed motion. Let us assume this field being potential, in the general case it can be unsteady. Denote its gradient by **j**, and its potential by U. We shall choose the coordinate system $O^\circ x^\circ y^\circ z^\circ$ so, that $O^\circ x^\circ$ axis is anti-parallel to vector j; axes $O^\circ y^\circ$ and $O^\circ z^\circ$ have arbitrary orientation with respect to the body (see Fig. 2.1.).

Let us denote the domain occupied by the liquid, by Q, the wetted surface of the cavity (including ribs' surface) by S, and free surface of the liquid in unperturbed motion by Σ. Then, with our assumptions, Σ is a plane perpendicular to vector j, the following expressions being true

$$\mathbf{j} = \mathbf{j}(t) = \nabla U, \quad U = -jx^{\circ}, \tag{2.1}$$

where $j(t) \geq j_{\min} > 0$. Choose the coordinate origin O position so that $O^{\circ}x^{\circ}$ axis runs over the mass centre of a flat plate coinciding with Σ.

The second coordinate system $Oxyz$ is a body related one, the body performing perturbed motion. The latter is considered the motion, with respect to the coordinate system $O^{\circ}x^{\circ}y^{\circ}z^{\circ}$, caused by initial deviation of the body's generalized coordinates and velocities (those will be discussed below) and exciting forces and moments affecting the body in the coordinate system $O^{\circ}x^{\circ}y^{\circ}z^{\circ}$.

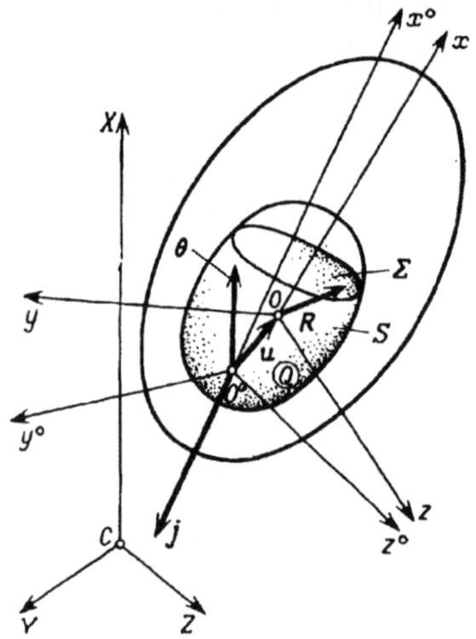

Fig. 2.1. A solid body with a cavity partially filled with liquid. Coordinate systems and main denominations.

Thus defined, the coordinate systems $Oxyz$ and $O^{\circ}x^{\circ}y^{\circ}z^{\circ}$ coincide in unperturbed motion.

We characterize perturbed motion by vectors of small displacement $\mathbf{u}(t) = (u_1, u_2, u_3)$ of the point O relative to O°; small rotation $\vec{\theta}(t) = (\theta_1, \theta_2, \theta_3)$ of coordinate system $Oxyz$ relative to $O^{\circ}x^{\circ}y^{\circ}z^{\circ}$, and infinite-dimensional vectors $\mathbf{s}(t) = (s_1, s_2, \ldots)$ or $\mathbf{s}(t)' = (s'_1, s'_2, \ldots)$, the components $s_n(t)$ and $s_n(t)'$ of which $(n = 1, 2, \ldots)$ are displacements of liquid free surface along $O^{\circ}x^{\circ}$ axis or Ox axis, respectively, at an arbitrary selected point of the unperturbed liquid free surface perimeter for its n-th natural oscillations mode. Vectors $\mathbf{s}$ and $\mathbf{s}'$ also differ in that the first one ($\mathbf{s}$) is related to the 'floating cover' surface Σ which remains perpendicular to $O^{\circ}x^{\circ}$ axis in perturbed motion, while the second one ($\mathbf{s}'$) is related to the 'fixed cover' surface Σ', which is perpendicular to Ox axis in perturbed motion.

Components of vectors $\mathbf{u}$, $\vec{\theta}$, $\mathbf{s}$ (or $\mathbf{s}'$) are assumed for generalized coordinates of the body – liquid system, and $\dot{\mathbf{u}}$, $\dot{\vec{\theta}}$, $\dot{\mathbf{s}}$ (or $\dot{\mathbf{s}}'$) are assumed for generalized velocities.

Suppose that all the generalized coordinates and generalized velocities are small quantities in the sense, that their squares and higher degrees may be neglected in comparison with linear terms. All the accelerations will be considered small in the sense of inequalities

$$|\ddot{\mathbf{u}}| \ll j_{\min}, \quad l\,|\ddot{\vec{\theta}}| \ll j_{\min}, \quad |\ddot{\mathbf{s}}| \ll j_{\min}, \quad n = 1, 2, \ldots, \tag{2.2}$$

where l is the characteristic size of the cavity.

The assumptions of ideal liquid and potential field of mass forces of unperturbed motion result, by virtue of Lagrange's theorem (see N.Ye.Kochin, I.A.Kibel, and

N.V.Roze [40]), in that liquid motion, being nonvortex for $t = 0$, remains nonvortex for any $t > 0$, and the flow around the ribs, including their sharp edges, can be considered nonstalling.

In the following section the assumption of ideal liquid is substituted by a weaker one, namely, the requirement for the liquid, to be low-viscous (Reynolds number is great enough); then for the presence of linear ribs with sharp edges the real picture appears closer to LV liquid vortex motion, resulting from vigorous eddying at the rib edges flowed around by real (even low-viscous) liquid. This model is discussed in the following sections.

2.1.2. Displacement potential and boundary-value problems for its components. In order to describe the kinematic picture of liquid motion let us introduce, following D.E.Okhotsimsky [72], displacement potential χ , for which velocity potential in the time derivative. For the potential of small displacements of fluid particles the same relations are valid as for velocity potential:

$$\mathbf{w} = \nabla \chi, \quad \Delta \chi = 0, \tag{2.3}$$

where $\mathbf{w}$ is the vector of fluid particles displacements, ∇ and Δ are Hamilton and Laplace operators, respectively

$$\nabla = \mathbf{i}_1 \frac{\partial}{\partial x} + \mathbf{i}_2 \frac{\partial}{\partial y} + \mathbf{i}_3 \frac{\partial}{\partial z}, \qquad \Delta = \frac{\partial^2}{\partial x^2} + \frac{\partial^2}{\partial y^2} + \frac{\partial^2}{\partial z^2}.$$

Liquid pressure is related to potentials U and χ by Lagrange – Cauchy integral (see N.Ye.Kochin, I.A.Kibel, and N.V.Roze [40]):

$$p = \rho \left(U - \frac{\partial^2 \chi}{\partial t^2} \right) + f(t), \tag{2.4}$$

where $f(t)$ is an arbitrary function of time.

Let us formulate boundary conditions for the function $\chi(x, y, z, t)$. At the wetted surface of the cavity we have the condition of solid walls impenetrability, and at the liquid free surface we have the kinematic condition of coincidence of fluid particles' displacements and those of the media boundary and the dynamic conditions of constant pressure. In virtue of assumptions of infinitesimal displacements of the body and fluid particles in their perturbed motion, the boundary conditions can be referred to the liquid volume Q surface in unperturbed motion, i.e. the surfaces S and Σ (the latter is identical to Σ' in unperturbed motion).

In consideration of what is mentioned above and applying the familiar liquid free surface dynamic condition (see N.Ye.Kochin, I.A.Kibel, and N.V.Roze [40]) equivalent to constant pressure condition, we can bring the boundary conditions for displacement potential χ to the following form:

$$\left.\frac{\partial \chi}{\partial \nu}\right|_S = \left(\mathbf{u}, \vec{\nu}\right) + \left(\vec{\theta}, \mathbf{R} \times \vec{\nu}\right), \quad \left.\frac{\partial \chi}{\partial \nu}\right|_\Sigma = \left(\mathbf{u}, \vec{\nu}\right) + \xi(y, z, t),$$

$$\left[\frac{\partial^2 \chi}{\partial t^2} + j\xi\right]_{\Sigma} = 0, \tag{2.5}$$

where $\xi\,(y, z, t)$ is the liquid free surface disturbance in $O^\circ x^\circ y^\circ z^\circ$ coordinate system; y, z are point coordinates at the surface Σ; $\vec{\nu}$ is the unit vector normal to the surface $S + \Sigma$; homogeneity of the last condition of (2.5) is achieved by special choice of functions $f\,(t)$ in (2.4).

Present the displacement potential in two modifications in the form of expansions

$$\chi = \left(\mathbf{u}, \vec{\Phi}\right) + \left(\vec{\theta}, \vec{\Psi}\right) + \sum_{n=1}^{\infty} \varphi_n s_n\,, \tag{2.6a}$$

$$\chi = \left(\mathbf{u}, \vec{\Phi}\right) + \left(\vec{\theta}, \vec{\Psi}'\right) + \sum_{n=1}^{\infty} \varphi_n s_n'\,, \tag{2.6b}$$

where

$$\vec{\Phi} = \left\{\Phi_1,\ \Phi_2,\ \Phi_3\right\}, \quad \vec{\Psi} = \left\{\Psi_1,\ \Psi_2,\ \Psi_3\right\}, \quad \vec{\Psi}' = \left\{\Psi_1',\ \Psi_2',\ \Psi_3'\right\}$$

are harmonic vector functions whose components are displacement potentials for unity values of generalized coordinates specifying linear and angular body displacements; φ_n $(n = 1, 2, \ldots)$ are eigenfunctions of the boundary-value problem of liquid free oscillations in a stationary cavity in the field of mass forces with gradient j (see N.Ye.Kochin, I.A.Kibel, and N.V.Roze [40]). Functions $\vec{\Phi}$, $\vec{\Psi}$, $\vec{\Psi}'$, φ_n $(n = 1, 2, \ldots)$ are solutions of the following boundary-value problems formulated in G.S.Narimanov [70] and B.I.Rabinovich [76]:

$$\Delta \vec{\Phi} = 0, \qquad \left.\frac{\partial \vec{\Phi}}{\partial \nu}\right|_{S+\Sigma} = \vec{\nu}. \tag{2.7}$$

$$\Delta \vec{\Psi} = 0, \qquad \left.\frac{\partial \vec{\Psi}}{\partial \nu}\right|_{S} = \mathbf{R} \times \vec{\nu}, \qquad \left.\frac{\partial \vec{\Psi}}{\partial \nu}\right|_{\Sigma} = 0. \tag{2.8a}$$

$$\Delta \vec{\Psi}' = 0, \qquad \left.\frac{\partial \vec{\Psi}'}{\partial \nu}\right|_{S+\Sigma} = \mathbf{R} \times \vec{\nu}. \tag{2.8b}$$

$$\Delta \varphi = 0, \qquad \left.\frac{\partial \varphi}{\partial \nu}\right|_{S} = 0, \qquad \left.\frac{\partial \varphi}{\partial \nu}\right|_{\Sigma} = \kappa\varphi, \tag{2.9}$$

where

$$\kappa = \kappa_1,\ \kappa_2,\ \ldots, \qquad \varphi = \varphi_1,\ \varphi_2,\ \ldots,$$

$$\left.\frac{\partial \varphi_n}{\partial \nu}\right|_{\Sigma} = \left.\frac{\partial \varphi_n}{\partial x}\right|_{\Sigma} = \psi_n\,, \quad \psi_n\,(y_s\,, z_s) = 1\,, \quad n = 1,\ 2,\ \ldots, \tag{2.10}$$

R is the radius vector of the surface $S+\Sigma$ point with the origin in the point O; and κ_n, φ_n $(n = 1, 2, ...)$ are eigenvalues and eigenfunctions of the homogeneous boundary-value problem (2.9). The last equality in (2.10) is the normalizing condition for functions ψ_n, and hence, for functions φ_n (y_s, z_s) are the coordinates of an arbitrary point of domain Σ perimeter. The normalizing condition supplies the generalized coordinates s_n and s'_n with the sense mentioned above.

According to N.Ye.Kochin, I.A.Kibel, and N.V.Roze [40], functions ψ_n are orthogonal at domain Σ and are orthogonal to constant, i.e. they satisfy the following conditions.

$$\int_{\Sigma} \psi_n \, dS = 0\,, \quad \int_{\Sigma} \psi_n \psi_m \, dS = \begin{cases} 0\,, & \text{for } m \neq n, \\ N_n^2\,, & \text{for } m = n. \end{cases} \tag{2.11}$$

The two modifications of expressions for χ correspond, as it is clear from (2.8a) and (2.8b), to the liquid wave motions coordinates counted off the 'floating cover' Σ (s_n), following [76], or off the 'fixed cover' $\Sigma'(s'_n)$ following G.S.Narimanov [70]. Both schemes are equally applicable, although they result in somewhat different forms of mathematical models of the body – liquid system perturbed motion.

Consider boundary-value Problems (2.7), (2.8a), and (2.8b). The solution of the first of them is straightforward. It has the form

$$\vec{\Phi} = \mathbf{R} + \mathbf{C}, \tag{2.12}$$

where **C** is an arbitrary constant vector, which can be assumed zero without loss of generality. Boundary-value Problems (2.8a) and (2.8b) cannot be solved in the general form (i.e. for any cavity); functions $\vec{\Psi}$ and $\vec{\Psi}'$ have some properties, however, that can be revealed without explicitly solving the problems.

Consider two locations O_1 and O_2 of the point O at the axis Ox, running through the area Σ mass centre. Write $\mathbf{R}_1$, $\mathbf{R}_2$ for radius vectors of the volume Q arbitrary point with the origin in O_1 and O_2 respectively; and $\mathbf{R}_{21}$ for the radius vector of the point O_2 with respect to O_1. It is easy to verify, that solutions $\vec{\Psi}_1$ and $\vec{\Psi}_2$ of the boundary-value Problem (2.8a) and $\vec{\Psi}'_1$ and $\vec{\Psi}'_2$ of the Problem (2.8b) corresponding to the points O_1 and O_2 are related by expressions

$$\vec{\Psi}_2 = \vec{\Psi}_1 + \mathbf{R}_2 \times \mathbf{R}_{21}, \quad \vec{\Psi}'_2 = \vec{\Psi}'_1 + \mathbf{R}_2 \times \mathbf{R}_{21}\,, \tag{2.13}$$

where $\vec{\Psi}_1$, $\vec{\Psi}_2$ and $\vec{\Psi}'_1$, $\vec{\Psi}'_2$ are the solutions of boundary-value Problems (2.8a) and (2.8b), when $\mathbf{R}_1$ and $\mathbf{R}_2$ are substituted for **R** in their right sides. It follows from (2.13), that difference $\vec{\Psi}' - \vec{\Psi}$ denoted by $\vec{\Phi}'$ does not depend on the coordinate origin:

$$\vec{\Phi}' = \vec{\Psi}' - \vec{\Psi} = \vec{\Psi}'_2 - \vec{\Psi}_2 = \vec{\Psi}'_1 - \vec{\Psi}_1\,. \tag{2.14}$$

Consider function $\vec{\Phi}'$ in detail. By definition, it is the solution of the following boundary-value problem (see (2.8a) and (2.8b)):

$$\Delta \vec{\Phi}' = 0, \quad \left. \frac{\partial \vec{\Phi}'}{\partial \nu} \right|_S = 0, \quad \left. \frac{\partial \vec{\Phi}'}{\partial \nu} \right|_\Sigma = \mathbf{R} \times \vec{\nu}. \tag{2.15}$$

Let us seek $\vec{\Phi}'$ in the form of an expansion in terms of functions φ_n with vector coefficients $\mathbf{a}_n$, unknown so far.

$$\vec{\Phi}' = \sum_{n=1}^{\infty} \mathbf{a}_n \varphi_n . \tag{2.16}$$

Assuming the possibility of termwise differentiation of this expansion, we obtain, that the infinite series in the right side of (2.16) is the solution of the boundary-value Problem (2.15), supposed the following condition is satisfied :

$$\left. \sum_{n=1}^{\infty} \mathbf{a}_n \psi_n \right|_\Sigma = \mathbf{R} \times \vec{\nu} = \mathbf{i}_2 z - \mathbf{i}_3 y , \tag{2.17}$$

where $\mathbf{i}_j$ ($j = 1, 2, 3$) are the unit vectors of the coordinate system $Oxyz$. Expression (2.17) yields the following equalities :

$$\mathbf{a}_n = \mathbf{i}_2 c_n - \mathbf{i}_3 d_n , \quad \sum_{n=1}^{\infty} c_n \psi_n = z , \quad \sum_{n=1}^{\infty} d_n \psi_n = y . \tag{2.18}$$

The problem, hence, consists in the expansion of functions y and z into generalized Fourier series in terms of functions ψ_n within domain Σ, that is possible due to (2.11), if functions y and z are orthogonal to constant within Σ domain, i.e.

$$\int_\Sigma y \, dS = \int_\Sigma z \, dS = 0. \tag{2.19}$$

These conditions are met because of special choice of the coordinate system ($O\overset{\circ}{x}{}^{\circ}$ axis runs through the mass centre of Σ). Using conditions of functions ψ_n orthogonality of (2.11) we obtain

$$c_n = \frac{C_n}{N_n^2}, \quad d_n = \frac{D_n}{N_n^2}, \quad C_n = \int_\Sigma z \, \psi_n \, dS , \quad D_n = \int_\Sigma y \, \psi_n \, dS ,$$

$$N_n^2 = \int_\Sigma \psi^2_n \, dS , \quad \mathbf{a}_n = \frac{\mathbf{A}_n}{N_n^2}, \quad \mathbf{A}_n = \int_\Sigma (\mathbf{R} \times \vec{\nu}) \, \psi_n \, dS . \tag{2.20}$$

Parseval's equalities for the generalized Fourier Series (2.18) allow us to obtain two more important expressions for the equatorial moment of inertia of the domain Σ (with respect to central axes) J_y and J_z :

$$J_y = \int_\Sigma z^2 \, dS = \sum_{n=1}^{\infty} \frac{C_n^2}{N_n^2}, \quad J_z = \int_\Sigma y^2 \, dS = \sum_{n=1}^{\infty} \frac{D_n^2}{N_n^2} . \tag{2.21}$$

Projecting the vector Equality (2.16) to the axes $O\overset{\circ}{x}{}^{\circ}$, $O\overset{\circ}{y}{}^{\circ}$, $O\overset{\circ}{z}{}^{\circ}$ and using (2.14) we obtain the relation of the components of functions $\vec{\Psi}'$ and $\vec{\Psi}$:

$$\Psi'_1 \equiv \Psi_1 , \quad \Psi'_2 = \Psi_2 + \sum_{n=1}^{\infty} c_n \varphi_n , \quad \Psi'_3 = \Psi_3 - \sum_{n=1}^{\infty} d_n \varphi_n . \tag{2.22}$$

Thus only two boundary-value problems are independent: (2.8a) and (2.9), or (2.8b) and (2.9); so it is enough to solve any couple, for instance, (2.8a) for $\vec{\Psi}$ and (2.9) for φ_n and we obtain all the sought functions. Numerous literature is devoted to solution of these problems and multiple results are obtained for domains of various configurations (see, e.g. [98], I.B.Bogoriad *et al.* [10], G.N.Mikishev and B.I.Rabinovich [63, 64], S.F.Feshchenko *et al.* [21], B.I.Rabinovich [76]).

2.1.3. Mathematical models of the system 'body – liquid'. These mathematical models are the equations of those systems perturbed motion for two ways of introducing generalized coordinates specifying liquid wave motion (s_n and s'_n). These equations can be obtained either as Lagrange's equations of the 2nd kind, or using theorems of the body – liquid system momentum and moment of momentum variation and conditions of liquid free surface pressure being constant. Let us consider the last method in detail.

For its implementation it is necessary to know the liquid perturbed motion absolute velocity field $\mathbf{v}_a (x, y, z, t)$ and relation of its particles' displacements and accelerations at the free surface for constant pressure ($\delta p = 0$). The appropriate expressions can be obtained from (2.3) – (2.9):

$$\mathbf{v}_a = \nabla (\dot{\mathbf{u}}, \vec{\Phi}) + \nabla (\dot{\vec{\theta}}, \vec{\Psi}) + \sum_{n=1}^{\infty} \nabla \varphi_n \dot{s}_n ,$$

$$\delta p = -\rho \left[(\ddot{\mathbf{u}}, \vec{\Phi}) + (\ddot{\vec{\theta}}, \vec{\Psi}) + \sum_{n-1}^{\infty} (\ddot{s}_n + \omega_n^2 s_n) \varphi_n \right]_\Sigma = 0. \tag{2.23a}$$

$$\mathbf{v}_a (x, y, z, t) = \nabla (\dot{\mathbf{u}}, \vec{\Phi}) + \nabla (\dot{\vec{\theta}}, \vec{\Psi}') + \sum_{n=1}^{\infty} \nabla \varphi_n \dot{s}_n = 0, \tag{2.23b}$$

$$\delta p = -\rho \left[(\ddot{\mathbf{u}}, \vec{\Phi}) + (\ddot{\vec{\theta}}, \vec{\Psi}') + j (\vec{\theta}, \mathbf{R} \times \vec{\nu}) + \sum_{n-1}^{\infty} (\ddot{s}'_n + \omega_n^2 s'_n) \varphi_n \right]_\Sigma = 0,$$

where $\omega_n^2 = j \kappa_n$ is the liquid n-th natural oscillation frequency square, ρ is the liquid mass density. Applying the theorem of variation of the body – liquid system momentum and moment of momentum with respect to point O and integrating over Σ domain the expression of δp multiplied by ψ_m function with an arbitrary subscript m, we obtain the following general set of equations of perturbed motion of the body containing liquid:

$$\frac{d}{dt} \left(\mathbf{K}^\circ + \rho \int_Q \mathbf{v}_a \, dQ \right) = \mathbf{P}, \quad \frac{d}{dt} \left[\mathbf{N}_O^\circ + \rho \int_Q (\mathbf{R} \times \mathbf{v}_a) \, dQ \right] = \mathbf{M}_O + \delta \mathbf{M}_O,$$

$$\int_\Sigma \delta p \psi_m \, dS = 0, \quad m = 1, 2, \ldots, \tag{2.24}$$

where $\mathbf{K}^\circ$ and $\mathbf{N}_O^\circ$ are momentum and moment of momentum of the solid with respect to point O ; $\mathbf{P}$ and $\mathbf{M}_O$ are the resultant force vector and the resultant moment with respect to the same point of the system of forces applied to the body; $\delta\mathbf{M}_O$ is the moment with respect to point O of the system of unperturbed motion mass forces associated with the body rotation and liquid free surface configuration change in perturbed motion.

Substituting (2.23a) and (2.23b) for appropriate expressions in (2.24) we obtain after some transformations the following two modifications of the body – liquid system mathematical models :

1. The 'floating cover' scheme (following [76]):

$$(m^\circ + m)\,\ddot{\mathbf{u}} - (\mathbf{L}^\circ + \mathbf{L}) \times \ddot{\vec{\theta}} + \left(\mathsf{L}, \ddot{\vec{\theta}}\right) + \sum_{n=1}^{\infty} \vec{\lambda}_n\, \ddot{s}_n = \mathbf{P}, \tag{2.25a}$$

$$\left(\mathsf{J}^\circ + \mathsf{J}, \ddot{\vec{\theta}}\right) + (\mathbf{L}^\circ + \mathbf{L}) \times \ddot{\mathbf{u}} + \left(\bar{\mathsf{L}}, \ddot{\mathbf{u}}\right) - (\mathbf{L}^\circ + \mathbf{L}) \times \mathbf{j} - \left(\bar{\mathsf{L}}, \mathbf{j}\right) + \sum_{n=1}^{\infty} \vec{\lambda}_{On}\, \ddot{s}_n = \mathbf{M}_O,$$

$$\mu_n\,(\ddot{s}_n + \omega_n^2 s_n) + (\vec{\lambda}_n, \ddot{\mathbf{u}}) + (\vec{\lambda}_{On}, \ddot{\vec{\theta}}) = 0, \qquad n = 1, 2, \ldots$$

2. The 'fixed cover' scheme (following G.S.Narimanov [70]):

$$(m^\circ + m)\,\ddot{\mathbf{u}} - (\mathbf{L}^\circ + \mathbf{L}) \times \ddot{\vec{\theta}} + \sum_{n=1}^{\infty} \vec{\lambda}_n\, \ddot{s}'_n = \mathbf{P}, \tag{2.25b}$$

$$\left(\mathsf{J}^\circ + \mathsf{J}', \ddot{\vec{\theta}}\right) + (\mathbf{L}^\circ + \mathbf{L}) \times \ddot{\mathbf{u}} - (\mathbf{L}^\circ + \mathbf{L}) \times \mathbf{j} + \sum_{n=1}^{\infty} \vec{\lambda}'_{On}\, \ddot{s}'_n - \sum_{n=1}^{\infty} (\vec{\lambda}_n \times \mathbf{j})\, s'_n = \mathbf{M}_O,$$

$$\mu_n\,(\ddot{s}'_n + \omega_n^2 s'_n) + (\vec{\lambda}_n, \ddot{\mathbf{u}}) + (\vec{\lambda}'_{On}, \ddot{\vec{\theta}}) - (\vec{\lambda}_n, \mathbf{j}) = 0, \qquad n = 1, 2, \ldots,$$

where $\left(\mathsf{L}, \ddot{\vec{\theta}}\right)$ and other similar symbols denote a right product of tensor L by vector $\ddot{\vec{\theta}}$; a bar over a letter denotes a conjugated tensor, in the first equation all the vectors should be projected to axes $O\mathring{x}^\circ, O\mathring{y}^\circ, O\mathring{z}^\circ$, and in the second one to axes Ox , Oy. Coefficients of Equations (2.25a) and (2.25b) are expressed by the following formulae (see [76]) :

Vectors and scalars:

$$\vec{\lambda}_n = \rho \int_S \varphi_n \vec{\nu}\, dS = \rho \int_\Sigma \mathbf{R}\, \frac{\partial \varphi_n}{\partial \nu}\, dS\,,$$

$$\vec{\lambda}_{On} = \rho \int_S (\mathbf{R} \times \vec{\nu})\, \varphi_n\, dS = \rho \int_\Sigma \Psi\, \frac{\partial \varphi_n}{\partial \nu}\, dS\,,$$

$$\vec{\lambda}'_{On} = \rho \oint\limits_{S+\Sigma} (\mathbf{R} \times \vec{\nu})\, \varphi_n\, dS = \rho \int\limits_{\Sigma} \Psi' \frac{\partial \varphi_n}{\partial \nu}\, dS =$$

$$= \vec{\lambda}_{On} + \frac{\rho}{\kappa_n} \int\limits_{\Sigma} (\mathbf{R} \times \vec{\nu})\, \psi_n\, dS = \vec{\lambda}_{On} + \frac{\vec{\lambda}_n \times \mathbf{i}_1}{\kappa_n}, \tag{2.26}$$

$$\mathbf{L} = \rho \int\limits_{Q} \mathbf{R}\, dQ, \qquad \mathbf{L}^{\circ} + \mathbf{L} = (m^{\circ} + m)\, \mathbf{R}_G,$$

$$\mu_n = \rho \int\limits_{\Sigma} \varphi_n \frac{\partial \varphi_n}{\partial \nu}\, dS = \rho\, \frac{N_n^2}{\kappa_n},$$

$$\omega_n^2 = j\, \kappa_n\,, \qquad \left.\frac{\partial \varphi_n}{\partial \nu}\right|_{\Sigma} = \psi_n\,, \qquad \varphi_n\, |_{\Sigma} = \frac{\psi_n}{\kappa_n}.$$

Tensors:

$$\mathsf{L} = \begin{Bmatrix} 0 & 0 & 0 \\ 0 & -\rho J_{yz} & \rho J_z \\ 0 & -\rho J_y & \rho J_{yz} \end{Bmatrix}, \quad \bar{\mathsf{L}} = \begin{Bmatrix} 0 & 0 & 0 \\ 0 & -\rho J_{yz} & -\rho J_y \\ 0 & \rho J_z & \rho J_{yz} \end{Bmatrix},$$

$$\mathsf{J} = \begin{Bmatrix} J_{11} & J_{12} & J_{13} \\ J_{21} & J_{22} & J_{23} \\ J_{31} & J_{32} & J_{33} \end{Bmatrix}, \qquad \mathsf{J}' = \begin{Bmatrix} J'_{11} & J'_{12} & J'_{13} \\ J'_{21} & J'_{22} & J'_{23} \\ J'_{31} & J'_{32} & J'_{33} \end{Bmatrix},$$

$$J_{ji} = J_{ij} = \rho \int\limits_{S} \frac{\partial \Psi_j}{\partial \nu} \Psi_i\, dS\,, \quad J'_{ji} = J'_{ij} = \rho \oint\limits_{S+\Sigma} \frac{\partial \Psi'_j}{\partial \nu} \Psi'_i\, dS\,, \tag{2.27}$$

$$\rho J_y = \rho \int\limits_{\Sigma} z^2\, dS = \sum_{n=1}^{\infty} \frac{\lambda_{n3}^2}{\kappa_n \mu_n}, \qquad \rho J_z = \rho \int\limits_{\Sigma} y^2\, dS = \sum_{n=1}^{\infty} \frac{\lambda_{n2}^2}{\kappa_n \mu_n},$$

$$\rho J_{yz} = \rho \int\limits_{\Sigma} yz\, dS.$$

Here m°, $\mathbf{L}^{\circ}$, J° are the solid's mass, vector of static moments, and tensor of inertia; $\mathbf{R}$ is the radius vector of the mass centre of the body with liquid solidified in unperturbed state, the origin of the radius vector being in point O; J' is N.Joukowski's tensor of associated moments of inertia of liquid for the cavity with 'fixed cover'; J is the tensor of inertia, equivalent to J', for the cavity with 'floating cover'; J_y, J_z, $J_{yz} = J_{zy}$ are equatorial and centrifugal moments of domain Σ with respect to its central axes (see (2.21)). Equations similar to (2.25b) were obtained in a number of works (see, e.g. N.N.Moiseyev and V.V.Rumiantsev, H.A.Abramson (Ed.) [66,1]).

Mathematical Models (2.25a) and (2.25b) are somewhat different in form, but they are equivalent in essence. Each of them can be transformed to another with non-singular linear transformation.

Let us represent function $\xi\ (y, z, t\ )$ characterizing perturbed free surface of liquid in $O^{\circ}x^{\circ}y^{\circ}z^{\circ}$ coordinate system in two equivalent forms according to (2.5) − (2.9) :

$$\xi = \sum_{n=1}^{\infty} \psi_n s_n. \tag{2.28a}$$

$$\xi = \left(\vec{\theta}, \mathbf{R} \times \vec{\nu}\right)\Big|_{\Sigma} + \sum_{n=1}^{\infty} \psi_n s'_n. \tag{2.28b}$$

Using (2.28), (2.17), and (2.20) we obtain

$$\sum_{n=1}^{\infty} \psi_n s_n = \left(\vec{\theta}, \mathbf{R} \times \vec{\nu}\right) + \sum_{n=1}^{\infty} \psi_n s'_n = \sum_{n=1}^{\infty} \left[\left(\vec{\theta}, \mathbf{a}_n\right) + s'_n\right] \psi_n,$$

which, after inclusion of (2.26), yields

$$s_n = s'_n + \left(\vec{\theta}, \mathbf{a}_n\right) = s'_n + \frac{\rho}{\mu_n \kappa_n}\left[\vec{\theta}, \int_{\Sigma} (\mathbf{R} \times \vec{\nu}\,) \varphi_n \, dS\right] =$$

$$= s'_n - \frac{1}{\mu_n \omega_n^2}\left(\vec{\lambda}_n \times \mathbf{j}, \vec{\theta}\,\right) = s'_n - \frac{1}{\mu_n \omega_n^2}\left(\vec{\lambda}_n, \mathbf{j}\right). \tag{2.29}$$

Let us, further, consider vectors $\left(\mathsf{L}, \ddot{\vec{\theta}}\,\right)$ and $\left(\bar{\mathsf{L}}, \ddot{\mathbf{u}} - \mathbf{j}\right)$ which, after some simple transformations, can be represented in the form of the following expansions :

$$\left(\mathsf{L}, \ddot{\vec{\theta}}\,\right) = -\rho \sum_{j=1}^{3} \mathbf{i}_j \int_{\Sigma} \left(\mathbf{R} \times \vec{\nu}, \ddot{\vec{\theta}}\,\right) \Phi_j \, dS = \sum_{n=1}^{\infty} \frac{\vec{\lambda}_n}{\omega_n^2 \mu_n}\left(\vec{\lambda}_n \times \mathbf{j}, \ddot{\vec{\theta}}\,\right), \tag{2.30}$$

$$\left(\bar{\mathsf{L}}, \ddot{\mathbf{u}} - \mathbf{j}\right) = -\rho \sum_{j=1}^{3} (\mathbf{i}_j, \ddot{\mathbf{u}} - \mathbf{j}) \int_{\Sigma} \left(\mathbf{R} \times \vec{\nu}\right) \Phi_j \, dS = \sum_{n=1}^{\infty} \frac{(\vec{\lambda}_n, \ddot{\mathbf{u}} - \mathbf{j})}{\omega_n^2 \mu_n}\left(\vec{\lambda}_n \times \mathbf{j}\right).$$

Derivation of Expressions (2.29) and (2.30) is the central moment when transforming Equations (2.25a) to (2.25b). Further operations are reduced to simple substitutions that we do not show.

Consider some possible ways of reducing Equations (2.25a) and (2.25b). Let the body and the cavity have a common plane Oxy of geometric and mass symmetry, which is the main central plane. In this case $J_{yz} = J_{zy} = 0$ and, besides that, $z_G = 0$. Let us introduce, besides the system's mass centre G, points F_1 and F_2 lying in Gx axis, which coordinates are defined as follows.

$$x_{F_1} = x_G + \frac{\rho J_z}{\overset{\circ}{m} + m}, \qquad y_{F_1} = y_G, \qquad z_{F_1} = z_G = 0,$$

$$x_{F_2} = x_G + \frac{\rho J_y}{\overset{\circ}{m} + m}, \qquad y_{F_2} = y_G, \qquad z_{F_2} = z_G = 0. \tag{2.31}$$

These points are metacentres of the body – liquid system (see [76]). They play the same role in the 'floating cover' Model (2.25a), as the mass centre G of the body with solidified liquid plays in the 'fixed cover' Model (2.25b). Both metacentres F_1 and F_2 lie in Gx axis, upper than mass centre G (in the direction of Gx axis). In the particular case of $J_y = J_z$ (e.g. for a cavity of revolution) (2.31) yields

$$x_{F1} = x_{F2} = x_F = x_G + \frac{\rho J_y}{m^\circ + m}. \tag{2.32}$$

Let us introduce two coordinate systems $F_1x_1y_1z_1$ and $F_2x_2y_2z_2$ for Model (2.25a), axes F_1x_1 and F_2x_2 coinciding and F_1y_1 parallel to F_2y_2, and F_1z_1 parallel to F_2z_2, and corresponding coordinate systems $F_1^\circ x_1^\circ y_1^\circ z_1^\circ$ and $F_2^\circ x_2^\circ y_2^\circ z_2^\circ$, and for Model (2.25b) coordinate system $Gxyz$ with origin in mass centre G and $G^\circ x^\circ y^\circ z^\circ$ respectively (the assumption of symmetry plane Oxy existence is irrelevant here). Then Models (2.25a) and (2.25b) are reduced and transformed to the following (see G.S.Narimanov, B.I.Rabinovich [70,76]):

$$(m^\circ + m)\,\ddot{\mathbf{u}} + \sum_{n=1}^{\infty} \vec{\lambda}_n\, \ddot{s}_n = \mathbf{P},$$

$$\left(\mathsf{J}^\circ + \mathsf{J}, \ddot{\vec{\theta}}\right) + \sum_{n=1}^{\infty} \vec{\lambda}_{Fjn}\, \ddot{s}_n = \mathbf{M}_{Fj},$$

$$\mu_n\left(\ddot{s}_n + \omega_n^2 s_n\right) + \left(\vec{\lambda}_n, \ddot{\mathbf{u}}\right) + \left(\vec{\lambda}_{Fjn}, \ddot{\vec{\theta}}\right) = 0, \qquad n = 1, 2, \ldots \tag{2.33a}$$

$$(m^\circ + m)\,\ddot{\mathbf{u}} + \sum_{n=1}^{\infty} \vec{\lambda}_n\, \ddot{s}'_n = \mathbf{P},$$

$$\left(\mathsf{J}^{\circ\prime} + \mathsf{J}', \ddot{\vec{\theta}}\right) + \sum_{n=1}^{\infty} \left[\vec{\lambda}'_{Gn}\, \ddot{s}'_n - \left(\vec{\lambda}_n \times \mathbf{j}\right) s'_n\right] = \mathbf{M}_G,$$

$$\mu_n\left(\ddot{s}'_n + \omega_n^2 s'_n\right) + \left(\vec{\lambda}_n, \ddot{\mathbf{u}} - \mathbf{j}\right) + \left(\vec{\lambda}'_{Gn}, \ddot{\vec{\theta}}\right) = 0, \qquad n = 1, 2, \ldots \tag{2.33b}$$

Because of the symmetry plane existence, each of sets of Equations (2.33) splits into two independent subsets, one of which describes motion in the symmetry plane, for

$$J_{23} = J_{32} = J'_{23} = J'_{32} = 0, \qquad J_{13} = J_{31} = J'_{13} = J'_{31} = 0.$$

The principal moment of perturbing forces in the right side of (2.33a) should be referred either to point F_1 or to point F_2 (to F_1 for the subset describing motion in the symmetry plane).

With this we conclude our short excursus to equations of dynamics of a solid with a cavity partially filled with liquid performing potential motion. The following sections are devoted to generalization of the adjacent mathematical models for the case of LV liquid vortex motion in cavities having inner damping ribs, with perturbed motion of the solid.

2.2. Velocity Field of LV Liquid Vortex Motion in a Moving Cavity

Consider the problem of perturbed motion of a solid with cavity having 'narrow' inner ribs, radial or ring-shaped ones, the statement of the problem being the same as in Article 2.1. Denote the number of ribs by K. Assume the liquid to be low-viscous now. For characteristic dimensionless parameters, introduce Strouhal number and Reynolds number and maximum relative width of the ribs defined as follows :

$$\mathrm{Sh} = \frac{\omega_F l}{V_F^*} = \frac{V_F}{V_F^*}, \quad \mathrm{Re} = \frac{\omega_F l^2}{\nu_F}, \quad \tilde{b} = \frac{b_{\max}}{l}, \tag{2.34}$$

where $V_F = \omega_F l$ is velocity of rib vibrations, relative to liquid, in the direction of normal to the rib's surface for perturbed motion with frequency ω_F; V_F^* is the value of velocity V_F related to vigorous vortex formation at the rib edge; l is the cavity characteristic size (for an axisymmetric cavity, for instance, it is the maximum diameter $2r_0$); ν_F is the liquid kinematic viscosity coefficient; $b_{\max}$ is the rib maximum width. Consider the liquid motion for great Reynolds number values, 'intermediate' Strouhal number values, and small relative width of ribs:

$$\mathrm{Re} \gg 1, \quad \mathrm{Sh} \le \mathrm{Sh}^* \sim 1, \quad \tilde{b} \ll 1. \tag{2.35}$$

Use once more coordinate systems $O^\circ x^\circ y^\circ z^\circ$ and $Oxyz$. The body's perturbed motion is the motion of coordinate system $Oxyz$ with respect to $O^\circ x^\circ y^\circ z^\circ$. Denote pole O velocity and the body's angular velocity of this motion by $\mathbf{V}_O$ and $\vec{\omega}$ respectively. Besides Cartesian coordinate system $Oxyz$, introduce cylindric coordinate system $Oxr\theta$, which Or axis, coinciding with Oy, is oriented relative to the cavity in such a way, that it coincides with the normal to the line of surface S intersection with a coaxial rib, nearest to free surface S.

Let us define more exactly the terms 'radial' and 'ring-shaped' ribs. Assume that both are line surfaces, generating lines of which coincide with inner normals to the wetted surface S . The line of surface S intersection with a 'radial' rib is the line of its intersection with the plane parallel to Ox axis and running through the liquid unperturbed free surface Σ mass centre. The line of surface S intersection with a 'ring-shaped' rib is the line of surface intersection with the plane perpendicular to Ox axis.

For the character of unperturbed and perturbed motion we keep the hypotheses formulated in Article 2.1, particularly, we assume $|\mathbf{V}_O|$ and $|\vec{\omega}|$ small and assume the concept of 'fixed cover' coinciding with unperturbed surface of liquid, but unlike Article 2.1, we omit further the 'prime' sign at all functions.

Because of great Reynolds numbers we neglect vortex motion of liquid in the wall boundary layer. Assume, that the most general case of vortex motion is stalling of vortexes from ribs' edges and, hence, nonuniform vorticity of the entire liquid mass. It is the case, the general mathematical model of vortex motions, considered below, refers to. In contrast to B.I.Rabinovich [77] and B.I.Rabinovich and V.M.Rogovoy [81], besides uniform vorticity independent from domain Q points coordinates, we consider additional components varying from point to point, i.e. 'higher harmonics' of the vortex field.

For this purpose, following [78], we apply Boubnov – Galerkin technique, the coordinate functions being particular solutions of ideal liquid hydrodynamics equations, satisfying continuity equation and the condition of impenetrability of boundary $S + \Sigma$ of the domain filled with liquid.

Present the field of absolute liquid velocities $\mathbf{v}_a$ in the form of

$$\mathbf{v}_a = \mathbf{V}_O + \vec{\omega} \times \mathbf{R} + \mathbf{v}, \quad \mathbf{v} = \sum_{k=1}^{2N} \mathbf{v}_k + \sum_{n=1}^{\infty} \dot{s}_n \nabla \varphi_n, \tag{2.36}$$

where $\mathbf{v}$ is the field of relative velocities; φ_n are the eigenfunctions of the boundary-value Problem (2.9) considering nonvortex nonstalling flow around ribs; $\mathbf{v}_k$ are velocity field components associated with forementioned coordinate functions; N is some integer that is chosen each time based on heuristic considerations.

Introduce the following set of functions for $\mathbf{v}_k$:

$$\mathbf{v}_k = \vec{\Omega}_k \times \mathbf{R}_k - \nabla \left(\vec{\Omega}_k, \vec{\Psi}_k\right), \tag{2.37}$$

where $\vec{\Omega}_k$ are vector functions of time, and $\mathbf{R}_k$ and $\vec{\Psi}_k$ are vector functions of coordinates of domain Q points, that satisfy the following equations and boundary conditions:

$$\operatorname{div} \left(\vec{\Omega}_k \times \mathbf{R}_k\right) = 0, \quad \Delta\vec{\Psi}_k = 0, \quad \frac{\partial\vec{\Psi}_k}{\partial\vec{\nu}} = \mathbf{R}_k \times \vec{\nu}, \tag{2.38}$$

where $\vec{\nu}$ is the unit vector normal to surface $S + \Sigma$, which includes in this case the ribs surface. We choose functions $\mathbf{R}_k$ so, that they belong to two families, as follows. For $1 \le k \le N$ they correspond to the vortex field 'adjusted' to radial ribs principally (nonclosed vortex lines with ends resting against surface $S + \Sigma$), and for $N + 1 \le k \le 2N$ they correspond to vortex field 'adjusted' to coaxial ring-shaped ribs principally (closed vortex rings). The following expressions refer to it:

For $1 \le k \le N$:

$$\vec{\Omega}_k = \sum_{j=1}^{3} \mathbf{i}_j\Omega_{kj}(t), \quad \mathbf{R}_k = \mathbf{i}_1 X_k(x) + \mathbf{i}_2 Y_k(y) + \mathbf{i}_3 Z_k(z), \tag{2.39a}$$

$$\mathbf{R}_1 \equiv \mathbf{R} = \mathbf{i}_1 x + \mathbf{i}_2 y + \mathbf{i}_3 z, \quad \vec{\Psi}_1 \equiv \vec{\Psi} = \mathbf{i}_1\Psi_1 + \mathbf{i}_2\Psi_2 + \mathbf{i}_3\Psi_3,$$

$$\operatorname{div}\left(\vec{\Omega}_k \times \mathbf{R}_k\right) = \frac{\partial}{\partial x}(\Omega_{k2}Z_k - \Omega_{k3}Y_k) + \\ + \frac{\partial}{\partial y}(\Omega_{k3}X_k - \Omega_{k1}Z_k) + \frac{\partial}{\partial z}(\Omega_{k1}Y_k - \Omega_{k2}X_k) \equiv 0,$$

$$\operatorname{curl}\left(\vec{\Omega}_k \times \mathbf{R}_k\right) = \mathbf{i}_1\Omega_{k1}(Y_k' + Z_k') + \mathbf{i}_2\Omega_{k2}(Z_k' + X_k') + \mathbf{i}_3\Omega_{k3}(X_k' + Y_k'),$$

$$\operatorname{curl}\left(\vec{\Omega}_1 \times \mathbf{R}_1\right) = 2\vec{\Omega}\,.$$

For $N + 1 \leq k \leq 2N$:

$$\vec{\Omega}_k = \mathbf{i}_x\Omega_{kx}(t) + \mathbf{i}_r\Omega_{kr}(t) + \mathbf{i}_\theta\,\Omega_{k\theta}\,(t), \tag{2.39b}$$

$$\mathbf{R}_k = \frac{r_0}{r}\,[\mathbf{i}_x X_k(x) + \mathbf{i}_r R_k(r) + \mathbf{i}_\theta\, T_k(\theta)], \qquad \mathbf{R}_{N+1} \equiv \frac{r_0}{r}\,\mathbf{R},$$

$$\operatorname{div}\left(\vec{\Omega}_k \times \mathbf{R}_k\right) = \frac{\partial}{\partial x}\left[\frac{r_0}{r}\,(\Omega_{kr}T_k - \Omega_{k\theta}R_k)\right] +$$

$$+ \frac{r_0}{r}\frac{\partial}{\partial r}\,(\Omega_{k\theta}X_k - \Omega_{kx}T_k) + \frac{1}{r}\frac{\partial}{\partial \theta}\left[\frac{r_0}{r}\,(\Omega_{kx}R_k - \Omega_{kr}X_k)\right] \equiv 0,$$

$$\operatorname{curl}\left(\vec{\Omega}_k \times \mathbf{R}_k\right) = \mathbf{i}_x\,\frac{r_0}{r}\left(R_k' + \frac{1}{r}\,T_k'\right)\Omega_{kx} + \mathbf{i}_r\,\frac{r_0}{r}\left(X_k' + \frac{1}{r}\,T_k'\right)\Omega_{kr} +$$

$$+ \mathbf{i}_\theta\left[\frac{r_0}{r}\,X_k' + \frac{d}{dr}\left(\frac{r_0}{r}R_k\right)\right]\Omega_{k\theta}\,,$$

$$\operatorname{curl}\left(\vec{\Omega}_{N+1} \times \mathbf{R}_{N+1}\right) = \frac{r_0}{r}\,\vec{\Gamma}\,,$$

where r_0 is the characteristic radial size of the cavity; $\mathbf{i}_1$, $\mathbf{i}_2$, $\mathbf{i}_3$, $\mathbf{i}_x \equiv \mathbf{i}_1$, $\mathbf{i}_r$, $\mathbf{i}_\theta$ are unit vectors of coordinate systems $Oxyz$ and $Oxr\theta$; $X_k(x)$, $Y_k(y)$, $Z_k(z)$, $R_k(r)$, $T_k(\theta)$ are linearly independent functions, continuous together with their first derivatives. 'Prime' sign denotes derivative with respect to the appropriate argument. Function $\vec{\Psi}_1 \equiv \vec{\Psi}$ is the solution of boundary-value Problem (2.8b) allowing for nonvortex nonstalling flow around ribs.

One can easily see that velocity field $\mathbf{v}_a(x, y, z, t)$ represented by Expressions (2.36), (2.37) satisfies the continuity equation in domain Q and boundary conditions at the surface $S + \Sigma$:

$$\operatorname{div}\mathbf{v}_a = 0, \quad \int_Q \operatorname{div}\mathbf{v}_a\,dQ = \sum_{k=1}^{2N}\ \oint_{S+\Sigma}\left(\vec{\Omega}_k \times \mathbf{R}_k, \vec{\nu}\right)\,dS =$$

$$= \sum_{k=1}^{2N}\left(\vec{\Omega}_k, \oint_{S+\Sigma}\left(\mathbf{R}_k \times \vec{\nu}\right)\,dS\right) \equiv 0 \rightarrow \oint_{S+\Sigma}\left(\mathbf{R}_k \times \vec{\nu}\right)\,dS = 0,$$

$$v_{a\nu}\Big|_S = V_{O\nu} + (\vec{\omega}, \mathbf{R} \times \vec{\nu}\,), \qquad v_{a\nu}\Big|_S = V_{O\nu} + (\vec{\omega}, \mathbf{R} \times \vec{\nu}\,) + \sum_{n=1}^{\infty}\dot{s}_n\psi_n,$$

where $v_{a\nu}$ and $V_{O\nu}$ are the projections of adjacent velocities to the direction of outer normal to surface $S + \Sigma$ of domain Q. A consequence is satisfaction of boundary-value Problem (2.38) solvability condition for arbitrary subscript k value.

Expression (2.37) for $\mathbf{v}_k$ can be presented in another equivalent form:

$$\mathbf{v}_k = \operatorname{curl}\mathbf{A}_k - \nabla\left(\vec{\Omega}_k, \vec{\Psi}_k\right), \tag{2.37a}$$

where $\operatorname{curl}\mathbf{A}_k = \vec{\Omega}_k \times \mathbf{R}_k$.

Consider the expression curl $\mathbf{v}_k$ for $1 \le k \le N$:

$$\text{curl } \mathbf{v}_k = \text{curl curl } \mathbf{A}_k = \nabla \text{ div } \mathbf{A}_k - \Delta \mathbf{A}_k.$$

Suppose, that we have managed to construct vector potential $\mathbf{A}_k$ satisfying the condition div $\mathbf{A}_k = 0$. Then

$$\text{curl } \mathbf{v}_k = - \Delta \mathbf{A}_k. \tag{2.40}$$

Introduce new denominations for vector $\mathbf{A}_k$ components A_{kj} (j=1, 2, 3):

$$A_{k1} = - \Omega_{k1}\, \alpha_{k1}, \quad A_{k2} = - \Omega_{k2}\, \alpha_{k2}, \quad A_{k3} = - \Omega_{k3}\, \alpha_{k3}. \tag{2.41}$$

Then (2.37a), (2.40), and (2.41) yield Poisson's equations for functions of space coordinates α_{kj} (j=1, 2, 3) :

$$\Delta\, \alpha_{k1} = Y_k'(y) + Z_k'(z)\,, \quad \Delta\, \alpha_{k2} = Z_k'(z) + X_k'(x)\,,$$
$$\Delta\, \alpha_{k3} = X_k'(x) + Y_k'(y)\,. \tag{2.42}$$

Assume

$$\alpha_{k1} = \alpha_{k1}\,(y, z)\,, \quad \alpha_{k2} = \alpha_{k2}\,(z, x)\,, \quad \alpha_{k3} = \alpha_{k3}\,(x, y)\,.$$

Laplace operators Δ in (2.42) become two-dimensional in this case, and we can easily obtain the following particular solutions of Equations (2.42):

$$\begin{aligned} \alpha_{k1} &= \int Y_k\,(y)\, dy + \int Z_k\,(z)\, dz\,, \quad \alpha_{11} = \frac{1}{2}\,(y^2 + z^2)\,, \\ \alpha_{k2} &= \int Z_k\,(z)\, dz + \int X_k\,(x)\, dx\,, \quad \alpha_{22} = \frac{1}{2}\,(z^2 + x^2)\,, \\ \alpha_{k3} &= \int X_k\,(x)\, dx + \int Y_k\,(y)\, dy\,, \quad \alpha_{33} = \frac{1}{2}\,(x^2 + y^2)\,. \end{aligned} \tag{2.43}$$

It follows from (2.43) that vector $\mathbf{A}_k$ divergence is equal to zero, that was postulated above:

$$\text{div } \mathbf{A}_k = - \left(\Omega_{k1} \frac{\partial \alpha_{k1}}{\partial x} + \Omega_{k2} \frac{\partial \alpha_{k2}}{\partial y} + \Omega_{k3} \frac{\partial \alpha_{k3}}{\partial z} \right) \equiv 0.$$

Consider symmetric tensors of rank 2, A_k and A_k':

$$\mathsf{A}_k = \left\{ \overset{k}{\alpha}_{ji} \right\}, \quad \mathsf{A}_k' = \left\{ \overset{k}{\alpha}_{ji}{}' \right\}, \quad \overset{k}{\alpha}_{ji} = \overset{k}{\alpha}_{ji}{}' = 0 \qquad \text{for } i \ne j,$$
$$i, j = 1, 2, 3, \tag{2.44}$$
$$\overset{k}{\alpha}_{jj} = \alpha_{kj}\,, \quad \overset{k}{\alpha}_{jj}{}' = \alpha_{kj}' = \Delta\alpha_{kj}\,, \quad j = 1, 2, 3,$$

where $\Delta\alpha_{kj}$ and α_{kj} are defined by Expressions (2.42), (2.43).

With right products of tensors A_k and A_k' by vector $\vec{\Omega}_k$ we can represent expressions for $\mathbf{A}_k$, $\mathbf{v}_k$, and curl $\mathbf{v}_k$ in the following form

$$\mathbf{A}_k = - \left(\mathsf{A}_k, \vec{\Omega}_k\right), \quad \mathbf{v}_k = - \left[\text{curl} \left(\mathsf{A}_k, \vec{\Omega}_k\right) + \nabla \left(\vec{\Psi}_k, \vec{\Omega}_k\right)\right], \tag{2.45}$$

$$\text{curl } \mathbf{v}_k = \left(\mathsf{A}'_k,\ \vec{\Omega}_k\right).$$

Suppose now $N + 1 \leq k \leq 2N$. Expression (2.40) in cylindric coordinates is not valid now, but Expressions (2.41) and (2.45) are valid, if elements of tensors A_k and A'_k are defined as follows:

$$\overset{k}{\alpha}_{ji} = \overset{k}{\alpha}{}'_{ji} = 0 \quad \text{for} \quad i \neq j, \quad i, j = 1, 2, 3, \tag{2.46}$$

$$\alpha_{k1} = \int \frac{r_0}{r} R_k(r)\, dr + r_0 \int T_k(\theta)\, d\theta\,, \quad \alpha'_{k1} = \frac{r_0}{r}\left(R'_k + \frac{1}{r} T'_k\right),$$

$$\alpha_{k2} = \frac{r_0}{r} \int X_k(x)\, dx + r_0 \int T_k(\theta)\, d\theta\,, \quad \alpha'_{k2} = \frac{r_0}{r}\left(X'_k + \frac{1}{r} T'_k\right),$$

$$\alpha_{k3} = \frac{r_0}{r} \int X_k(x)\, dx + \frac{r_0}{r} \int R_k(r)\, dr\,, \quad \alpha'_{k3} = \frac{r_0}{r} X'_k + \frac{d}{dr}\left(\frac{r_0}{r} R'_k\right).$$

Now we can present Expression (2.36) for absolute velocity of liquid in the following two equivalent forms:

$$\mathbf{v}_a = \mathbf{V}_O + \vec{\omega} \times \mathbf{R} + \sum_{k=1}^{2N} \left[\vec{\Omega}_k \times \mathbf{R}_k - \nabla(\vec{\Omega}_k, \vec{\Psi}_k)\right] + \sum_{n=1}^{\infty} \dot{s}_n \nabla \varphi_n. \tag{2.47a}$$

$$\mathbf{v}_a = \mathbf{V}_O - \text{curl}\,(\mathsf{A}_1, \vec{\omega}) - \sum_{k=1}^{2N} \left[\text{curl}(\mathsf{A}_k, \vec{\Omega}_k) + \nabla(\vec{\Psi}_k, \vec{\Omega}_k)\right] + \sum_{n=1}^{\infty} \dot{s}_n \nabla \varphi_n. \tag{2.47b}$$

The first representation is preferable from the point of view of further reductions, and the second one is in better agreement with conventional division of a vector field to potential and solenoidal components.

Absolute liquid velocity curl and its volume mean component are defined, due to (2.40) and (2.47). by expressions:

$$\text{curl } \mathbf{v}_a = 2(\vec{\omega} + \vec{\Omega}) + \sum_{k=2}^{2N} (\mathsf{A}'_k, \vec{\Omega}_k)\,,$$

$$\frac{1}{Q} \int_Q \text{curl } \mathbf{v}_a\, dQ = 2(\vec{\omega} + \vec{\Omega}) + \frac{1}{Q} \sum_{k=2}^{2N} \int_Q \text{curl}\,(\vec{\Omega}_k \times \mathbf{R}_k)\, dQ =$$

$$= 2(\vec{\omega} + \vec{\Omega}) - \frac{1}{Q} \sum_{k=2}^{2N} \oint_{S+\Sigma} (\vec{\Omega}_k \times \mathbf{R}_k) \times \vec{\nu}\, dS\,. \tag{2.48}$$

2.3. Generalized Forces and Derivatives of Liquid Kinetic Energy with Respect to Generalized Velocities

Let us consider the expression of hydrodynamic forces $\mathbf{F}_\nu$ acting on unit length of a rib, perpendicular to its surface, the force associated with vortex formation at the edges. The following expression, suggested by V.M.Rogovoi [86], is in good agreement with experimental data by G.N.Mikishev [62, 64], G.H.Keulegan and L.H.Carpenter [36], and J.W. Miles [65] in the considered range of Strouhal numbers :

$$\mathbf{F}_\nu = \rho k_s b^{3/2} \varphi(x') \left(\frac{|\mathbf{v}_\nu|}{\pi} \right)^{1/2} \mathbf{V}_\nu , \qquad k_s = 1.1\, k_s^* , \tag{2.49}$$

where $\mathbf{v}_\nu$ is relative liquid velocity $\mathbf{v}^\circ$ component normal to rib surface, the relative velocity being calculated in points of the medium line of the rib, as if there were no ribs (functions $\vec{\Psi}_k^\circ$, $\vec{\varphi}_n^\circ$, $\mathbf{R}_k^\circ$), and $\mathbf{V}_\nu$ is its counterpart, according to the following expressions:

$$\mathbf{v}_\nu = (\mathbf{v}^\circ, \vec{\nu})\,|_L\, \vec{\nu} = \left[\sum_{k=1}^{2N} \left(\vec{\Omega}_k , \ \mathbf{R}_k^\circ \times \vec{\nu} - \frac{\partial \vec{\Psi}_k^\circ}{\partial \nu} \right) + \sum_{n=1}^{\infty} \dot{s}_n \frac{\partial \varphi_n^\circ}{\partial \nu} \right] \Bigg|_L \vec{\nu},$$

$$\mathbf{V}_\nu = \int_{-\infty}^{t} \frac{\partial \mathbf{v}_\nu}{\partial \tau} \bigg|_L \frac{d\tau}{\sqrt{t-\tau}} , \tag{2.50}$$

$\vec{\nu}$ is the unit vector normal to the rib surface, making an acute angle with $\mathbf{v}^\circ$, ρ is mass density of liquid, b is the rib width, k_s^* is an empiric constant introduced in [62, 77]; $k_s = 2.71$ for antisymmetric oscillations, $k_s = 4.1$ for symmetric oscillations in the case of ring-shaped ribs, $k_s = 4.57$ for antisymmetric oscillations in the case of radial ribs,

$$\varphi(x') = 1 + \alpha e^{-\beta x'}$$

is an empiric function, $\alpha = 0.4$, $\beta = 0.131$ for axisymmetric cavities (x' is the coordinate counted off the free surface towards the bottom, related to the rib width). If there is no free surface $s_n \equiv 0$.

It is shown in [79], that expressions under sum sign with subscript k in (2.36) and (2.50) do not depend on the choice of pole O, i.e. the system of hydrodynamic forces associated with vortex motion of liquid can be reduced to a couple. Hence, vortex motions of liquid do not affect translational motion of the body and do not depend on this motion, that makes it possible to assume $\mathbf{V}_O \equiv 0$ further in this section.

For efficient use of the mathematical model of vortex motions of liquid, while constructing equations of the body – liquid system perturbed motion, it is necessary to have the expressions of generalized forces associated with vortex motion of liquid and expressions of liquid kinetic energy derivatives with respect to generalized velocities ω_j, Ω_{kj} ($k = 1, 2, ..., 2N$; $j = 1, 2, 3$) and $\dot{s}_n$ ($n = 1, 2, ...$).

Denote generalized forces corresponding to generalized velocities $\vec{\Omega}_k$ and $\dot{s}_n$ by $\mathbf{M}_k$ and P_n, liquid kinetic energy for $\mathbf{V}_O = 0$ by T_F, and variation of liquid total energy per time unit by $\dot{W}_F$. Write expressions $\dot{W}_F$, $\frac{\partial T_F}{\partial \omega_j}$, $\frac{\partial T_F}{\partial \Omega_{kj}}$, $\frac{\partial T_F}{\partial \dot{s}_n}$ in two equivalent forms taking into consideration (2.36), (2.39), (2.8b), (2.9), (2.37) and applying Green and Gauss – Ostrogradsky formulae.

Form (a) :

$$\dot{W}_F = -\frac{\rho k_s b^{3/2}}{\sqrt{\pi}} \int_{L_\mu} \sqrt{|\mathbf{v}_\nu|}\, (\mathbf{V}_\nu, \mathbf{v}_\nu)\, \varphi(x')\, ds,$$

$$\frac{\partial T_F}{\partial \omega_j} = \rho \int_Q \left(\frac{\partial \mathbf{v}_a}{\partial \omega_j}, \mathbf{v}_a\right) dQ = \rho \sum_{i=1}^{3} \left\{(\omega_i + \Omega_{1i}) \int_Q (\mathbf{i}_i \times \mathbf{R}, \mathbf{i}_j \times \mathbf{R})\, dQ - \right.$$

$$- \Omega_{1i} \oint_{S+\Sigma} \frac{\partial \Psi_j}{\partial \nu} \Psi_i\, dS + \sum_{k=2}^{2N} \Omega_{ki} \left[\left. \int_Q (\mathbf{i}_i \times \mathbf{R}_k, \mathbf{i}_j \times \mathbf{R})\, dQ - \oint_{S+\Sigma} \frac{\partial \Psi_j}{\partial \nu} \Psi_{ki}\, dS \right]\right\} +$$

$$+ \rho \sum_{n=1}^{\infty} \dot{s}_n \oint_{S+\Sigma} (\mathbf{R} \times \vec{\nu})_j\, \varphi_n\, dS,$$

$$\frac{\partial T_F}{\partial \Omega_{1j}} = \rho \int_Q \left(\frac{\partial \mathbf{v}_a}{\partial \Omega_{1j}}, \mathbf{v}_a\right) dQ =$$

$$= \rho \sum_{i=1}^{3} \left\{(\omega_i + \Omega_{1i}) \left[\int_Q (\mathbf{i}_i \times \mathbf{R}, \mathbf{i}_j \times \mathbf{R})\, dQ - \oint_{S+\Sigma} \frac{\partial \Psi_j}{\partial \nu} \Psi_i\, dS\right] + \right.$$

$$\left. + \sum_{k=2}^{2N} \Omega_{ki} \left[\int_Q (\mathbf{i}_i \times \mathbf{R}_k, \mathbf{i}_j \times \mathbf{R})\, dQ - \oint_{S+\Sigma} \frac{\partial \Psi_j}{\partial \nu} \Psi_{ki}\, dS\right]\right\}, \tag{2.51a}$$

$$\frac{\partial T_F}{\partial \Omega_{kj}} = \rho \int_Q \left(\frac{\partial \mathbf{v}_a}{\partial \Omega_{kj}}, \mathbf{v}_a\right) dQ =$$

$$= \sum_{l=1}^{2N} \Omega_{li} \left[\int_Q (\mathbf{i}_i \times \mathbf{R}, \mathbf{i}_j \times \mathbf{R}_k)\, dQ - \oint_{S+\Sigma} \frac{\partial \Psi_{kj}}{\partial \nu} \Psi_{li}\, dS\right],$$

$$\frac{\partial T_F}{\partial \dot{s}_n} = \rho \int_Q \left(\frac{\partial \mathbf{v}_a}{\partial \dot{s}_n}, \mathbf{v}_a\right) dQ = \rho \left[\dot{s}_n \int_\Sigma \frac{\partial \varphi_n}{\partial \nu} \varphi_n\, dS + \sum_{i=1}^{3} \omega_i \oint_{S+\Sigma} \frac{\partial \Psi_i}{\partial \nu} \varphi_n\, dS\right],$$

$$j = 1, 2, 3, \quad k = 1, 2, 3, \ldots, 2N, \quad n = 1, 2, \ldots$$

Form (b) :

$$\dot{W}_F = \sum_{k=1}^{2N} (\mathbf{M}_k, \vec{\Omega}_k) + \sum_{n=1}^{\infty} P_n \dot{s}_n =$$

$$= \sum_{k=1}^{2N} \left[- \sum_{l=1}^{2N} ((\mathsf{B}_{kl}, \vec{\Theta}_l), \vec{\Omega}_k) + \sum_{n=1}^{\infty} S_n (\vec{\gamma}_{kn}, \vec{\Omega}_k) \right] +$$

$$+ \sum_{n=1}^{\infty} \left[\sum_{k=1}^{2N} (\vec{\gamma}_{kn}, \vec{\Theta}_k) - \mu_n \sum_{m=1}^{\infty} \beta_{nm} S_m \right] \dot{s}_n ,$$

$$\vec{\Theta}_k (t) = \int_{-\infty}^{t} \frac{\dot{\vec{\Omega}}_k(\tau)\, d\tau}{\sqrt{t-\tau}}, \quad S_n (t) = \int_{-\infty}^{t} \frac{\ddot{s}_n(\tau)\, d\tau}{\sqrt{t-\tau}}, \tag{2.51b}$$

$$\frac{\partial T_F}{\partial \omega_j} = (\mathsf{J}^{(0)}, \vec{\omega})_j + (\mathsf{J}_1^*, \vec{\Omega}_1)_j + \sum_{l=2}^{2N} (\mathsf{J}_{1l}, \vec{\Omega}_l)_j + \sum_{n=1}^{\infty} \vec{\lambda}_{Onj}\, \dot{s}_n ,$$

$$\frac{\partial T_F}{\partial \Omega_{1j}} = (\mathsf{J}_1^*, \vec{\Omega}_1)_j + (\mathsf{J}_1^*, \vec{\omega})_j + \sum_{l=2}^{2N} (\mathsf{J}_{1l}, \vec{\Omega}_l)_j ,$$

$$\frac{\partial T_F}{\partial \Omega_{kj}} = \sum_{l=2}^{2N} (\mathsf{J}_{kl}, \vec{\Omega}_l)_j , \qquad \frac{\partial T_F}{\partial \dot{s}_n} = \mu_n \dot{s}_n + (\vec{\lambda}_{Onj}, \vec{\omega}) ,$$

$$j = 1, 2, 3, \qquad k = 1, 2, 3, ..., 2N, \qquad n = 1, 2, ...,$$

where L_μ is the contour formed by the μ -th rib middle line; B_{kl}, $\mathsf{J}^{(0)}$, J_1^*, J_{kl} are the tensors of rank 2; $\vec{\gamma}_{kn}$, $\vec{\lambda}_{On}$ - are vectors; μ_n, β_{nm} - are scalars;

$$\mathsf{B}_{kl} = \left\{ \beta_{ji}^{kl} \right\}, \quad \mathsf{J}_1^* = \mathsf{J}^{(0)} - \mathsf{J} = \left\{ J_{ji}^* \right\}, \quad \mathsf{J}^{(0)} = \left\{ J_{ji}^{(0)} \right\},$$
$$\mathsf{J} = \left\{ J_{ji} \right\}, \quad \mathsf{J}_{kl} = \left\{ J_{ji}^{kl} \right\}, \tag{2.52}$$
$$\vec{\beta}_{kn} = \sum_{j=1}^{3} \mathbf{i}_j \beta_{knj} , \quad \vec{\lambda}_{On} = \sum_{j=1}^{3} \mathbf{i}_j \lambda_{Onj} ,$$

$$j = 1, 2, 3, \qquad k = 1, 2, 3, ..., 2N, \qquad n = 1, 2, \ldots$$

Tensor $\mathsf{J}^{(0)}$ is the solidified liquid's tensor of inertia; tensor J is N.Joukowski's tensor of the liquid's associated moments of inertia calculated considering existing baffles with potential nonstall flow around them.

Tensors elements and vectors components in (2.52) are expressed by the following formulae obtained from comparison of Expressions (2.51a) and (2.51b) with identical right sides.

$$\beta_{ji}^{kl} = \beta_{ij}^{lk} =$$

$$= \frac{\rho k_s}{\sqrt{\pi}} \sum_{\mu=1}^{K} \int_{L_\mu} \sqrt{|\mathbf{v}_\nu|} \left[- (\mathbf{R}_l^\circ \times \vec{\nu})_i + \frac{\partial \Psi_{li}^\circ}{\partial \nu} \right] \left[- (\mathbf{R}_k^\circ \times \vec{\nu})_j + \frac{\partial \Psi_{kj}^\circ}{\partial \nu} \right] b^{3/2} \varphi(x')\, ds,$$

$$\gamma_{knj} = \frac{\rho k_s}{\sqrt{\pi}} \sum_{\mu=1}^{K} \int_{L_\mu} \sqrt{|\mathbf{v}_\nu|} \left[- (\mathbf{R}_k^\circ \times \vec{\nu})_j + \frac{\partial \Psi_{kj}^\circ}{\partial \nu} \right] \frac{\partial \varphi_n^\circ}{\partial \nu} b^{3/2} \varphi(x')\, ds,$$

(2.53a)

$$\beta_{nm} = \beta_{mn} = \frac{\rho k_s}{\mu_n \sqrt{\pi}} \sum_{\mu=1}^{K} \int_{L_\mu} \sqrt{|\mathbf{v}_\nu|} \, \frac{\partial \overset{\circ}{\varphi}_n}{\partial \nu} \frac{\partial \overset{\circ}{\varphi}_m}{\partial \nu} b^{3/2} \varphi(x') \, ds,$$

$$i, j = 1, 2, 3, \qquad k, l = 1, 2, 3, \ldots, 2N, \qquad m, n = 1, 2, \ldots,$$

$$J_{ij}^{*} = J_{ji}^{*} = J_{ij}^{(0)} - J_{ij}, \qquad J_{ij}^{(0)} = J_{ji}^{(0)} = \rho \int_{Q} (\mathbf{R} \times \mathbf{i}_i, \mathbf{R} \times \mathbf{i}_j) \, dQ,$$

$$J_{ij} = J_{ji} = \rho \oint_{S+\Sigma} \frac{\partial \Psi_{kj}}{\partial \nu} \Psi_j \, dS,$$

$$J_{ji}^{kl} = J_{ij}^{lk} = \rho \left[\int_{Q} (\mathbf{i}_i \times \mathbf{R}_l, \mathbf{i}_j \times \mathbf{R}_k) \, dQ - \oint_{S+\Sigma} \frac{\partial \Psi_{kj}}{\partial \nu} \Psi_{li} \, dS \right], \tag{2.53b}$$

$$\mu_n = \rho \int_{\Sigma} \frac{\partial \varphi_n}{\partial \nu} \varphi_n \, dS = \rho \, \kappa_n \int_{\Sigma} \varphi_n^2 \, dS,$$

$$\lambda_{Onj} = \rho \oint_{S+\Sigma} (\mathbf{R} \times \vec{\nu})_j \, \varphi_n \, dS = \rho \oint_{S+\Sigma} \frac{\partial \Psi_{kj}}{\partial \nu} \varphi_n \, dS.$$

Potentials $\overset{\circ}{\varphi}_n$ and $\overset{\circ}{\Psi}_{kj}$ refer to potential motion of liquid in the cavity without ribs, i.e. they are equivalent to φ_n and Ψ_{kj} from Article 2.2.

Expressions (2.53a) let us determine generalized forces corresponding to generalized velocities $\vec{\Omega}_k$, $\dot{s}_n$, and (2.53b) determine kinetic energy derivatives with respect to their components.

Let us introduce now, in addition to the previous, the following orthogonality conditions

$$\int_{Q} (\mathbf{i}_i \times \mathbf{R}_l, \mathbf{i}_j \times \mathbf{R}_k) \, dQ - \oint_{S+\Sigma} \frac{\partial \Psi_{kj}}{\partial \nu} \Psi_{li} \, dS = 0, \tag{2.54a}$$

$$i, j = 1, 2, 3, \qquad k = 2, 3, \ldots, 2N, \qquad l = 1, 2, \ldots, k-1,$$

that, if obeyed, result in

$$\mathrm{J}_{kl} \equiv 0 \quad \text{for} \quad l \neq k,$$

$$\mathrm{J}_{kk} = \mathrm{J}_k^{*} = \left\{ J_{ji}^{kk} \right\}, \qquad k, l = 2, 3, \ldots, 2N, \qquad i, j = 1, 2, 3. \tag{2.54b}$$

Each k value corresponds to $9C_k^{(2)}$ Equations (2.54a), where $C_k^{(2)}$ is the number of combinations of k elements taken 2 at a time.

Conditions (2.54a) can be satisfied by appropriate choice of functions $X_k(x)$, $Y_k(y)$, $Z_k(z)$, $R_k(r)$, $T_k(\theta)$. One of the opportunities consists in polynomial representation of X_k, Y_k, Z_k, R_k, and a part of Fourier series expansion of $T_k(\theta)$ in terms of sines (as $T_k(0) = 0$) the coefficients being unknown so far:

$$X_k(x) = x + \sum_{\nu=2}^{M_k} A_{k\nu} x^\nu, \quad Y_k(y) = y + \sum_{\nu=2}^{M_k} B_{k\nu} y^\nu, \quad Z_k(z) = z + \sum_{\nu=2}^{M_k} C_{k\nu} z^\nu,$$

$$M_k = 3C_k^{(2)} + 1, \quad C_k^{(2)} = \frac{k!}{2(k-1)!}, \quad k = 2, 3, ..., N, \tag{2.55}$$

$$X_k(x) = x + \sum_{\nu=2}^{M_{k-N}} A_{k\nu} x^\nu, \quad R_k(r) = x + \sum_{\nu=2}^{M_{k-N}} D_{k\nu} r^\nu,$$

$$T_k(\theta) = \sum_{\nu=2}^{M_{k-N}} E_{k\nu} \sin \nu\,\theta, \quad M_{k-N} = 3C_{k-N+1}^{(2)}, \quad C_{k-N}^{(2)} = \frac{(k-N)!}{2(k-2-N)!}.$$

Properties of polynomials and sines allow to meet the conditions of Functions (2.55) continuity and linear independence formulated above. On the other side, for each k value we now have at our disposal the same number of coefficients, as many Equations (2.54a) correspond to the given k value (e.g. for $k = 2$, $9C_2^{(2)} = 9$ equations for 9 coefficients $A_{22}, A_{23}, A_{24}, B_{22}, B_{23}, B_{24}, C_{22}, C_{23}, C_{24}$, for $k = 3$, $9C_3^{(2)} = 27$ equations for 27 coefficientsA_{32}, ..., B_{32}, ..., C_{32}, ..., and so on). That enables determination of all the coefficients $A_{k\nu}$, $B_{k\nu}$, $C_{k\nu}$ of Expressions (2.55) from Equations (2.54a). It can be easily seen that the process of determination of those coefficients is recurrent and consists at each step in solution of a set of nonhomogeneous equations of order 9 $(k-1)$. These sets of equations are solvable (with the exception of a special case when the determinant is zero).

We assume further, that all the auxiliary boundary-value problems are solved, and systems of Functions (2.55) are constructed in such a way that Conditions (2.54a) are obeyed for all values of $k = 2, 3, ..., 2N$. Then to construct the mathematical model of liquid vortex motion we have only to calculate Coefficients (2.53), except $J_{ji}^{kl} = J_{ij}^{lk}$ which are zero for all considered values of indices $i, j, k, l, k \neq l$.

Two asymptotic models are further considered parallel to the general mathematical model of vortex motions of liquid :

a) Uniform vorticity all over the considered volume ($k = 1$ in (2.39a)). Axisymmetric oscillations of liquid in a cavity of revolution correspond to vorticity inversely proportional to radius vector of the cylindric coordinate system ($k = N + 1$ in (2.39b)).

b) Vorticity is localized in the vicinity of rib edges, and in the rest of the volume liquid motion is nonvortex. This model corresponds to the one most widely used (see, e.g. [76]).

2.4. The Mathematical Model of a Solid with a Cavity Containing Vortex Low-Viscous Liquid.

We proceed now to consider the mathematical models of the system 'body vortex LV liquid'. We now consider translational motion of the body, not only rotational one as in Article 2.3, and make use of Lagrange's equations of kind 2:

$$\frac{d}{dt}\frac{\partial T}{\partial \dot{q}_j} - \frac{\partial T}{\partial q_j} + \frac{\partial U}{\partial q_j} = Q_j, \qquad j = 1, 2, \ldots, \tag{2.56}$$

where T is kinetic energy of body liquid system, U is potential energy, Q_j are generalized forces, are generalized coordinates. Let us also use the general Equations (2.25b) written for the ideal liquid case and liquid kinetic energy derivatives with respect to generalized velocities, and generalized forces obtained in Article 2.3. We obtain the following set of equations describing perturbed motion of a solid with a cavity partially filled with LV liquid, allowing for vortex motion of the latter.

$$(m^\circ + m)\,\dot{\mathbf{V}}_O + \dot{\vec{\omega}} \times (\mathbf{L}^\circ + \mathbf{L}) + \sum_{n=1}^{\infty} \vec{\lambda}_n \ddot{s}_n = \mathbf{P},$$

$$(\mathrm{J}^\circ + \mathrm{J}^{(0)}, \dot{\vec{\omega}}) + (\mathbf{L}^\circ + \mathbf{L}) \times \dot{\mathbf{V}}_O - (\mathbf{L}^\circ + \mathbf{L}) \times \mathbf{j} +$$
$$+ (\mathrm{J}_1^*, \dot{\vec{\Omega}}_1) + \sum_{n=1}^{\infty} [\vec{\lambda}_{On} \ddot{s}_n - (\vec{\lambda}_n \times \mathbf{j})\, s_n] = \mathbf{M}_O, \tag{2.57}$$

$$(\mathrm{J}_k^*, \dot{\vec{\Omega}}_k + \delta_k \dot{\vec{\omega}}) + \sum_{l=1}^{2N} \left(\mathrm{B}_{kl}, \int_{-\infty}^{t} \frac{\dot{\vec{\Omega}}_l(\tau)\, d\tau}{\sqrt{t-\tau}} \right) - \sum_{n=1}^{\infty} \vec{\gamma}_{kn} \int_{-\infty}^{t} \frac{\ddot{s}_n(\tau)\, d\tau}{\sqrt{t-\tau}} = 0,$$

$$\mu_n (\ddot{s}_n + \omega_n^2 s_n) + (\vec{\lambda}_n, \dot{\mathbf{V}}_O) + (\vec{\lambda}_{On}, \dot{\vec{\omega}}) - (\vec{\lambda}_n, \mathbf{j}) +$$
$$+ \mu_n \sum_{m=1}^{\infty} \beta_{nm} \int_{-\infty}^{t} \frac{\ddot{s}_m(\tau)\, d\tau}{\sqrt{t-\tau}} - \sum_{k=1}^{2N} \left(\vec{\gamma}_{kn}, \int_{-\infty}^{t} \frac{\dot{\Omega}_l(\tau)\, d\tau}{\sqrt{t-\tau}} \right) = 0,$$

$$k = 1, 2, 3, \ldots, 2N, \quad n = 1, 2, \ldots, \quad \delta_k = 0 \text{ for } k \neq 1, \ \delta_1 = 1,$$

where $\mathbf{P}$ and $\mathbf{M}_O$ are the resultant vector and resultant moment (with respect to point O) of the system of external forces, $\mathbf{L}^\circ$ and $\mathbf{L}$ are vectors of static moments (with respect to point O) of the solid body and 'fixed - cover' solidified liquid; J° is the tensor of inertia of the solid body.

Consider now some particular cases.

2.4.1. The general model of liquid vortex motion.

2.4.1a. Special choice of the coordinate origin. Equations (2.57) correspond to arbitrary pole O selection. If we place the pole to the mass centre G of the system "body – 'fixed cover' solidified liquid" (2.33b), total static moment $\mathbf{L}^\circ + \mathbf{L}$ becomes zero, and Equations (2.57) are transformed to the following:

$$(m^\circ + m)\,\dot{\mathbf{V}} + \sum_{n=1}^{\infty} \vec{\lambda}_n \ddot{s}_n = \mathbf{P},$$

$$(\mathrm{J}^\circ + \mathrm{J}^{(0)}, \dot{\vec{\omega}}) + (\mathrm{J}_1^*, \dot{\vec{\Omega}}_1) + \sum_{n=1}^{\infty} [\vec{\lambda}_{Gn} \ddot{s}_n - (\vec{\lambda}_n \times \mathbf{j})\, s_n] = \mathbf{M}_G,$$

$$(\mathrm{J}_k^*, \dot{\vec{\Omega}}_k + \delta_k \dot{\vec{\omega}}) + \sum_{l=1}^{2N} \left(\mathrm{B}_{kl}, \int_{-\infty}^{t} \frac{\dot{\vec{\Omega}}_l(\tau)\, d\tau}{\sqrt{t-\tau}} \right) - \sum_{n=1}^{\infty} \vec{\gamma}_{kn} \int_{-\infty}^{t} \frac{\ddot{s}_n(\tau)\, d\tau}{\sqrt{t-\tau}} = 0, \tag{2.58}$$

$$\mu_n(\ddot{s}_n + \omega_n^2 s_n) + (\vec{\lambda}_n, \dot{\mathbf{V}}) + (\vec{\lambda}_{Gn}, \dot{\vec{\omega}}) - (\vec{\lambda}_n, \mathbf{j}) +$$

$$+ \mu_n \sum_{m=1}^{\infty} \beta_{nm} \int_{-\infty}^{t} \frac{\ddot{s}_m(\tau)\, d\tau}{\sqrt{t-\tau}} - \sum_{k=1}^{2N} \left(\vec{\gamma}_{kn}, \int_{-\infty}^{t} \frac{\dot{\vec{\Omega}}_k(\tau)\, d\tau}{\sqrt{t-\tau}} \right) = 0,$$

$$k = 1, 2, 3, \ldots, 2N, \quad n = 1, 2, \ldots, \quad \delta_k = 0 \quad \text{for} \quad k \neq 1, \quad \delta_1 = 1.$$

where $\mathbf{V}$ is point G velocity, and all coefficients of Equations (2.58), depending on the pole O location, are referred to the new pole G, as well as $\mathbf{M}_G$.

2.4.1b. Cavity completely filled with LV liquid. In this case wave motions of liquid are absent, and we are to assume $s_n \equiv 0$ in Equations (2.57). As a result we obtain

$$(m^\circ + m)\,\dot{\mathbf{V}}_O + \vec{\omega} \times (\mathbf{L}^\circ + \mathbf{L}) = \mathbf{P},$$

$$(\mathrm{J}^\circ + \mathrm{J}^{(0)}, \dot{\vec{\omega}}) + (\mathbf{L}^\circ + \mathbf{L}) \times \dot{\mathbf{V}}_O - (\mathbf{L}^\circ + \mathbf{L}) \times \mathbf{j} + (\mathrm{J}_1, \dot{\vec{\Omega}}_1) = \mathbf{M}_O,$$

$$(\mathrm{J}_k^*, \dot{\vec{\Omega}}_k + \delta_k \dot{\vec{\omega}}) + \sum_{l=1}^{2N} \left(\mathrm{B}_{kl}, \int_{-\infty}^{t} \frac{\dot{\vec{\Omega}}_l(\tau)\, d\tau}{\sqrt{t-\tau}} \right) = 0, \tag{2.59}$$

$$k = 1, 2, 3, \ldots, 2N, \quad \delta_k = 0 \quad \text{for} \quad k \neq 1, \quad \delta_1 = 1,$$

This set of equations describes motion of a solid with a cavity completely filled with LV liquid, vorticity of which being quite considerable.

2.4.1c. Cavity completely filled with high-viscous liquid with $\nu_F = \infty$. HV liquid with infinite viscosity ($\nu_F = \infty$) behaves like a solid, and in this sense it is an analogue of a medium with infinite electrical resistance. In this case relative velocity of liquid is identically zero, particularly, there are no wave motions. Equations (2.57) convert, hence, to the ordinary equations of solid body dynamics. To see this assume $\vec{\Omega}_k \equiv 0$ $(k = 1, 2, 3, \ldots, 2N)$, $s_n \equiv 0$ $(n = 1, 2, \ldots)$ in (2.57) and obtain

$$(m^\circ + m)\,\dot{\mathbf{V}}_O + \vec{\omega} \times (\mathbf{L}^\circ + \mathbf{L}) = \mathbf{P},$$

$$(\mathrm{J}^{\circ}+\mathrm{J}^{(0)}, \dot{\vec{\omega}}) + (\mathbf{L}^{\circ}+\mathbf{L}) \times \dot{\mathbf{V}}_O - (\mathbf{L}^{\circ}+\mathbf{L}) \times \mathbf{j} = \mathbf{M}_O. \qquad (2.60)$$

These are familiar equations describing 'body solidified liquid' system motion.

It is to be noted that, if vortex motion of the whole liquid mass in the absolute coordinate system is not considered, the passing to the limit from the adjacent equations to the solid body dynamics equations, demonstrated above, is impossible.

2.4.2. Liquid motion for the case of developed stalled flow around ribs with known space distribution of absolute velocities. Suppose that the picture of stalled flow around ribs is obtained somehow (e.g. empirically), and liquid absolute velocities field outside stalling zones can be represented in the following form

$$\mathbf{v}_a = \mathbf{V}_O + \vec{\omega} \times \mathbf{R} + \mathbf{v}, \quad \mathbf{v} = \vec{\Omega}^* \times \mathbf{R}^* - \nabla(\vec{\Omega}^*, \vec{\Psi}^*) + \sum_{n=1}^{\infty} \dot{s}_n \nabla \varphi_n. \qquad (2.61)$$

Here $\vec{\Omega}^* = \vec{\Omega}^*(t)$ is the new generalized velocity characterizing vortex motion of the whole liquid mass in the integral sense. Velocity field space distribution in Q domain (stalling zones in the ribs vicinity excluded) is specified by function $\mathbf{R}^*(x, y, z)$, which can be represented, like $\mathbf{R}_k$ (2.39a), in the form of

$$\mathbf{R}^* = \mathbf{i}_1 X^*(x) + \mathbf{i}_2 Y^*(y) + \mathbf{i}_3 Z^*(z), \qquad (2.62)$$

where $X^*(x)$, $Y^*(y)$, $Z^*(z)$ - are known functions continuous together with their first derivatives in Q domain; and $\vec{\Psi}^*$ - is a harmonic vector function being the solution of the following boundary-value problem

$$\Delta \vec{\Psi}^* = 0, \quad \left.\frac{\partial \vec{\Psi}^*}{\partial \nu}\right|_{S+\Sigma} = \mathbf{R}^* \times \vec{\nu}, \qquad (2.63)$$

similar to (2.38). Equations of perturbed motion can be obtained following the same scheme as (2.57).

2.4.2a. General equations. Special choice of coordinate origin. The body liquid system mass centre G chosen for the coordinate origin, we obtain equations similar to (2.58)

$$(m^{\circ}+m)\dot{\mathbf{V}} + \sum_{n=1}^{\infty} \vec{\lambda}_n \ddot{s}_n = \mathbf{P},$$

$$(\mathrm{J}^{\circ}+\mathrm{J}^{(0)}, \dot{\vec{\omega}}) + (\mathrm{J}^*, \dot{\vec{\Omega}}^*) + \sum_{n=1}^{\infty} [\vec{\lambda}_{Gn} \ddot{s}_n - (\vec{\lambda}_n \times \mathbf{j}) s_n] = \mathbf{M}_G, \qquad (2.64)$$

$$(\mathrm{J}^{**}, \dot{\vec{\Omega}}^*) + (\bar{\mathrm{J}}^*, \dot{\vec{\omega}}) + \left(\mathrm{B}^*, \int_{-\infty}^{t} \frac{\dot{\vec{\Omega}}^*(\tau)\, d\tau}{\sqrt{t-\tau}}\right) - \sum_{n=1}^{\infty} \vec{\gamma}_n^* \int_{-\infty}^{t} \frac{\ddot{s}_n(\tau)\, d\tau}{\sqrt{t-\tau}} = 0,$$

$$\mu_n(\ddot{s}_n + \omega_n^2 s_n) + (\vec{\lambda}_n, \dot{\mathbf{V}}) + (\vec{\lambda}_{Gn}, \dot{\vec{\omega}}) - (\vec{\lambda}_n, \mathbf{j}) +$$

$$+\mu_n \sum_{m=1}^{\infty} \beta_{nm} \int_{-\infty}^{t} \frac{\ddot{s}_m(\tau)\, d\tau}{\sqrt{t-\tau}} - \left(\vec{\gamma}_n^{\,*}, \int_{-\infty}^{t} \frac{\dot{\vec{\Omega}}^*(\tau)\, d\tau}{\sqrt{t-\tau}}\right) = 0,$$

$$n = 1, 2, \ldots,$$

in which, compared to (2.58), included are, besides generalized velocity $\vec{\Omega}$, new tensors $\underline{\mathsf{B}}^*$, J^{**}, J^* and conjugated $\bar{\mathsf{J}}^*$, and vectors $\vec{\gamma}_n^{\,*}$. Elements of those tensors and vector $\vec{\gamma}_n^{\,*}$ components, unlike (2.53), are specified by the following formulae:

$$\beta_{ij}^* = \beta_{ji}^* =$$

$$= \frac{\rho k_s}{\sqrt{\pi}} \sum_{\mu=1}^{K} \int_{L_\mu} \sqrt{|\mathbf{v}_\nu|} \left[-(\mathbf{R}^* \times \vec{\nu})_i + \frac{\partial \overset{\circ}{\Psi}{}_i^{*}}{\partial \nu}\right]\left[-(\mathbf{R}^* \times \vec{\nu})_j + \frac{\partial \overset{\circ}{\Psi}{}_j^{*}}{\partial \nu}\right] b^{3/2} \varphi(x')\, ds,$$

$$\gamma_{nj}^* = \frac{\rho k_s}{\sqrt{\pi}} \sum_{\mu=1}^{K} \int_{L_\mu} \sqrt{|\mathbf{v}_\nu|} \left[-(\mathbf{R}^* \times \vec{\nu})_j + \frac{\partial \overset{\circ}{\Psi}{}_j^{*}}{\partial \nu}\right] \frac{\partial \overset{\circ}{\varphi}{}_n^{*}}{\partial \nu} b^{3/2} \varphi(x')\, ds, \tag{2.65}$$

$$J_{ij}^{**} = J_{ji}^{**} = \rho \int_Q (\mathbf{R}^* \times \mathbf{i}_i\,,\, \mathbf{R}^* \times \mathbf{i}_j)\, dQ - \rho \oint_{S+\Sigma} \frac{\partial \Psi_i^*}{\partial \nu} \Psi_j^*\, dS\,,$$

$$J_{ji}^{*} = \rho \int_Q (\mathbf{R}^* \times \mathbf{i}_i\,,\, \mathbf{R} \times \mathbf{i}_j)\, dQ - \rho \oint_{S+\Sigma} \frac{\partial \Psi_j}{\partial \nu} \Psi_i^*\, dS\,,$$

where potentials $\overset{\circ}{\Psi}{}_j^{*}$, $\overset{\circ}{\varphi}{}_n^{*}$ are components of a harmonic vector function being the solution of boundary-value Problem (2.63) for the case of absence of ribs.

2.4.2b. C a v i t y c o m p l e t e l y f i l l e d w i t h l i q u i d . For the same coordinate origin choice, as in 2.4.2a., assumed $s_n \equiv 0$ in Equations (2.64), we obtain the following set of equations:

$$(m^\circ + m)\, \dot{\mathbf{V}} = \mathbf{P}, \quad (\mathsf{J}^\circ + \mathsf{J}^{(0)}, \dot{\vec{\omega}}) + (\mathsf{J}_1^*, \dot{\vec{\Omega}}^*) = \mathbf{M}_G, \tag{2.66}$$

$$(\mathsf{J}^{**}, \dot{\vec{\Omega}}^*) + (\mathsf{J}^*, \dot{\vec{\omega}}) + \left(\mathsf{B}^*, \int_{-\infty}^{t} \frac{\dot{\vec{\Omega}}^*(\tau)\, d\tau}{\sqrt{t-\tau}}\right) = 0.$$

These equations are very close in structure to Equations (1.65), i.e. to the mathematical model of eddy currents in HECF elements.

2.4.3. Asymptotical models of liquid vortex motion.

2.4.3a. L i q u i d v o r t i c i t y u n i f o r m l y d i s t r i b u t e d i n Q v o l u m e . Suppose that vortex field, which appeared in Q domain resulting from stalling of vortices from rib edges, is uniform all over Q volume, that corresponds to $\vec{\Omega}_k \equiv 0$ in (2.57) for $k = 2, 3, \ldots, 2N$. Introduce now the same denominations as in [77]: $\vec{\Omega}_1 \equiv \vec{\Omega}$; $\mathsf{J}_1^{*\prime} \equiv \mathsf{J}^*$; $\mathsf{B}_1 \equiv \mathsf{B}$. Then Equations (2.57) convert to the following ones:

$$(m^\circ + m)\,\dot{\mathbf{V}}_O + \dot{\vec{\omega}} \times (\mathbf{L}^\circ + \mathbf{L}) + \sum_{n=1}^{\infty} \vec{\lambda}_n \ddot{s}_n = \mathbf{P},$$

$$(\mathrm{J}^\circ + \mathrm{J}^{(0)}, \dot{\vec{\omega}}) + (\mathbf{L}^\circ + \mathbf{L}) \times \dot{\mathbf{V}}_O - (\mathbf{L}^\circ + \mathbf{L}) \times \mathbf{j} +$$

$$+ (\mathrm{J}^*, \dot{\vec{\Omega}}) + \sum_{n=1}^{\infty} [\vec{\lambda}_{On} \ddot{s}_n - (\vec{\lambda}_n \times \mathbf{j})\, s_n] = \mathbf{M}_O, \qquad (2.67)$$

$$(\mathrm{J}^*, \dot{\vec{\Omega}}) + (\mathrm{J}^*, \dot{\vec{\omega}}) + \left(\mathrm{B}, \int_{-\infty}^{t} \frac{\dot{\vec{\Omega}}(\tau)\, d\tau}{\sqrt{t-\tau}}\right) - \sum_{n=1}^{\infty} \vec{\gamma}_n \int_{-\infty}^{t} \frac{\ddot{s}_n(\tau)\, d\tau}{\sqrt{t-\tau}} = 0,$$

$$\mu_n (\ddot{s}_n + \omega_n^2 s_n) + (\vec{\lambda}_n, \dot{\mathbf{V}}_O) + (\vec{\lambda}_{On}, \dot{\vec{\omega}}) - (\vec{\lambda}_n, \mathbf{j}) +$$

$$+ \mu_n \sum_{m=1}^{\infty} \beta_{nm} \int_{-\infty}^{t} \frac{\ddot{s}_m(\tau)\, d\tau}{\sqrt{t-\tau}} - \left(\vec{\gamma}_n, \int_{-\infty}^{t} \frac{\dot{\vec{\Omega}}(\tau)\, d\tau}{\sqrt{t-\tau}}\right) = 0.$$

These equations are close in concept to Equations (14) of [77, p. 110, 111] and have the same form.

2.4.3b. Absolute motion of liquid in Q domain, nonvortex outside ribs sharp edges vicinity. This case corresponds to $\vec{\Omega} + \vec{\omega} \equiv 0$ in (2.67). Eliminate $\vec{\Omega}$ from the second equation with the help of the third one, then assume $\vec{\Omega} = -\vec{\omega}$. We obtain then equations which correspond to potential motion of liquid in Q domain with the exeption of small vicinity of ribs edges:

$$(m^\circ + m)\,\dot{\mathbf{V}}_O + \dot{\vec{\omega}} \times (\mathbf{L}^\circ + \mathbf{L}) + \sum_{n=1}^{\infty} \vec{\lambda}_n \ddot{s}_n = \mathbf{P},$$

$$(\mathrm{J}^\circ + \mathrm{J}^{(0)}, \dot{\vec{\omega}}) + (\mathbf{L}^\circ + \mathbf{L}) \times \dot{\mathbf{V}}_O - (\mathbf{L}^\circ + \mathbf{L}) \times \mathbf{j} + \sum_{n=1}^{\infty} [\vec{\lambda}_{On} \ddot{s}_n - (\vec{\lambda}_n \times \mathbf{j})\, s_n] +$$

$$+ \left(\mathrm{B}, \int_{-\infty}^{t} \frac{\dot{\vec{\omega}}(\tau)\, d\tau}{\sqrt{t-\tau}}\right) + \sum_{n=1}^{\infty} \vec{\gamma}_n \int_{-\infty}^{t} \frac{\ddot{s}_n(\tau)\, d\tau}{\sqrt{t-\tau}} = \mathbf{M}_O, \qquad (2.68)$$

$$\mu_n (\ddot{s}_n + \omega_n^2 s_n) + (\vec{\lambda}_n, \dot{\mathbf{V}}_O) + (\vec{\lambda}_{On}, \dot{\vec{\omega}}) - (\vec{\lambda}_n, \mathbf{j}) +$$

$$+ \mu_n \sum_{m=1}^{\infty} \beta_{nm} \int_{-\infty}^{t} \frac{\ddot{s}_m(\tau)\, d\tau}{\sqrt{t-\tau}} + \left(\vec{\gamma}_n, \int_{-\infty}^{t} \frac{\dot{\vec{\omega}}(\tau)\, d\tau}{\sqrt{t-\tau}}\right) = 0,$$

$$n = 1, 2, \ldots$$

The set of Equations (2.68) is equivalent to (2.115) of [76], if we consider generalized coordinates s_n in (2.68) corresponding to 'floating cover'. The transformation of one set of equations to the other will be considered in detail further on, in Article 2.5, for the example of an axially symmetric cavity.

2.4.4. Ideal liquid. Nonvortex absolute motion of the whole liquid mass (an analogue of a magnet with a superconducting core (1.69)). This case is realized for $B \equiv 0,\ \vec{\gamma}_n \equiv 0,\ \beta_{nm} \equiv 0\ (n,\ m = 1, 2, \ldots)$ in (2.68). The adjacent equations have the form

$$(m^\circ + m)\,\dot{\mathbf{V}}_O + \dot{\vec{\omega}} \times (\mathbf{L}^\circ + \mathbf{L}) + \sum_{n=1}^{\infty} \vec{\lambda}_n \ddot{s}_n = \mathbf{P},$$

$$(\mathrm{J}^\circ + \mathrm{J}, \dot{\vec{\omega}}) + (\mathbf{L}^\circ + \mathbf{L}) \times \dot{\mathbf{V}}_O - (\mathbf{L}^\circ + \mathbf{L}) \times \mathbf{j} + \\ + \sum_{n=1}^{\infty} [\vec{\lambda}_{On} \ddot{s}_n - (\vec{\lambda}_n \times \mathbf{j})\, s_n] = \mathbf{M}_O, \tag{2.69}$$

$$\mu_n (\ddot{s}_n + \omega_n^2 s_n) + (\vec{\lambda}_n, \dot{\mathbf{V}}_O) + (\vec{\lambda}_{On}, \dot{\vec{\omega}}) - (\vec{\lambda}_n, \mathbf{j}) = 0, \quad n = 1, 2, \ldots$$

The set of equations (2.69) coincides, up to denominations, with the general equations of perturbed motion of a solid with cavity partially filled with ideal liquid, that were first derived by G.S.Narimanov [68, 70].

2.5. Axially Symmetric Body. Particular Cases of General Equations of Perturbed Motion

Consider a body characterized by axial symmetry with respect to Ox axis, having a cavity with radial and ring shape 1 inner ribs, the walls of the cavity being a surface of revolution. The set of Equations (2.57) splits in this case into four independent subsets of scalar equations describing perturbed motion in planes $O^\circ x^\circ y^\circ$ and $O^\circ x^\circ z^\circ$, rotation about Ox axis and motion along $O^\circ x^\circ$ axis.

If there are only ring shape ribs in the cavity, we are interested in the motion in planes $O^\circ x^\circ y^\circ$ and $O^\circ x^\circ z^\circ$ and along the longitudinal axis $O^\circ x^\circ$. In the first two cases we ought to keep both sets of coordinate functions describing vortex motion, introduced in Article 2.2, i.e. assume $1 \le k \le 2N$. In the last case, corresponding to axisymmetric oscillations of liquid, we ought to keep only the second set of coordinate functions, i.e. $N + 1 \le k \le 2N$, except for functions $T_k(\theta)$ (See Art. 8.6). If there are only radial ribs in the cavity, we are interested in the motion in planes $O^\circ x^\circ y^\circ$ and $O^\circ x^\circ z^\circ$ and rotation about Ox axis. We can confine to the first set of coordinate functions only, i.e. assume $1 \le k \le 2N$.

Let us write down, for instance, equations of motion in $O^\circ x^\circ z^\circ$ plane and rotation about Ox axis for the case of radial ribs, introducing new denominations.

Motion in $O^\circ x^\circ z^\circ$ plane and rotation around Ox axis:

$$\Omega_{11} = \Gamma_1, \quad \Omega_{12} = \Omega_1, \quad \Omega_{k1} = \Gamma_k, \quad \Omega_{k2} = \Omega_k,$$

$$V_{O3} = V_O, \quad \omega_1 = \gamma, \quad \omega_2 = \omega = \dot{\psi},$$

$$J_{11} = I, \quad J_{11}^{\circ} = I^{\circ}, \quad J_{11}^{(0)} = I^{(0)}, \quad J_{11}^{*} = I_1^{*}, \quad J_{11}^{kk} = I_k^{*}, \tag{2.70}$$

$$J_{22} = J, \quad J_{22}^{\circ} = J^{\circ}, \quad J_{22}^{(0)} = J^{(0)}, \quad J_{22}^{*} = J_1^{*}, \quad J_{22}^{kk} = J_k^{*},$$

$$I_1^{*} = I^{(0)} - I, \quad J_1^{*} = J^{(0)} - J, \quad \beta_{11}^{kl} = \beta_{kl}^{*}, \quad \beta_{22}^{kl} = \overset{\circ}{\beta}_{kl}, \quad \gamma_{kn2} = \gamma_{nk},$$

where J and I are N.Joukowski's associated moments of inertia of liquid, calculated for a cavity with ribs and 'fixed cover' substituting the liquid free surface.

Omitting some indices which are insignificant now, we obtain the following independent sets of equations.

1. Motion in $O^{\circ}x^{\circ}z^{\circ}$ plane:

$$(m^{\circ} + m)\dot{V}_O - (m^{\circ} + m)x_G\dot{\omega} + \sum_{n=1}^{\infty} \lambda_n \ddot{s}_n = P_z,$$

$$(J^{\circ} + J^{(0)})\,\dot{\omega} - (m^{\circ} + m)\,x_G\dot{V}_O + J^{*}\dot{\Omega}_1 - (m^{\circ} + m)\,x_G j\,\psi +$$

$$+ \sum_{n=1}^{\infty} (\lambda_{On}\,\ddot{s}_n + \lambda_{nj}\,s_n) = M_{Oy}, \tag{2.71}$$

$$J_1^{*}\,(\dot{\Omega}_1 + \dot{\omega}) + \sum_{n=1}^{N} \overset{\circ}{\beta}_{1k} \int_{-\infty}^{t} \frac{\dot{\Omega}_k(\tau)\,d\tau}{\sqrt{t-\tau}} - \sum_{n=1}^{\infty} \gamma_{n1} \int_{-\infty}^{t} \frac{\ddot{s}_n(\tau)\,d\tau}{\sqrt{t-\tau}} = 0,$$

$$J_k^{*}\,\dot{\Omega}_k + \sum_{n=1}^{N} \overset{\circ}{\beta}_{kl} \int_{-\infty}^{t} \frac{\dot{\Omega}_l(\tau)\,d\tau}{\sqrt{t-\tau}} - \sum_{n=1}^{\infty} \gamma_{nk} \int_{-\infty}^{t} \frac{\ddot{s}_n(\tau)\,d\tau}{\sqrt{t-\tau}} = 0,$$

$$\mu_n\,(\,\ddot{s}_n + \omega_n^2 s_n\,) + \lambda_n\,\dot{V}_O + \lambda_{On}\,\dot{\omega} + \lambda_{nj}\,\psi +$$

$$+\,\mu_n \sum_{m=1}^{\infty} \beta_{nm} \int_{-\infty}^{t} \frac{\ddot{s}_m(\tau)\,d\tau}{\sqrt{t-\tau}} - \sum_{k=1}^{N} \gamma_{nk} \int_{-\infty}^{t} \frac{\dot{\Omega}_k(\tau)\,d\tau}{\sqrt{t-\tau}} = 0,$$

$$\omega = \dot{\psi}, \quad k = 2, 3, \ldots, N, \quad n = 1, 2, \ldots$$

2. Rotation around Ox axis:

$$(I^{\circ} + I^{(0)})\,\dot{\gamma} + I_1^{*}\,\dot{\Gamma}_1 = M_{Ox},$$

$$I_1^{*}\,(\dot{\Gamma}_1 + \dot{\gamma}) + \sum_{k=1}^{N} \beta_{1k}^{*} \int_{-\infty}^{t} \frac{\dot{\Gamma}_k(\tau)\,d\tau}{\sqrt{t-\tau}} = 0, \tag{2.72}$$

$$I_k^{*}\,\dot{\Gamma}_k + \sum_{l=1}^{N} \beta_{kl}^{*} \int_{-\infty}^{t} \frac{\dot{\Gamma}_l(\tau)\,d\tau}{\sqrt{t-\tau}} = 0, \quad k = 2, 3, \ldots, N.$$

Further on we discuss Equations (2.71) and (2.72) in a more detail.

In the considered case of a cavity of revolution all non-diagonal elements of tensors $\mathbf{J}_k^{*}$ in (2.54b) are identically zero. Besides that, all the integrals over Q domain and over the surface $S + \Sigma$, involved in the expressions of those tensors' diagonal elements, become zero after substituting into them terms of Series (2.55) with even powers of y and z. Functions $X_k(x)$, $Y_k(y)$, $Z_k(z)$ of (2.55) can therefore be expressed in the form

$$X_k(x) = x + \sum_{\nu=2,3,\ldots}^{C_k^{(2)}+1} A_{k\nu}\, x^{\nu}, \qquad Y_k(y) = y + \sum_{\nu=2,3,\ldots}^{C_k^{(2)}+1} B_{k\nu}\, y^{2\nu-1},$$

$$Z_k(z) = z + \sum_{\nu=2,3,\ldots}^{C_k^{(2)}+1} C_{k\nu}\, z^{2\nu-1}, \qquad C_k^{(2)} = \frac{k!}{2(k-2)!}, \tag{2.73}$$

$$k = 2, 3, \ldots, N.$$

In the first-order approximation $X_1 = x$, $Y_1 = y$, $Z_1 = z$. . The number of Equations (2.54a) will now be $3C_k^{(2)}$.

Let us consider Equations (2.71) of second-order approximation for example. Omitting irrelevant indices we obtain from (2.73)

$$X_2(x) = x + A\,x^2, \qquad Y_2(y) = y + B\,y^3, \qquad Z_2(z) = z + C\,z^3. \tag{2.74}$$

Substituting Expressions (2.74) into (2.54a) and considering axial symmetry of the cavity, we come to a set of three nonhomogeneous equations

$$a_{12}\,B + a_{13}\,C = -\,I^*, \qquad a_{21}\,B + a_{23}\,C = -\,J^*,$$

$$a_{31}\,A + a_{32}\,B = -\,J^*, \tag{2.75}$$

where

$$a_{12} = \int_Q y^4\,dQ - \oint_{S+\Sigma} \Psi_1\, y^3\, \nu_z\, dS,$$

$$a_{13} = \int_Q z^4\,dQ + \oint_{S+\Sigma} \Psi_1\, z^3\, \nu_y\, dS = a_{12},$$

$$a_{21} = \int_Q x^3\,dQ + \oint_{S+\Sigma} \Psi_2\, x^2\, \nu_z\, dS, \tag{2.76}$$

$$a_{31} = \int_Q x^3\,dQ - \oint_{S+\Sigma} \Psi_3\, x^2\, \nu_y\, dS = a_{21},$$

$$a_{23} = \int_Q z^4\,dQ - \oint_{S+\Sigma} \Psi_2\, z^3\, \nu_x\, dS,$$

$$a_{32} = \int_Q y^4\,dQ + \oint_{S+\Sigma} \Psi_3\, y^3\, \nu_x\, dS = a_{23}.$$

Equations (2.75) with coefficients specified by Expressions (2.76), have the following solution:

$$B = C = -\frac{\Gamma^*}{2a_1}, \qquad A = \frac{1}{a_2}\left(-J^* + \frac{a_3\Gamma^*}{2a_1}\right), \tag{2.77}$$

where

$$a_1 = a_{12} = a_{13}, \qquad a_2 = a_{21} = a_{31}, \qquad a_3 = a_{32} = a_{23}.$$

It we suppose the ribs not radial, but coaxial, with the axis coinciding with Ox, we obtain

$$\Gamma^* = 0, \qquad B = C = 0, \qquad A = -\frac{J^*}{a_2}. \tag{2.78}$$

In this case vorticity changes linearly with respect to x.

If we suppose $\Omega_k \equiv 0$, $\gamma_{kl} \equiv 0$, $\beta_{nk} \equiv 0$, $\beta_{nm} \equiv 0$ $(k, l, n, m = 1, 2, \ldots)$ in Equations (2.71), those equations convert to the following ones:

$$(m^\circ + m)\dot{V} - (m^\circ + m)x_G\dot{\omega} + \sum_{n=1}^{\infty} \lambda_n \ddot{s}_n = P_z,$$

$$(J^\circ + J)\,\dot{\omega} - (m^\circ + m)\,x_G\dot{V} - (m^\circ + m)\,x_G j\,\psi + \sum_{n=1}^{\infty} (\lambda_{On}\,\ddot{s}_n + \lambda_{nj}\,s_n) = M_{Oy}, \tag{2.79}$$

$$\mu_n\,(\,\ddot{s}_n + \omega_n^2 s_n\,) + \lambda_n\,\dot{V} + \lambda_{On}\,\dot{\omega} + \lambda_{nj}\,\psi = 0,$$

$$n = 1, 2, \ldots, \quad \omega = \dot{\psi}$$

These are a particular case of (2.25b), corresponding to the axisymmetric body's motion in the yawing plane. They coincide, up to denominations, with those obtained by G.S.Narimanov [70].

2.6. Axisymmetric Body. Coefficients Conversion Formulae

Let us now derive the formulae connecting all the coefficients of equations resulting from (2.71) ('fixed cover' scheme), corresponding to uniform vorticity of liquid all over the volume ($\Omega_k \equiv 0$ for $k \neq 1$), and coefficients of the equations for the considered particular case of the 'floating cover' scheme. These formulae simplify usage of numerical values of those coefficients (both theoretical and experimental ones), presented in literature for the 'floating cover' scheme equations of perturbed motion (see [76]).

Let us introduce once more the temporary denomination 'prime' for coefficients and generalized coordinates corresponding to the 'fixed cover' scheme (the same denominations keep for the quantities invariant of the selected scheme). Besides that, we omit subscript '1' at Ω_1 and at coefficients γ_{n1}, J_1^* and change $\overset{\circ}{\beta}_{11}$ to β.

Equations (2.71) for $\Omega_k \equiv 0$ $(k = 2, 3, \ldots, N)$ gain the following form:

$$(m^\circ + m)\dot{V} - (m^\circ + m)x_G\dot{\omega} + \sum_{n=1}^{\infty} \lambda_n \ddot{s}'_n = P_z,$$

$$(J^\circ + J' + J^{*\prime})\,\dot{\omega} + J^{*\prime}\dot{\Omega} - (m^\circ + m)\,x_G\dot{V} -$$

$$- (m^\circ + m)\,x_G j\,\psi + \sum_{n=1}^{\infty} (\lambda'_{On}\ddot{s}'_n + \lambda_n j\, s'_n) = M_{Oy}, \qquad (2.80)$$

$$J^{*\prime}\,\dot{\Omega} + J^{*\prime}\dot{\omega} + \beta' \int_{-\infty}^{t} \frac{\dot{\Omega}(\tau)\,d\tau}{\sqrt{t-\tau}} - \sum_{n=1}^{\infty} \gamma'_n \int_{-\infty}^{t} \frac{\ddot{s}_n(\tau)'\,d\tau}{\sqrt{t-\tau}} = 0,$$

$$\mu_n\,(\ddot{s}'_n + \omega_n^2\, s'_n) + \lambda_n\dot{V} + \lambda'_{On}\dot{\omega} + \lambda_n j\,\psi +$$

$$+ \mu_n \sum_{m=1}^{\infty} \beta_{nm} \int_{-\infty}^{t} \frac{\ddot{s}_m(\tau)'\,d\tau}{\sqrt{t-\tau}} - \gamma'_n \int_{-\infty}^{t} \frac{\dot{\Omega}(\tau)\,d\tau}{\sqrt{t-\tau}} = 0, \qquad n = 1, 2, \ldots$$

Here J' is the associated moment of inertia of liquid (see (2.52)). Let us now write down two equivalent expressions of perturbed free surface $\xi\,(y, z, t)$ in $O^\circ x^\circ y^\circ z^\circ$ coordinate system.

In this case, distinct from (2.28), the expressions are as follows:

$$\xi\,(y, z, t) = z\,(\psi + \vartheta) + \sum_{n=1}^{\infty} s_n\psi_n, \qquad (2.81a)$$

$$\xi\,(y, z, t) = z\,\psi + \sum_{n=1}^{\infty} s'_n\,\psi_n, \qquad (2.81b)$$

where $\dot{\vartheta} = \Omega$, $\dot{\psi} = \omega$, $\psi_n = \frac{\partial\varphi_n}{\partial x}\Big|_{\Sigma}$ – is a complete set of functions corresponding to forms of natural oscillations of liquid in the cavity. Those functions are orthogonal on domain Σ , (associated with the unperturbed free surface of liquid). Expressions (2.18), (2.20), and (2.26) result in the following expansion:

$$z = \sum_{n=1}^{\infty} \frac{\lambda_n\,\psi_n}{\mu_n\,\kappa_n}, \qquad \kappa_n = \frac{\omega_n^2}{j}\,. \qquad (2.82)$$

Using it we find from (2.81a) and (2.81b)

$$s'_n = s_n + \frac{\lambda_n\vartheta}{\mu_n\kappa_n}. \qquad (2.83)$$

Substituting this expression into the first Equation (2.80), considering Formulae (2.27) and (2.32),

$$\sum_{n=1}^{\infty} \frac{\lambda_n^2}{\mu_n\,\kappa_n} = \rho J_y\,, \qquad x_F = x_G + \frac{\rho J_y}{m^\circ + m}\,, \qquad (2.84)$$

where J_y is the equatorial moment of inertia of Σ domain, x_F – is the coordinate of metacentre, we obtain the following equation:

$$(m^\circ + m)\dot{V} - (m^\circ + m)x_F\dot{\omega} + \rho J_z(\dot{\omega} + \dot{\Omega}) + \sum_{n=1}^{\infty} \lambda_n \ddot{s}_n = P_z. \tag{2.85}$$

Further, substituting $J^{*\prime}\,\dot{\Omega}$ from the third Equation (2.80) and $j\lambda_n = \dfrac{\lambda_n\omega_n^2}{\kappa_n}$ from the fourth one into the second Equation (2.80), and considering (2.83) and (2.84) we obtain

$$\left[J^\circ + J' - 2\sum_{n=1}^{\infty}\frac{\lambda_n}{\mu_n\kappa_n}\left(\lambda'_{On} - \frac{\lambda_n}{\kappa_n}\right)\right]\dot{\omega} + \sum_{n=1}^{\infty}\frac{\lambda_n}{\mu_n\kappa_n}\left(\lambda'_{On} - \frac{\lambda_n}{\kappa_n}\right)(\dot{\omega} + \dot{\Omega}) -$$

$$- (m^\circ + m)x_F\dot{V} - (m^\circ + m)x_G j\,\psi + \sum_{n=1}^{\infty}\left(\lambda'_{On} - \frac{\lambda_n}{\kappa_n}\right)\ddot{s}_n -$$

$$- \left(\beta' - 2\sum_{n=1}^{\infty}\frac{\gamma'_n\lambda_n}{\mu_n\kappa_n} + \sum_{n=1}^{\infty}\frac{\lambda_n}{\kappa_n}\sum_{m=1}^{\infty}\frac{\beta_{nm}\lambda_m}{\mu_m\kappa_m}\right)\int_{-\infty}^{t}\frac{\dot{\Omega}(\tau)\,d\tau}{\sqrt{t-\tau}} +$$

$$+ \sum_{n=1}^{\infty}\gamma'_n\int_{-\infty}^{t}\frac{\ddot{s}_n(\tau)\,d\tau}{\sqrt{t-\tau}} - \sum_{n=1}^{\infty}\frac{\lambda_n}{\kappa_n}\sum_{m=1}^{\infty}\beta_{nm}\int_{-\infty}^{t}\frac{\ddot{s}_m(\tau)\,d\tau}{\sqrt{t-\tau}} = 0. \tag{2.86}$$

Finally, using the substituion of (2.83) into the fourth Equation (2.80), we can present the latter in the form of

$$\mu_n(\ddot{s}_n + \omega_n^2 s_n) + \frac{\lambda_n}{\kappa_n}(\dot{\omega} + \dot{\Omega}) + \lambda_n j(\vartheta + \psi) +$$

$$+ \lambda_n\dot{V} + \left(\lambda'_{On} - \frac{\lambda_n}{\kappa_n}\right)\dot{\omega} + \mu_n\sum_{m=1}^{\infty}\beta_{nm}\int_{-\infty}^{t}\frac{\ddot{s}_m(\tau)\,d\tau}{\sqrt{t-\tau}} -$$

$$- \left(\gamma'_n - \sum_{m=1}^{\infty}\beta_{nm}\frac{\lambda_m}{\kappa_m}\right)\int_{-\infty}^{t}\frac{\dot{\Omega}(\tau)\,d\tau}{\sqrt{t-\tau}} = 0. \tag{2.87}$$

Interchanging subscripts n and m in the last term of (2.86) and considering symmetry of coefficients $\beta_{nm} = \beta_{mn}$, we obtain from Equations (2.85) - (2.87) all together the following set of equations corresponding to the 'floating cover' scheme:

$$(m^\circ + m)\dot{V} - (m^\circ + m)x_F\dot{\omega} + \rho J_y(\dot{\omega} + \dot{\Omega}) + \sum_{n=1}^{\infty}\lambda_n\ddot{s}_n = P_z,$$

$$(J^\circ + J)\,\dot{\omega} - (m^\circ + m)\,x_F\dot{V} - (m^\circ + m)\,x_F j\,\psi + \sum_{n=1}^{\infty}\lambda_{On}\ddot{s}_n +$$

$$+ \sum_{n=1}^{\infty}\frac{\lambda_{On}\lambda_n}{\mu_n\kappa_n}(\dot{\omega} + \dot{\Omega}) - \beta\int_{-\infty}^{t}\frac{\dot{\Omega}(\tau)\,d\tau}{\sqrt{t-\tau}} + \sum_{n=1}^{\infty}\gamma_n\int_{-\infty}^{t}\frac{\ddot{s}_n(\tau)\,d\tau}{\sqrt{t-\tau}} = M_{Oy},$$

$$\mu_n(\ddot{s}_n + \omega_n^2 s_n) + \frac{\lambda_n}{\kappa_n}(\dot{\omega} + \dot{\Omega}) + \lambda_n j(\vartheta + \psi) + \lambda_n\dot{V} + \lambda_{On}\dot{\omega} +$$

$$+\mu_n \sum_{m=1}^{\infty} \beta_{nm} \int_{-\infty}^{t} \frac{\ddot{s}_m(\tau)\,d\tau}{\sqrt{t-\tau}} - \gamma_n \int_{-\infty}^{t} \frac{\dot{\Omega}(\tau)\,d\tau}{\sqrt{t-\tau}} = 0. \quad n = 1, 2, \ldots. \tag{2.88}$$

where

$$\lambda_{On} = \lambda'_{On} - \frac{\lambda_n}{\kappa_n}, \qquad \lambda'_{On} = \lambda_{On} + \frac{\lambda_n}{\kappa_n},$$

$$J = J' - 2\sum_{n=1}^{\infty} \frac{\lambda'_{On}\lambda_n}{\mu_n\kappa_n} + \sum_{n=1}^{\infty} \frac{\lambda_n^2}{\mu_n\kappa_n^2},$$

$$J' = J + 2\sum_{n=1}^{\infty} \frac{\lambda'_{On}\lambda_n}{\mu_n\kappa_n} + \sum_{n=1}^{\infty} \frac{\lambda_n^2}{\mu_n\kappa_n^2},$$

$$\gamma_n = \gamma'_n - \sum_{m=1}^{\infty} \frac{\beta_{nm}\lambda_m}{\kappa_m}, \qquad \gamma'_n = \gamma_n + \sum_{m=1}^{\infty} \frac{\beta_{nm}\lambda_m}{\kappa_m}, \tag{2.89}$$

$$\beta = \beta' - 2\sum_{n=1}^{\infty} \frac{\gamma'_n\lambda_n}{\mu_n\kappa_n} + \sum_{n=1}^{\infty} \frac{\lambda_n}{\kappa_n} \sum_{m=1}^{\infty} \frac{\beta_{nm}\lambda_m}{\mu_m\kappa_m},$$

$$\beta' = \beta + 2\sum_{n=1}^{\infty} \frac{\gamma_n\lambda_n}{\mu_n\kappa_n} + \sum_{n=1}^{\infty} \frac{\lambda_n}{\mu_n\kappa_n} \sum_{m=1}^{\infty} \frac{\beta_{nm}\lambda_m}{\mu_m\kappa_m}.$$

Here J is the associated moment of inertia of liquid, calculated from the 'floating cover' scheme, according to [76].

The set of Equations (2.88) is completed with the third of Equations (2.80), but to change over to the equations corresponding to the asymptotic Model (2.68), the third Equation (2.80) must be substituted by the following finite expression and initial condition

$$\Omega + \omega = 0, \qquad (\vartheta + \psi)\Big|_{t=0} = 0, \tag{2.90}$$

which corresponds to volume-uniform vorticity of liquid in its absolute motion and its time integral being identically zero. Expressions (2.90) substituted into (2.87), we obtain the following simplified set of equations

$$(m^\circ + m)\dot{V} - (m^\circ + m)x_F\dot{\omega} + \sum_{n=1}^{\infty} \lambda_n \ddot{s}_n = P_z,$$

$$(J^\circ + J)\dot{\omega} - (m^\circ + m)x_F\dot{V} - (m^\circ + m)x_F j\psi + \sum_{n=1}^{\infty} \lambda_{On}\ddot{s}_n +,$$

$$+\beta \int_{-\infty}^{t} \frac{\dot{\omega}(\tau)\,d\tau}{\sqrt{t-\tau}} + \sum_{n=1}^{\infty} \gamma_n \int_{-\infty}^{t} \frac{\ddot{s}_n(\tau)\,d\tau}{\sqrt{t-\tau}} = M_{Oy},$$

$$\mu_n(\ddot{s}_n + \omega_n^2 s_n) + \lambda_n\dot{V} + \lambda_{On}\dot{\omega} +$$

$$+\mu_n \sum_{m=1}^{\infty} \beta_{nm} \int_{-\infty}^{t} \frac{\ddot{s}_m(\tau)\,d\tau}{\sqrt{t-\tau}} + \gamma_n \int_{-\infty}^{t} \frac{\dot{\omega}(\tau)\,d\tau}{\sqrt{t-\tau}} = 0. \quad n = 1, 2, \ldots, \tag{2.91}$$

This set of equations is identical to that obtained for the axisymmetric domain from (2.68) and from Expressions (2.115) of [76] for the yawing plane motion.

The following considerations are to be noted in association with the obtained results.

1. Change-over to the mathematical model, considering kinetic energy of liquid vortex motion for volume-uniform vorticity, does not require calculation of some new coefficients in addition to those already used in the model (2.115) of [76].

2. All the coefficients of (2.89), depending on the way of introducing generalized coordinates, which describe wave motions of liquid (counted from the 'fixed cover' or 'floating cover'), split into two groups. Coefficients of the first group, associated with the ideal liquid potential motion, depend on the coordinate origin choice, while coefficients of the second group, associated with the liquid vortex motion, do not depend on it, because coefficients γ_n and β are independent of coordinate origin, that is proved in [79] (and also in [76]).

3. The first three Formulae (2.89) coincide with (2.134) of [76] (the formula for λ'_{On} – is a particular case of (2.26)). The last three Formulae (2.89) are a generalization of Formulae (2.135) of [76], to which they reduce, supposed $\beta_{nm} = 0$ for $n \neq m$; $\beta_{nm} = \beta_n$, for $m = n$.

Let us present an example of transformation, according to Formulae (2.82), of coefficients corresponding to the cavity of the shape of a rectangular circular cylinder without ribs having r_0 radius (liquid depth h). Formulae (2.103) of [63] yield for this case

$$\kappa_n = \frac{\xi_n}{r_0} \tanh(\xi_n \bar{h}), \qquad \mu_n = \frac{\rho\,\pi\, r_0^3 (\xi_n^2 - 1)}{2\,\xi_n^3 \tanh(\xi_n \bar{h})},$$

$$\lambda_n = \frac{\rho\,\pi\, r_0^3}{\xi_n^2}, \qquad \lambda_{On} = -\frac{\rho\,\pi\, r_0^4}{\xi_n^2}\left[\tilde{x}_0 + \bar{h} + \frac{2 - \cosh(\xi_n \bar{h})}{\xi_n \sinh(\xi_n \bar{h})}\right],$$

$$J = \rho\,\pi\, r_0^5 \left[\frac{1}{3}\left((\tilde{x}_0 + \bar{h})^3 - \tilde{x}_0^3\right) + 2\sum_{n=1}^{\infty} \frac{5\cosh(\xi_n \bar{h}) - 4}{\xi_n^3 (\xi_n^2 - 1)\sinh(\xi_n \bar{h})}\right], \tag{2.92}$$

where $\bar{h} = \dfrac{h}{r_0}$ – is dimensionless depth of liquid, $\tilde{x} = \dfrac{x}{r_0}$; $\tilde{x}_0 = \dfrac{x_0}{r_0}$; x_0 – is the distance from the metacentre to the cavity bottom, ξ_n – are roots of the equation $J'_1(\xi) = 0$, where $J_1(\xi)$ is Bessel's function of the 1-st kind, 1-st order ($\xi_1 = 1{,}8412$; $\xi_2 = 5{,}3315$; $\xi_3 = 8{,}5363$; $\xi_4 = 11{,}7060$).

Simple transformation of (2.88) and (2.92) result in

$$\lambda'_{On} = -\frac{\rho\pi r_0^4}{\xi_n^2}\left(\tilde{x}_0 + \bar{h} - \frac{2}{\xi_n}\tanh\frac{\xi_n \bar{h}}{2}\right), \tag{2.93}$$

$$J' = \rho\pi r_0^5 \left[\frac{1}{3}\left((\tilde{x}_0 + \bar{h})^3 - \tilde{x}_0^3\right) - \frac{3}{4}\bar{h} - 16\sum_{n=1}^{\infty} \frac{\tanh\dfrac{\xi_n \bar{h}}{2}}{\xi_n^3 (\xi_n^2 - 1)}\right],$$

The first of the expressions coincides for $\tilde{x}_0 = -\frac{\tilde{h}}{2}$ with that obtained by G.S.Narimanov [70], and the second one with that obtained by N.Joukowski [33].

2.7. Axisymmetric Body. Derivation of Equations of the Body – Liquid System Perturbed Motion from the Variational Principle.

Suppose that the cavity and the body are characterized by axial symmetry with respect to O_x axis. Consider the problem of a body with an axisymmetric cavity containing liquid with the help of the variational principle, namely with Ritz – Trefftz method. We apply this method as an alternative to the method used above, which is in fact Boubnov – Galerkin method. We can consider the expression for kinetic energy of the vortex liquid only, for the structure of generalized forces associated with liquid interaction with ribs remains the same.

Let us confine to the class of liquid motions, being antisymmetric with respect to Oxy andOxz planes. As it is clear from above, one of the central moments is approximation of that part of the field of relative velocities $\mathbf{v}$, which is associated with liquid vortex motion and satisfies the following equation and boundary conditions:

$$\operatorname{div} \mathbf{v} = 0, \qquad (\mathbf{v}, \vec{\nu}) \,|_{S+\Sigma} = 0\,, \tag{2.94}$$

where $\vec{\nu}$ is the unit vector of outer normal to the surface $S + \Sigma$ including the surface of internal ribs too. Let us represent that field of velocities in the form

$$\mathbf{v} = \vec{\Omega} \times \mathbf{R}^* - \kappa \nabla (\vec{\Omega}, \vec{\Psi}^*), \tag{2.95}$$

where κ is a parameter unknown so far, $\mathbf{R}^*(x, y, z)$, $\vec{\Psi}^*(x, y, z)$ – are the functions to be determined.

Introduce kinetic energy of liquid corresponding to field of velocities (2.95), which, the cavity's symmetry being considered, can be expressed in a quadratic form

$$T_F = \frac{\rho}{2} \int_Q (\mathbf{v}, \mathbf{v})\, dQ = \tag{2.96}$$

$$= \frac{\rho}{2} \sum_{j=1}^{3} \Omega_j^2 \left[\int_Q (\mathbf{i}_j \times \mathbf{R}^*)^2\, dQ + \kappa^2 \int_Q (\nabla \Psi_j^*)^2\, dQ - 2\kappa \int_Q (\nabla \Psi_j^*, \mathbf{i}_j \times \mathbf{R}^*)\, dQ \right]$$

(ρ is mass density of liquid). Consider the functionals being the coefficients of this quadratic form:

$$T_j = \frac{1}{2} \left[\int_Q (\mathbf{i}_j \times \mathbf{R}^*)^2\, dQ + \kappa^2 \int_Q (\nabla \Psi_j^*)^2\, dQ - 2\kappa \int_Q (\nabla \Psi_j^*, \mathbf{i}_j \times \mathbf{R}^*)\, dQ \right],$$

$$j = 1, 2, 3. \tag{2.97}$$

Suppose function $\mathbf{R}^*$ is chosen so, that the first term in Expression (2.95) satisfies Equation (2.94). Setting the first variation of the functionals zero, we obtain the following equations for determination of stationary values (minima in this case) T_j:

$$\delta T_j = \int_Q (\mathbf{i}_j \times \mathbf{R}^*, \mathbf{i}_j \times \delta\mathbf{R}^*)\, dQ + \kappa^2 \int_Q (\nabla \Psi_j^*, \nabla (\delta\Psi_j^*))\, dQ -$$

$$- \kappa \int_Q (\nabla \Psi_j^*, \mathbf{i}_j \times \delta\mathbf{R}^*)\, dQ - \kappa \int_Q (\mathbf{i}_j \times \mathbf{R}^*, \nabla \delta \Psi_j^*)\, dQ = 0,$$

or, after applying Green and Gauss – Ostrogradsky formulae,

$$\int_Q (\mathbf{i}_j \times \mathbf{R}^*, \mathbf{i}_j \times \delta\mathbf{R}^*)\, dQ - \kappa \oint_{S+\Sigma} (\delta\mathbf{R}^* \times \vec{\nu})_j\, \Psi_j^*\, dS -$$

$$- \kappa^2 \int_Q \Delta\Psi_j^*\, \delta\Psi_j^*\, dQ + \kappa \oint_{S+\Sigma} \left[\kappa \frac{\partial \Psi_j^*}{\partial \nu} - (\mathbf{R}^* \times \vec{\nu})_j \right] \delta\Psi_j^*\, dS = 0,$$

$$j = 1, 2, 3, \tag{2.98}$$

where ∇ and Δ are three-dimensional Hamilton and Laplace operators. Suppose variations $\delta\mathbf{R}^*$ and $\delta\Psi_j^*$ are independent, that corresponds to the conception of sequential (stepwise) construction of $\mathbf{R}^*$ and $\vec{\Psi}^*$ functions (see Article 2.3). Then, considering arbitrary character and independence of $\delta\Psi_j^*$ variations in Q domain and at its surface $S + \Sigma$, and arbitrary character of variation $\delta\mathbf{R}^*$, we obtain the following equations and boundary conditions from (2.98):

$$\Delta \Psi_j = 0, \qquad \left.\frac{\partial \Psi_j}{\partial \nu}\right|_{S + \Sigma} = (\mathbf{R}^* \times \vec{\nu})_j. \tag{2.99}$$

$$\int_Q (\mathbf{i}_j \times \mathbf{R}^*, \mathbf{i}_j \times \delta\mathbf{R}^*)\, dQ - \oint_{S+\Sigma} (\delta\mathbf{R}^* \times \vec{\nu})_j\, \Psi_j^*\, dS = 0, \tag{2.100}$$

where $\Psi_j = \kappa\, \Psi_j^* \quad (j = 1, 2, 3)$.

The condition of functionals T_j minimum, i.e. $\delta T_j = 0$, is thus equivalent to the total of boundary-value Problems (2.99) and auxiliary Equations (2.100). On the other side, if (2.99) is satisfied and if $\operatorname{div} (\vec{\Omega} \times \mathbf{R}^*) = 0$, Expression (2.95) is the solution of boundary-value Problem (2.94). We show later on, that for a certain structure of expressions of $\mathbf{R}^*$ and $\delta\mathbf{R}^*$ Equations (2.100) are equivalent to Equations (2.54b).

Let us now consider a direct method of solving the boundary-value problems, we are interested in, namely, by minimization of functionals T_j with Ritz method, that corresponds to the conception of simultaneous functions $\mathbf{R}^*$ and $\vec{\Psi}^*$ construction. Represent those functions by the following expansions:

$$\mathbf{R}^* = \sum_{k=1}^{N} (\mathrm{R}_k, \mathbf{E}_k), \qquad \vec{\Psi}^* = \sum_{k=1}^{N} (\mathbf{E}_k \times \vec{\psi}_k), \tag{2.101}$$

where R_k, $\vec{\psi}_k$, $\mathbf{E}_k$ are symmetric tensors of rank 2 and vectors respectively

$$\mathsf{R}_k = \left\{ \rho_{ij}^k \right\}, \qquad \rho_{ij}^k = 0 \quad \text{for} \quad i \neq j,$$
$$\rho_{11}^k = \xi_k(x), \qquad \rho_{22}^k = \eta_k(y), \qquad \rho_{33}^k = \zeta_k(z),$$
$$\vec{\psi}_k = \mathbf{i}_1 \psi_{1k}(x, y, z) + \mathbf{i}_2 \psi_{2k}(x, y, z) + \mathbf{i}_3 \psi_{3k}(x, y, z)\,, \tag{2.102}$$
$$\mathbf{E}_k = \mathbf{i}_1 A_k + \mathbf{i}_2 B_k + \mathbf{i}_3 C_k\,.$$

Here

$$\xi_k(x), \quad \eta_k(y), \quad \zeta_k(z), \quad \psi_{jk}(x, y, z), \qquad j = 1, 2, 3; \qquad k = 1, 2, \ldots, N$$

are sets of linearly independent functions having the necessary completeness, considering antisymmetry of the sought velocity field with respect to Oxy and Oxz planes; functions ψ_{jk} are to take into account the singularities due to ribs' existence, and N is the number of coordinate functions used. It is obvious that the structure of functions $\mathbf{R}^*(x, y, z)$ in (2.101), (2.102) ensures compliance with the equation $\operatorname{div}(\vec{\Omega} \times \mathbf{R}^*) = 0$.

If we additionally require ψ_{jk} functions to be harmonic, Equations (2.99) will be satisfied. The modification of Ritz method with such choice of coordinate functions is known in literature as Trefftz method.

The coefficients of function Sequences (2.101) minimizing Functionals (2.97) for any N value are determined, following Ritz – Trefftz technique, from the following equations

$$\frac{\partial T_j}{\partial \mathbf{e}_l} = 0, \qquad j = 1, 2, 3, \qquad l = 1, 2, \ldots, N, \tag{2.103}$$

where $\mathbf{e}_l$ is the unit vector, $\mathbf{e} = \mathbf{i}_1 + \mathbf{i}_2 + \mathbf{i}_3$, which is added to $\mathbf{E}_l$ vector before differentiation (subscript l comes from it). The derivative with respect to $\mathbf{e}_l$ vector is calculated according to a known formula:

$$\frac{\partial T_j}{\partial \mathbf{e}_l} = \lim_{\varepsilon \to 0} \frac{1}{\varepsilon} \left[T_j(\mathbf{E}_l + \mathbf{e}) - T_j(\mathbf{E}_l) \right]. \tag{2.104}$$

Using it we obtain from (2.97), (2.101)

$$\frac{\partial T_j}{\partial \mathbf{e}_l} = \sum_{k=1}^{N} \Bigg\{ \int_Q (\mathbf{i}_j \times (\mathsf{R}_k, \mathbf{E}_k), \mathbf{i}_j \times (\mathsf{R}_l, \mathbf{e}))\, dQ -$$
$$- \kappa \left[\int_Q (\nabla (\mathbf{E}_k \times \vec{\psi}_k)_j, \mathbf{i}_j \times (\mathsf{R}_l, \mathbf{e}))\, dQ + \int_Q (\nabla (\mathbf{e} \times \vec{\psi}_l)_j, \mathbf{i}_j \times (\mathsf{R}_k, \mathbf{E}_k))\, dQ \right] +$$
$$+ \kappa^2 \int_Q (\nabla (\mathbf{E}_l \times \vec{\psi}_k)_j, \nabla (\mathbf{e} \times \vec{\psi}_l)_j)\, dQ \Bigg\}, \qquad j = 1, 2, 3, \qquad l = 1, 2, \ldots, N.$$

Setting the right side equal to zero, according to (2.103), and applying Gauss – Ostrogradsky and Green formulae, with consideration of equations, we obtain the following set of equations of Trefftz method.

$$\sum_{k=1}^{N} \left\{ \int_{Q} (\mathbf{i}_j \times (\mathbf{R}_k, \mathbf{E}_k), \mathbf{i}_j \times (\mathbf{R}_l, \mathbf{e}))\, dQ - \right.$$

$$- \kappa \left[\oint_{S+\Sigma} (\mathbf{E}_k \times \vec{\psi}_k)_j ((\mathbf{R}_l, \mathbf{e}) \times \vec{\nu})\, dS + \oint_{S+\Sigma} (\mathbf{e} \times \vec{\psi}_l)_j ((\mathbf{R}_k, \mathbf{E}_k) \times \vec{\nu})_j\, dS \right] +$$

$$\left. + \kappa^2 \oint_{S+\Sigma} (\mathbf{e} \times \vec{\psi}_l)_j \frac{\partial}{\partial \nu} (\mathbf{E}_k \times \vec{\psi}_k)_j\, dS \right\} = 0, \tag{2.105}$$

$$j = 1, 2, 3; \qquad l = 1, 2, \ldots, N.$$

Consider the first group of Equations (2.105) corresponding to $j = 1$:

$$\sum_{k=1}^{N} \kappa \left[B_l \left(\kappa \oint_{S+\Sigma} \frac{\partial \psi_{3k}}{\partial \nu} \psi_{3l}\, dS - \oint_{S+\Sigma} \eta_k \psi_{3l} \nu_z\, dS \right) - \right.$$

$$\left. - C_l \left(\kappa \oint_{S+\Sigma} \frac{\partial \psi_{2k}}{\partial \nu} \psi_{2l}\, dS - \oint_{S+\Sigma} \zeta_k \psi_{2l} \nu_y\, dS \right) \right] +$$

$$+ \sum_{k=1}^{N} \left[- \kappa \left(B_k \oint_{S+\Sigma} \eta_l \psi_{3k} \nu_z\, dS - C_k \oint_{S+\Sigma} \zeta_l \psi_{2k} \nu_y\, dS \right) + B_k \int_{Q} \eta_k \eta_l\, dQ + C_k \int_{Q} \zeta_k \zeta_l\, dQ \right] = 0,$$

$$l = 1, 2, \ldots, N. \tag{2.106}$$

We are going to show that (2.106) can yield equations corresponding to (2.54a) with $j = 1$. Let us instead of harmonic functions ψ_{2k}, ψ_{3k} satisfying arbitrary boundary conditions, introduce functions $\Psi_{2k} = \kappa \psi_{2k}$, $\Psi_{3k} = \kappa \psi_{3k}$ being the solutions of boundary-value problems

$$\Delta \Psi_{2k}^{(1)} = 0, \qquad \left. \frac{\partial \Psi_{2k}^{(1)}}{\partial \nu} \right|_{S+\Sigma} = \zeta_k \nu_y, \tag{2.107}$$

$$\Delta \Psi_{3k}^{(1)} = 0, \qquad \left. \frac{\partial \Psi_{3k}^{(1)}}{\partial \nu} \right|_{S+\Sigma} = \eta_k \nu_z,$$

(superscript 1 corresponds to $j = 1$). Functions $\Psi_{2k}^{(1)}$ and $\Psi_{3k}^{(1)}$ assumed known, we can reduce the set of Equations (2.106) to the following one:

$$\sum_{k=1}^{N} \left[B_k \left(- \oint_{S+\Sigma} \Psi_{3k}^{(1)} \eta_l \nu_z\, dS + \int_{Q} \eta_k \eta_l\, dQ \right) + \right.$$

$$\left. + C_k \left(\oint_{S+\Sigma} \Psi_{2k}^{(1)} \zeta_l \nu_y\, dS + \int_{Q} \zeta_k \zeta_l\, dQ \right) \right] = 0, \tag{2.108}$$

$$l = 1, 2, \ldots, N - 1.$$

This set of equations is, in turn, equivalent up to denominations to Equation (2.100) for $j = 1$ (for $N \to \infty$) and to the set of equations, to which (2.54a) is reduced in the considered case. This result is valid for $j = 2, 3$ also, that can be proved using functions $\Psi_{3k}^{(2)}, \Psi_{1k}^{(2)}$ $(j = 2)$ and $\Psi_{2k}^{(3)}, \Psi_{1k}^{(3)}$ $(j = 3)$. Let

$$N = 2, \quad k = 1, 2, \quad l = 1, \quad \xi_1 = x, \quad \eta_1 = y, \quad \zeta_1 = z, \quad A_1 = B_1 = C_1 = 1.$$

Then (2.107) yields

$$B_2\left(- \oint_{S+\Sigma} \Psi_{32}^{(1)}\, y\, \nu_z\, dS + \int_Q \eta_2\, y\, dQ \right) + C_2 \left(\oint_{S+\Sigma} \Psi_{22}^{(1)}\, z\, \nu_y\, dS + \int_Q \zeta_2\, z\, dQ \right) =$$

$$= \oint_{S+\Sigma} (\Psi_{31}^{(1)}\, y\, \nu_z - \Psi_{21}^{(1)}\, z\, \nu_y)\, dS - \int_Q (y^2 + z^2)\, dQ \tag{2.109}$$

and similar equations for $j = 2, 3$ enabling determination of coefficients A_2, B_2, C_2. Sequential coefficients for $k = 3, 4, \ldots, N$; $l = 2, 3, \ldots, N - 1$ are found from a recurrent procedure described in Article 2.3.

Thus, applying more rigid Constraints (2.107) to coordinate functions ψ_{jl}, than Trefftz technique requires, we can reconstruct the results obtained above with Boubnov – Galerkin method with special coordinate functions selection.

Let us now return to general Trefftz method Equations (2.105). For the considered case of cavity of revolution we can confine to the values $j = 1, 2$ of subscript (rotation about Ox axis and motion in $O^\circ x^\circ z^\circ$ plane). Taking into account axial symmetry of the cavity, which results in

$$C_l \equiv B_l, \qquad \oint_{S+\Sigma} \frac{\partial \psi_{3l}}{\partial \nu}\, \psi_{3k}\, dS = \oint_{S+\Sigma} \frac{\partial \psi_{2l}}{\partial \nu}\, \psi_{2k}\, dS, \tag{2.110}$$

we obtain the following two independent sets of equations:

$$(\kappa\, \mathsf{B}' + \mathsf{B})\, \mathbf{B} = 0,$$

$$(-\kappa^2\, \mathsf{A}'' + \kappa \mathsf{A}' + \mathsf{A})\, \mathbf{A} = (-\kappa^2\, \mathsf{C}'' + \kappa\, \mathsf{C}' - \mathsf{C})\, \mathbf{B}, \tag{2.111}$$

where B', B, A'', A', A, C'', C', C are symmetric square matrices of size $N \times N$:

$$\mathsf{B}' = [b'_{lk}], \qquad \mathsf{B} = [b_{lk}],$$

$$\mathsf{A}'' = [a''_{lk}], \qquad \mathsf{A}' = [a'_{lk}], \qquad \mathsf{A} = [a_{lk}], \tag{2.112}$$

$$\mathsf{C}'' = [c''_{lk}], \qquad \mathsf{C}' = [c'_{lk}], \qquad \mathsf{C} = [c_{lk}],$$

$\mathbf{A}$ and $\mathbf{B}$ are N -dimensional vectors. Elements of Matrices (2.112) are expressed by the formulae:

$$b'_{lk} = b'_{kl} = \oint_{S+\Sigma} [(\psi_{2k}\, \zeta_l + \psi_{2l}\, \zeta_k)\, \nu_y - (\psi_{3k} \eta_l + \psi_{3l}\, \eta_k)\, \nu_z]\, dS\,, \tag{2.113}$$

$$b_{lk} = b_{kl} = \int_Q (\eta_l\, \eta_k + \zeta_l\, \zeta_k)\, dQ\,,$$

$$a''_{lk} = a''_{kl} = \oint_{S+\Sigma} \frac{\partial \psi_{3l}}{\partial \nu}\, \psi_{3k}\, dS\,, \qquad c''_{lk} = c''_{kl} = \oint_{S+\Sigma} \frac{\partial \psi_{1l}}{\partial \nu}\, \psi_{1k}\, dS\,,$$

$$a'_{lk} = a'_{kl} = \oint_{S+\Sigma} (\psi_{3k}\,\xi_l + \psi_{3l}\,\xi_k)\,\nu_z\,dS\,,$$

$$c'_{lk} = c'_{kl} = \oint_{S+\Sigma} (\psi_{1k}\,\zeta_l + \psi_{1l}\,\zeta_k)\,\nu_x\,dS\,,$$

$$a_{lk} = a_{kl} = \int_Q \xi_l\,\xi_k\,dQ\,, \qquad c_{lk} = c_{kl} = \int_Q \zeta_k\,\zeta_l\,dQ\,.$$

The symmetry of matrices B , B′ causes all the eigenvalues κ_n ($n = 1, 2, ..., N$) to be real. They correspond to N eigenvectors $\mathbf{B}^{(n)}$ determined up to a normalizing factor. So we can assume $B_1^{(n)} = 1$ ($n = 1, 2, ..., N$) . With $\mathbf{B}^{(n)}$ known, we can find the solution of the nonhomogeneous matrix Equation (2.111), i.e. N vectors $\mathbf{A}^{(n)}$. Since $\mathbf{C}^{(n)} \equiv \mathbf{B}^{(n)}$, we thus determine $\mathbf{E}_k^{(n)}$ vectors and eigenfunctions $\mathbf{R}_n^*$ and $\vec{\Psi}_n^*$ of the boundary-value problem (2.94):

$$\mathbf{R}_n^* = \sum_{k=1}^{N} (\mathrm{R}_k, \mathbf{E}_k^{(n)}), \qquad \vec{\Psi}_n^* = \sum_{k=1}^{N} (\mathbf{E}_k^{(n)} \times \vec{\psi}_k). \tag{2.114}$$

Its general solution can now be presented, using (2.95), by an expansion with respect to eigenfunctions (2.114):

$$\mathbf{v} = \sum_{n=1}^{N} \left[\vec{\Omega}_n \times \mathbf{R}_n^* - \kappa_n \nabla (\vec{\Omega}_n, \vec{\Psi}_n^*)\right], \tag{2.115}$$

which structure is identical to (2.36), (2.37), if we omit from (2.36) the terms corresponding to wave motions of liquid and keep only the first N coordinate functions. Generalized angular velocities $\vec{\Omega}_n$ in (2.115) are certainly different from (2.37), since velocity field $\mathbf{v}\,(x, y, z, t)$ is now expanded to a series with respect to another set of functions.

Contrast of those expansions is displayed particularly, in the fact, that function $\mathbf{R}_1 = \mathbf{i}_1 x + \mathbf{i}_2 y + \mathbf{i}_3 z$ is not a member of the set of Functions (2.115) any longer, and so they are not orthogonal to $\mathbf{R} \equiv \mathbf{R}_1$ in the sense of (2.54a). On the other hand, if certain conditions are met, the minimizing Sequences (2.114) converge in energy. In other words, with $N \to \infty$ $T_j \to T_{j\min} > 0$ ($j = 1, 2, 3$) and left sides of Equations (2.100) tend to zero. For no finite N value, however, can Equations (2.100) equivalent to (2.54b) be satisfied. Thus, unlike the orthogonalization procedure of Article 2.3, for any finite N we can only speak of approximate satisfaction of orthogonality conditions for Functions (2.114) in the sense of (2.54b).

The question of the character of convergence of minimizing Sequences (2.114) with $N \to \infty$ for various sets of coordinate functions requires a special analysis. We assume that in each case it is possible to choose such values N_1 and $N < N_1$, that N orthogonality conditions in the sense of (2.54b) are satisfied with the required accuracy for the first coordinate functions. In this case it is easy to obtain a new form of equations of the body – liquid system perturbed motion. These are equivalent to (2.71), (2.72), but are of somewhat different structure due to absence of orthogonality in the sense of (2.54a). Let us write down these equations assuming $x_G = 0$ for simplification.

1. Motion in $O^\circ x^\circ z^\circ$ plane

$$(m^\circ + m)\,\dot{V} + \sum_{n=1}^{\infty} \lambda_n \ddot{s}_n = P_z\,,$$

$$(\overset{\circ}{J} + J^{(0)})\,\dot{\omega} + \sum_{k=1}^{N} J_k^*\,\dot{\Omega}_k + \sum_{n=1}^{\infty} (\lambda_{Gn}\ddot{s}_n + \lambda_n j\, s_n) = M_{Gy},$$

$$J_k^{**}\,\dot{\Omega}_k + J_k^*\,\dot{\omega} + \sum_{l=1}^{N} \overset{\circ}{\beta}_{kl} \int_{-\infty}^{t} \frac{\dot{\Omega}_l(\tau)\,d\tau}{\sqrt{t-\tau}} - \sum_{n=1}^{\infty} \gamma_{nk}^* \int_{-\infty}^{t} \frac{\ddot{s}_n(\tau)\,d\tau}{\sqrt{t-\tau}} = 0, \tag{2.116}$$

$$\mu_n(\ddot{s}_n + \omega_n^2 s_n) + \lambda_n \dot{V} + \lambda_{Gn}\dot{\omega} + \lambda_n j\,\psi +$$

$$+ \mu_n \sum_{m=1}^{\infty} \beta_{nm} \int_{-\infty}^{t} \frac{\ddot{s}_m(\tau)\,d\tau}{\sqrt{t-\tau}} - \sum_{k=1}^{N} \gamma_{nk}^* \int_{-\infty}^{t} \frac{\dot{\Omega}_k(\tau)\,d\tau}{\sqrt{t-\tau}} = 0\,,$$

$$k = 1, 2, .., N, \qquad n = 1, 2, \ldots$$

2. Rotation around Ox axis

$$(\overset{\circ}{I} + I^{(0)})\,\dot{\gamma} + \sum_{k=1}^{N} I_k^*\,\dot{\Gamma}_k = M_{Gx}\,,$$

$$I_k^{**}\,\dot{\Gamma}_k + I_k^*\,\dot{\gamma} + \sum_{l=1}^{N} \beta_{kl}^* \int_{-\infty}^{t} \frac{\dot{\Gamma}_l(\tau)\,d\tau}{\sqrt{t-\tau}} = 0\,, \qquad k = 1, 2, \ldots, N\,. \tag{2.117}$$

Expressions for coefficients of Equations (2.116) and (2.117) can be obtained from (2.65) after changing Ψ_j, $\overset{\circ}{\Psi}_j^*$ and $\mathbf{R}^*$, $\overset{\circ}{\mathbf{R}}^*$ to Ψ_{jk}^*, $\overset{\circ}{\Psi}_{jk}^*$ and $\mathbf{R}_k^*$, $\overset{\circ}{\mathbf{R}}_k^*$ respectively, where $\overset{\circ}{\Psi}_j^*$; $\overset{\circ}{\Psi}_{jk}^*$, $\overset{\circ}{\mathbf{R}}_k^*$ correspond to absence of ribs:

$$\overset{\circ}{\beta}_{kl} = \overset{\circ}{\beta}_{lk} = \beta_{kl2}^*\,, \qquad \beta_{kl}^* = \beta_{lk}^* = \beta_{kl1}^*\,,$$

$$\beta_{klj}^* = \frac{\rho\,k_s}{\sqrt{\pi}} \sum_{\mu=1}^{K} \int_{L_\mu} \sqrt{|\mathbf{v}_\nu|} \left[-\left(\overset{\circ}{\mathbf{R}}_k^* \times \vec{\nu}\right)_j + \frac{\partial \overset{\circ}{\Psi}{}_{jk}^*}{\partial \nu}\right] \left[-\left(\overset{\circ}{\mathbf{R}}_l^* \times \vec{\nu}\right)_j + \frac{\partial \overset{\circ}{\Psi}{}_{jl}^*}{\partial \nu}\right] b^{3/2}\varphi(x')\,ds, \tag{2.118}$$

$$\gamma_{nk}^* = \frac{\rho\,k_s}{\sqrt{\pi}} \sum_{\mu=1}^{K} \int_{L_\mu} \sqrt{|\mathbf{v}_\nu|} \left[-\left(\overset{\circ}{\mathbf{R}}_k^* \times \vec{\nu}\right)_2 + \frac{\partial \overset{\circ}{\Psi}{}_{2k}^*}{\partial \nu}\right] \frac{\partial \overset{\circ}{\varphi}_n}{\partial \nu}\, b^{3/2}\varphi(x')\,ds,$$

$$\beta_{nm} = \beta_{mn} = \frac{\rho\,k_s}{\sqrt{\pi}} \sum_{\mu=1}^{K} \int_{L_\mu} \sqrt{|\mathbf{v}_\nu|}\, \frac{\partial \overset{\circ}{\varphi}_n}{\partial \nu} \frac{\partial \overset{\circ}{\varphi}_m}{\partial \nu}\, b^{3/2}\varphi(x')\,ds,$$

$$J_k^* = J_{2k}^*, \qquad J_k^{**} = J_{2k}^{**}, \qquad I_k^* = J_{1k}^*, \qquad I_k^{**} = J_{1k}^{**},$$

$$J^*_{jk} = \rho \int\limits_Q (\mathbf{R}^*_k \times \mathbf{i}_j, \mathbf{R} \times \mathbf{i}_j)\, dQ - \rho \oint\limits_{S+\Sigma} \frac{\partial \Psi^*_{jk}}{\partial \nu} \left(\mathbf{R} \times \vec{\nu}\right)_j dS,$$

$$J^{**}_{jk} = \rho \int\limits_Q (\mathbf{R}^*_k \times \mathbf{i}_j)^2\, dQ - \rho \oint\limits_{S+\Sigma} \frac{\partial \Psi^*_{jk}}{\partial \nu} \left(\mathbf{R}^*_k \times \vec{\nu}\right)_j dS,$$

$$j = 1, 2; \qquad k = 1, 2, \ldots, N.$$

It is obvious that for $N = 1$ Equations (2.116) and (2.117) coincide with the corresponding scalar equations resulting from (2.64), if we take into account axial symmetry of the body and of the cavity in (2.64).

2.8. Spacecraft with LPRE Stabilized in Attitude

We now proceed to practical applications of the mathematical models obtained above. Let us consider a spacecraft with LPRE, having N tanks with inner ribs (assumed radial for definiteness). The spacecraft is stabilized in attitude and has a mass and geometry symmetry, the symmetry axis coinciding with its longitudinal axis. Suppose the spacecraft is moving at the boost phase of the flight in vacuum. Let the drive be due to one LPRE in cardanic suspension for stabilization in the pitch and yaw planes. Roll stabilization is performed by four special vernier engines with nozzles oriented in tangential direction.

Let us schematize the considered vehicle in its perturbed motion in the stabilization planes by a solid, absolutely rigid body with liquid filling, keeping valid all the assumptions on the character of its unperturbed and perturbed motion, formulated in Articles 2.1 and 2.2.

Consider the perturbed motion in two planes: yaw plane $G^\circ x^\circ z^\circ$ (an analogue of the pitch plane $G^\circ x^\circ y^\circ$ for an axisymmetric vehicle) and roll plane Gyz, based on the mathematical Models (2.71) and (2.72). In those models we confine (as in most applied investigations) to consideration of the first antisymmetric form of wave motions in each plane. We ignore all the higher harmonics of vortex motion of liquid keeping only vorticity uniformly distributed over the domain. Subscript n now corresponds not to the oscillation form number, but to the cavity number counted from the tail of the vehicle to the head; N is the quantity of cavities.

Let us use $G^\circ x^\circ y^\circ z^\circ$ and $Gxyz$ coordinate systems with the origin in the centre of mass of the vehicle with liquid solidified in the unperturbed state. We count all the coordinates along the longitudinal axis in an auxiliary body related coordinate system $Oxyz$ with the origin in the plane of the base arms of the body (see [76]).

2.8.1. Motion in the yaw plane $G^\circ x^\circ z^\circ$. We use generalized coordinates ζ, ψ, ϑ_n, related to generalized velocities V, ω, Ω_n by the expressions $\dot{\zeta} = V$, $\dot{\psi} = \omega$, $\dot{\vartheta}_n = \Omega_n$, and symbols J^*_n, β^*_n $(n = 1, 2, \ldots, N)$ for elements of tensors J^*_n, B_n corresponding to the yaw plane. Then with consideration of notes made above, we obtain from (2.71) the following set of equations of perturbed motion in the yaw plane:

$$(\overset{\circ}{m} + m)\ddot{\xi} + \sum_{n=1}^{N} \lambda_n \ddot{s}_n = P_z ,$$

$$(\overset{\circ}{J} + J^{(0)})\ddot{\psi} + \sum_{n=1}^{N} \left(J_n^* \ddot{\vartheta}_n + \lambda_{Gn} \ddot{s}_n + \lambda_n j \, s_n\right) = M_{Gy},$$

$$J_n^* (\ddot{\vartheta}_n + \ddot{\psi}) + \overset{\circ}{\beta}_n \int_{-\infty}^{t} \frac{\ddot{\vartheta}_n(\tau)\, d\tau}{\sqrt{t-\tau}} - \gamma_n \int_{-\infty}^{t} \frac{\ddot{s}_n(\tau)\, d\tau}{\sqrt{t-\tau}} = 0, \qquad (2.119)$$

$$\mu_n \left(\ddot{s}_n + \omega_n^2 s_n + \beta_n \int_{-\infty}^{t} \frac{\ddot{s}_n(\tau)\, d\tau}{\sqrt{t-\tau}}\right) + \lambda_n \ddot{\xi} + \lambda_{Gn} \ddot{\psi} + \lambda_n j \, \psi - \gamma_n \int_{-\infty}^{t} \frac{\ddot{\vartheta}(\tau)\, d\tau}{\sqrt{t-\tau}} = 0,$$

$$n = 1, 2, .., N.$$

To take into account specific features of the considered vehicle it is necessary to expand the right sides of Equations (2.119) to show explicitly the control forces and moments proportional to the small deflection angle of the LPRE axis from the SC longitudinal axis. We also have to join (2.119) with the equation of the regulator, whose actuator is the gimbaled LPRE.

Suppose the control law implemented by the regulator is constructed, as usual, based on two observed coordinates. These are angular deflections of the body from the programmed position and side shift of one of its sections. Let us ignore some minor terms associated with variable liquid mass and inertia of the LPRE, and refer coefficients of each equation to the coefficient at the highest-order derivative, as in [76]. Resulting is the following set of equations that can be considered the mathematical model of an automatically stabilized SC with LPRE at the boosted part of trajectory for perturbed motion in the yaw plane:

$$\ddot{\xi} + a_{\xi\psi}\psi + \sum_{n=1}^{N} a_{\xi s n} \ddot{s}_n + a_{\xi\delta}\delta_\psi = \widetilde{P}_z(t),$$

$$\ddot{\psi} + \sum_{n=1}^{N} \left(a_{\psi \vartheta n} \ddot{\vartheta}_n + a_{\psi\, sn} \ddot{s}_n + a'_{\psi\, sn} s_n\right) + a_{\psi\delta}\delta_\psi = \widetilde{M}_{Gy}(t),$$

$$\ddot{\vartheta}_n + \ddot{\psi} + \beta_{\vartheta n} \int_{-\infty}^{t} \frac{\ddot{\vartheta}_n(\tau)\, d\tau}{\sqrt{t-\tau}} + a_{\vartheta n s n} \int_{-\infty}^{t} \frac{\ddot{s}_n(\tau)\, d\tau}{\sqrt{t-\tau}} = 0, \qquad (2.120)$$

$$\ddot{s}_n + \omega_n^2 s_n + \beta_{sn} \int_{-\infty}^{t} \frac{\ddot{s}_n(\tau)\, d\tau}{\sqrt{t-\tau}} + a_{sn\xi}\ddot{\xi} +$$

$$+ a_{sn\psi} \ddot{\psi} + a'_{sn\psi} \psi + a_{sn\vartheta n} \int_{-\infty}^{t} \frac{\vartheta_n(\tau)\, d\tau}{\sqrt{t-\tau}} = 0,$$

$$\delta_\psi = \Im_1(\psi) - \Im_2(\zeta_S), \quad \zeta_S = \zeta - (x_S - x_G)\,\psi, \quad n = 1, 2, \ldots, N.$$

Here δ_ψ is the engine deflection angle in the yaw plane; ζ_S is displacement of the body section with x_S coordinate, containing the control devices; $\Im_1$ and $\Im_2$ are some linear-fractional (particularly, linear) operators characterizing the yaw plane control law. Coefficients and right sides of Equations (2.120) are expressed by the formulae

$$a_{\zeta\psi} = a_{\zeta\delta} = j, \quad a_{\zeta sn} = \frac{\lambda_n}{\overset{\circ}{m} + m'}, \quad j = \frac{P}{\overset{\circ}{m} + m'},$$

$$a_{\psi\delta} = \frac{P\,(x_G - x_Q)}{J^\circ + J^{(0)}}, \quad a_{\psi sn} = \frac{\lambda_{Gn}}{J^\circ + J^{(0)}}, \quad a'_{\psi sn} = \frac{\lambda_n j}{J^\circ + J^{(0)}},$$

$$a_{sn\zeta} = \frac{\lambda_n}{\mu_n}, \quad a_{\zeta n\psi} = \frac{\lambda_{Gn}}{\mu_n}, \quad a'_{sn\psi} = \frac{\lambda_n j}{\mu_n}, \tag{2.121}$$

$$\beta_{\vartheta n} = \frac{\overset{\circ}{\beta}_n}{J_n^*}, \quad \beta_{sn} = \beta_n, \quad a_{sn\vartheta n} = -\frac{\gamma_n}{\mu_n}, \quad a_{\vartheta n sn} = -\frac{\gamma_n}{J_n^*},$$

$$\tilde{P}_z(t) = \frac{P_z(t)}{\overset{\circ}{m} + m'}, \quad \tilde{M}_{Gy}(t) = \frac{M_{Gy}(t)}{J^\circ + J^{(0)}}.$$

Here x_G is the coordinate of the mass centre of the vehicle with solidified liquid; x_Q is the coordinate of the gimbaled LPRE spinning axis; P is the LPRE thrust; $P_z(t)$ and $M_{Gy}(t)$ are projections to $\overset{\circ}{G}\overset{\circ}{z}$ and Gy axes, respectively, of the resultant vector and resultant moment of the system of external perturbing forces with respect to G point.

2.8.2. Motion in the roll plane Gyz . In this case let us use generalized coordinates φ and χ_n connected with generalized velocities γ and Γ_n by the expressions $\gamma = \dot{\varphi}$; $\Gamma_n = \dot{\chi}_n$; and symbols for elements of tensors J_n^* and B_n corresponding to the roll plane. We then obtain from (2.72) the following set of equations of perturbed motion in the roll plane, describing the vehicle's rotation about its longitudinal axis (for small angles) with consideration of simplifications mentioned above:

$$(I^\circ + I^{(0)})\,\ddot{\varphi} + \sum_{n=1}^{N} I_n^*\,\ddot{\chi}_n = M_{Gx},$$

$$I_n^*\,(\ddot{\chi}_n + \ddot{\varphi}) + \beta_n^* \int_{-\infty}^{t} \frac{\dot{\chi}_n(\tau)\, d\tau}{\sqrt{t-\tau}} = 0, \quad n = 1, 2, \ldots, N. \tag{2.122}$$

Suppose that the control law in the roll plane is based on one observed coordinate, namely the roll angle. Let us transform Equations (2.122) to the form similar to (2.120). This brings us to the following mathematical model of automatically stabilized SC with LPRE, that describes its perturbed motion in the roll plane:

$$\ddot{\varphi} + \sum_{n=1}^{N} a_{\varphi\,\chi n}\ddot{\chi}_n + a_{\varphi\,\delta}\delta_\varphi = \widetilde{M}_{Gx}(t),$$

$$\ddot{\chi}_n + \ddot{\varphi} + \beta_{\chi n} \int_{-\infty}^{t} \frac{\ddot{\chi}_n(\tau)\, d\tau}{\sqrt{t-\tau}} = 0, \qquad \delta_\varphi = \Im_3(\varphi) \;\;, \qquad n = 1, 2, ..., N. \tag{2.123}$$

Here δ_φ is the generalized coordinate, to which the thrust of each vernier engine is proportional, those engines being the actuators of the vehicle roll stabilization system. $\Im_3$ is the linear-fractional (linear – in the particular case) operator symbolizing the roll plane stabilization algorithm. The coefficients and right side of Equations (2.123) are expressed by the formulae:

$$a_{\varphi\,\delta} = 4\left.\left(\frac{\partial P_1}{\partial \delta}\right)\right|_{\delta_\varphi = 0} r_Q\,, \qquad \widetilde{M}_{Gx} = \frac{M_{Gx}(t)}{I^{o} + I^{(0)}}\,,$$

$$a_{\varphi\,\chi n} = \frac{I_n^*}{I^{o} + I^{(0)}}\,, \qquad \beta_{\chi n} = \frac{\beta_n^*}{I_n^*}\,, \tag{2.124}$$

where $\left(\frac{\partial P_1}{\partial \delta}\right)_{\delta_\varphi = 0}$ is the gradient of a vernier engine power with respect to parameter δ_φ ; r_Q is the distance from the nozzle axis of each of the four vernier engines to the vehicle's longitudinal axis Gx ; $M_{Gx}(t)$ is the projection to axis of the perturbing forces resulting moment specified as a function of time with respect to point G .

2.9. Spacecraft with LPRE, Slowly Rotating around Its Longitudinal Axis

Up to this article throughout the chapter we discussed vehicles performing translational unperturbed motion. Let us now consider the case of a SC performing uniform rotation with angular velocity $\vec{\Omega}^\circ = \mathbf{i}_1\Omega^\circ$ about $O^\circ X^\circ$ axis of coordinate system $O^\circ X^\circ Y^\circ Z^\circ$, which coincides with $O^\circ x^\circ$ axis. Then absolute angular velocity $\vec{\Omega}$ of the body related coordinate system $Oxyz$ is

$$\vec{\Omega} = \vec{\Omega}^\circ + \vec{\omega}, \tag{2.125}$$

where $\vec{\omega}$ is the perturbed motion angular velocity of coordinate system $Oxyz$ with respect to $O^\circ x^\circ y^\circ z^\circ$; and $\vec{\Omega}^\circ$ is the angular velocity of $O^\circ x^\circ y^\circ z^\circ$ relative to $O^\circ X^\circ Y^\circ Z^\circ$, . We assume that liquid performing unperturbed motion rotates as a solid with the same angular velocity $\vec{\Omega}^\circ$.

Problems of dynamics of a rotating solid body with a cavity filled with liquid have been attracting researchers' attention for a long time. Fundamental investigations for the case of completely filled cavity have been published by S.L.Sobolev [90]. Quite a number of important results were obtained afterwards for bodies of revolution with ellipsoidal and cylindric cavities completely filled with liquid (see [30, 31, 88, 12]).

There is also a number of works on dynamics of gyroscopes with a cylindric cavity partially filled with liquid. The work by G.S.Narimanov **[69]** seems to be one of the first. All those works, however, concern a 'speedily rotating' body.

Another limiting case is considered below, namely the case of 'slow' rotation in the following sense: Froud's number corresponding to the angular velocity Ω° is substantially less then unity, i.e. the condition

$$\mathrm{Fr} = \frac{\Omega^{\circ 2}\, l}{j} << 1. \tag{2.126}$$

Here j is the acceleration modulus of the unperturbed motion mass forces. This means that for unperturbed state (uniform rotation of the body and liquid with the angular velocity Ω°) the liquid free surface differs negligibly from a plane perpendicular to O^*x^* axis. At the same time we assume that terms of the type of $\Omega^\circ \mid \mathbf{v} \mid$ ($\mathbf{v}$ is relative velocity of liquid) are first order infinitesimal. Note, that such assumptions, together with long-wave concept are conventionally used to solve the problem of liquid oscillations on rotating Earth surface (see, e.g. L.M.Brekhovskikh and V.V.Goncharov [13]).

Let us first construct the equations of perturbed motion of a 'slowly rotating' solid body with a cavity containing ideal liquid. This system differs from the considered SC in the absence of damping ribs only. To simplify the equations let us fair the pole to mass centre G° of the system 'body – liquid solidified in its unperturbed state'.

First write down system 'body – liquid' equations of motion, which are similar to the first two equations of (2.24), but consider the difference of the total time derivative $\frac{d}{dt}$ in the irrotational absolute coordinate system $G^\circ X^\circ Y^\circ Z^\circ$ from the local derivative $\frac{\partial}{\partial t}$ in coordinate system $Gxyz$. These equations have, as we know, the form

$$\frac{\partial}{\partial t}\left(\mathbf{K}^\circ + \rho \int_Q \mathbf{v}_a\, dQ\right) + \vec{\Omega} \times \left(\mathbf{K}^\circ + \rho \int_Q \mathbf{v}_a\, dQ\right) = \mathbf{P}, \tag{2.127}$$

$$\frac{\partial}{\partial t}\left[\mathbf{N}^\circ_G + \rho \int_Q (\mathbf{R} \times \mathbf{v}_a)\, dQ\right] + \vec{\Omega} \times \left[\mathbf{N}^\circ_G + \rho \int_Q (\mathbf{R} \times \mathbf{v}_a)\, dQ\right] =$$

$$= \mathbf{M}_G + \delta\, \mathbf{M}_G.$$

Here $\mathbf{K}^\circ$ and $\mathbf{N}^\circ_G$ are the momentum of the solid body and its moment of momentum with respect to point G; $\mathbf{P}$ and$\mathbf{M}_G$ are the resultant vector and the resultant moment of the system of external forces applied to the body; $\delta\, \mathbf{M}_G$ is the moment of unperturbed motion mass forces with respect to point G, those forces being associated with the body displacement and liquid free surface displacement in perturbed motion; $\mathbf{v}_a$ is the liquid absolute velocity. One more additional term is omitted in the left side of Equation (2.127), for it is a second-order infinitesimal.

Let us represent velocity field $\mathbf{v}_a$ by an expression differing from (2.36), (2.37) in that the role of $\vec{\omega}$ is now played by the absolute angular velocity $\vec{\Omega}$ of the body.

$$\mathbf{v}_a = \mathbf{V} + \vec{\Omega} \times (\mathbf{R} + \sum_{n=1}^{\infty} s_n \nabla \varphi_n) + \mathbf{v},$$

$$\mathbf{v} = \sum_{k=1}^{N} [\vec{\Omega}_k \times \mathbf{R}_k - \nabla(\vec{\Omega}_k, \vec{\Psi}_k)] + \sum_{n=1}^{\infty} \dot{s}_n \nabla \varphi_n, \quad (2.128)$$

where $\mathbf{V}$ is now, unlike (2.36), unperturbed motion velocity of the body-related point G with respect to point G° in the rotating coordinate system $G^\circ x^\circ y^\circ z^\circ$; $\mathbf{v}$ is relative velocity of liquid in $Gxyz$ coordinate system. Equations (2.127) can now be rewritten as follows, similar to (2.58) (if second-order infinitesimal quantities are ignored and inner ribs are taken into account):

$$(m^\circ + m)(\dot{\mathbf{V}} + \vec{\Omega}^\circ \times \mathbf{V}) + \sum_{n=1}^{\infty} [\lambda_n \ddot{s}_n + 2(\vec{\Omega}^\circ \times \vec{\lambda}_n)\dot{s}_n] = \mathbf{P}, \quad (2.129)$$

$$(\mathrm{J}^\circ + \mathrm{J}^{(0)}, \dot{\vec{\omega}}) + \vec{\Omega}^\circ \times (\mathrm{J}^\circ + \mathrm{J}^{(0)}, \vec{\omega}) - (\mathrm{J}^\circ + \mathrm{J}^{(0)}, \vec{\Omega}^\circ) \times \vec{\omega} + (\mathrm{J}_1^*, \dot{\vec{\Omega}}_1) +$$

$$+ \vec{\Omega}^\circ \times (\mathrm{J}_1^*, \vec{\Omega}_1) + \sum_{n=1}^{\infty} [\vec{\lambda}_{Gn}\ddot{s}_n + 2(\vec{\Omega}^\circ \times \vec{\lambda}_{Gn})\dot{s}_n - (\vec{\lambda}_n \times \mathbf{j}) s_n] = \mathbf{M}_G.$$

Consider now the equation of liquid relative motion in rotating coordinate system $Gxyz$. According to N.E.Kochin, I.A.Kibel, and N.V.Roze [40], these equations have the following form

$$\frac{\partial \mathbf{v}}{\partial t} + \dot{\mathbf{V}} + \vec{\Omega} \times \mathbf{V} + \dot{\vec{\omega}} \times \mathbf{R} + \vec{\Omega} \times (\vec{\Omega} \times \mathbf{R}) + 2\vec{\Omega} \times \mathbf{v} = -\frac{1}{\rho} \nabla p + \nabla U, \quad (2.130)$$

where ρ is liquid mass density; U is mass forces field potential in the nonrotating coordinate system $G^\circ X^\circ Y^\circ Z^\circ$ (i.e. for $\vec{\Omega}^\circ = 0$). All derivatives are taken, like in Equation (2.129), in the body-related coordinate system $Gxyz$.

As for the component $\vec{\Omega} \times (\vec{\Omega} \times \mathbf{R})$, it is transformed by the formulae

$$\frac{1}{2} \nabla (\vec{\Omega} \times \mathbf{R})^2 = (\vec{\Omega} \times \mathbf{R}, \nabla)(\vec{\Omega} \times \mathbf{R}) - 2\vec{\Omega} \times (\vec{\Omega} \times \mathbf{R})$$

and (up to first-order infinitesimal),

$$(\vec{\Omega} \times \mathbf{R}, \nabla)(\vec{\Omega} \times \mathbf{R}) = -(\vec{\Omega} \times \mathbf{R}) \times \vec{\Omega},$$

which yield the familiar equality

$$\vec{\Omega} \times (\vec{\Omega} \times \mathbf{R}) = -\frac{1}{2} \nabla (\vec{\Omega} \times \mathbf{R})^2 = \quad (2.131)$$

$$= -\frac{1}{2} \nabla (\vec{\Omega}^\circ \times \mathbf{R})^2 + \vec{\Omega}^\circ \times (\vec{\omega} \times \vec{\mathbf{R}}).$$

Equations (2.130) can now be rewritten as follows:

$$\frac{\partial \mathbf{v}}{\partial t} + \dot{\mathbf{V}} + \vec{\Omega}^{\circ} \times \mathbf{V} + \dot{\vec{\omega}} \times \mathbf{R} + \vec{\Omega}^{\circ} \times (\vec{\omega} \times \mathbf{R}) + 2\vec{\Omega} \times \mathbf{v} = -\frac{1}{\rho} \nabla p + \nabla U^{\circ} \quad (2.132)$$

Relative velocity field $\mathbf{v}\ (x, y, z, t)$ is specified by Expression (2.128), and U° by the following expression

$$U^{\circ} = -j x^{\circ} + \frac{1}{2} (\vec{\Omega}^{\circ} \times \mathbf{R})^2.$$

Condition (2.126) means a possibility to neglect the last term, i.e. set $U^{\circ} \approx U = -j x^{\circ}$. The vortex generating factor is now Coriolis force $\mathbf{F} = -2(\vec{\Omega}^{\circ} \times \mathbf{v})$.

Having carried out reductions similar to those of Article 2.3, one can obtain appropriate generalized forces and reduce equations of liquid motion to the form of (2.56). After joining them with Equations (2.129) we obtain the following complete set of equations of perturbed motion of the body – liquid system:

$$(m^{\circ} + m)(\dot{\mathbf{V}} + \vec{\Omega}^{\circ} \times \mathbf{V}) + \sum_{n=1}^{\infty} [\vec{\lambda}_n \ddot{s}_n + 2(\vec{\Omega}^{\circ} \times \vec{\lambda}_n) \dot{s}_n] = \mathbf{P},$$

$$(\mathsf{J}^{\circ} + \mathsf{J}^{(0)}, \dot{\vec{\omega}}) + \vec{\Omega}^{\circ} \times (\mathsf{J}^{\circ} + \mathsf{J}^{(0)}, \vec{\omega}) - (\mathsf{J}^{\circ} + \mathsf{J}^{(0)}, \vec{\Omega}^{\circ}) \times \vec{\omega} +$$

$$+ (\mathsf{J}_1^*, \dot{\vec{\Omega}}_1) + \vec{\Omega}^{\circ} \times (\mathsf{J}_1^*, \vec{\Omega}_1) +$$

$$+ \sum_{n=1}^{\infty} [\vec{\lambda}_{Gn} \ddot{s}_n + 2(\vec{\Omega}^{\circ} \times \vec{\lambda}_{Gn}) \dot{s}_n - (\vec{\lambda}_n \times \mathbf{j}) s_n] = \mathbf{M}_G. \quad (2.133)$$

$$(\mathsf{J}_k^*, \dot{\vec{\Omega}}_k + \delta_k (\dot{\vec{\omega}} + \vec{\Omega}^{\circ} \times \vec{\omega})) + \sum_{l=1}^{2N} (\mathsf{D}_{kl}, \vec{\Omega}_l) - \sum_{n=1}^{\infty} \vec{\alpha}_{kn} \dot{s}_n = 0,$$

$$\mu_n (\ddot{s}_n + \omega_n^2 s_n) + \mu_n \sum_{m=1}^{\infty} \alpha_{nm} \dot{s}_m + (\vec{\lambda}_n, \dot{\mathbf{V}} + \vec{\Omega}^{\circ} \times \mathbf{V}) +$$

$$+ \left(\vec{\lambda}_{Gn}, \dot{\vec{\omega}} + \vec{\Omega}^{\circ} \times \vec{\omega}\right) - \left(\vec{\lambda}_n, \mathbf{j}\right) + \sum_{k=1}^{2N} \left(\vec{\alpha}_{kn}, \vec{\Omega}_k\right) = 0,$$

$$k = 1, 2, \ldots, 2N, \quad n = 1, 2, \ldots, \quad \delta_1 = 1, \quad \delta_k = 0 \quad \text{for} \quad k \neq 1.$$

Here D_{kl}, $\vec{\alpha}_{kn}$ are tensors of rank 2 and vectors, respectively:

$$\mathsf{D}_{kl} = \left\{\delta_{ji}^{kl}\right\}, \quad \vec{\alpha}_{kn} = \sum_{j=1}^{3} \mathbf{i}_j \alpha_{knj}. \quad (2.134)$$

Tensors D_{kl} elements, vector $\vec{\alpha}_{kn}$ components, and scalar coefficients α_{nm} are expressed by formulae

$$\delta_{ji}^{kl} = -\delta_{ij}^{lk} = 2\rho\Omega^{\circ} \int_Q (\mathbf{f}_{li} \times \mathbf{f}_{kj}, \mathbf{i}_1)\, dQ,$$

$$\alpha_{knj} = -\,2\rho\Omega^\circ \int_Q (\nabla\varphi_n \times \mathbf{f}_{kj}, \mathbf{i}_1)\, dQ, \tag{2.135}$$

$$\alpha_{nm} = -\,\alpha_{mn} = \frac{2\rho\Omega^\circ}{\mu_n} \int_Q (\nabla\varphi_m \times \nabla\varphi_n, \mathbf{i}_1)\, dQ,$$

$$\mathbf{f}_{kj} = \mathbf{i}_j \times \mathbf{R}_k - \nabla\Psi_{kj}.$$

It is clear from (2.135), that new generalized forces associated with the body rotation with angular velocity $\vec{\Omega}^\circ$, which have appeared in Equations (2.133), are gyroscopic in structure, as it is appropriate.

Consider an important case of a completely filled cavity. Equations (2.133) convert to the following ones:

$$(m^\circ + m)(\dot{\mathbf{V}} + \vec{\Omega}^\circ \times \mathbf{V}) = \mathbf{P},$$

$$(\mathrm{J}^\circ + \mathrm{J}^{(0)}, \dot{\vec{\omega}}) + \vec{\Omega}^\circ \times (\mathrm{J}^\circ + \mathrm{J}^{(0)}, \vec{\omega}) - (\mathrm{J}^\circ + \mathrm{J}^{(0)}, \vec{\Omega}^\circ) \times \vec{\omega} +$$

$$+ (\mathrm{J}_1^*, \dot{\vec{\Omega}}_1) + \vec{\Omega}^\circ \times (\mathrm{J}_1^*, \vec{\Omega}_1) = \mathbf{M}_G, \tag{2.136}$$

$$\left(\mathrm{J}_k^*, \dot{\vec{\Omega}}_k + \delta_k (\dot{\vec{\omega}} + \vec{\Omega}^\circ \times \vec{\omega})\right) + \sum_{l=1}^{2N} (\mathrm{D}_{kl}, \vec{\Omega}_l) = 0,$$

$$k = 1, 2, \ldots, 2N, \quad \delta_1 = 1, \quad \delta_k = 0 \quad \text{for} \quad k \neq 1.$$

Condition (2.126) is now unnecessary, and the difference of 'slow' and 'fast' rotation about $G^\circ X^\circ$ axis vanishes. Therefore it is interesting to compare these equations for $\mathbf{V} \equiv 0$ with the corresponding equations for a fast rotating solid body with a cavity completely filled with ideal incompressible liquid, obtained by R.V.Rvalov and V.M.Rogovoy [88] (see also L.V.Dokuchaev [19]). The fundamental difference of these equations consists in the following. In [88] the orthogonal basis is the set of exact solenoidal particular solutions of the hydrodynamics problem for a body rotating with angular velocity $\vec{\Omega}^\circ$ about an axis motionless in the space, i.e. for $\vec{\omega} \equiv 0$. That is why in the ultimate set of equations (for $\vec{\omega} \neq 0$) there are no connections between equations for the corresponding generalized angular velocities, and these equations become independent for $\vec{\omega} \equiv 0$. The role of such a basis is played herein by the set of exact solenoidal particular solutions of the hydrodynamics problem for a non-rotating body, i.e. for $\vec{\Omega}^\circ \equiv 0$. Therefore in Equations (2.136) there are gyroscopic connections between all the equations for generalized angular velocities $\vec{\Omega}_k$ which have naturally another sense here, than in [88, 19], being not scalars, but vectors.

Let us now return to the general case described by Equations (2.133), and apply those equations to a slowly rotating SC with tanks having inner ring-shape or radial ribs. Obviously equations of perturbed motion of such an object can be obtained by superposition of Equations (2.57) and (2.133); they have the following form

$$(m^\circ + m)(\dot{\mathbf{V}} + \vec{\Omega}^\circ \times \mathbf{V}) + \sum_{n=1}^{\infty} [\vec{\lambda}_n \ddot{s}_n + 2(\vec{\Omega}^\circ \times \vec{\lambda}_n)\, \dot{s}_n] = \mathbf{P},$$

$$(\mathrm{J}^\circ + \mathrm{J}^{(0)}, \dot{\vec{\omega}}) + \vec{\Omega}^\circ \times (\mathrm{J}^\circ + \mathrm{J}^{(0)}, \vec{\omega}) - (\mathrm{J}^\circ + \mathrm{J}^{(0)}, \vec{\Omega}^\circ) \times \vec{\omega} +$$

$$+ (\mathrm{J}_1^*, \dot{\vec{\Omega}}_1) + \vec{\Omega}^\circ \times (\mathrm{J}_1^*, \vec{\Omega}_1) +$$

$$+ \sum_{n=1}^{\infty} [\vec{\lambda}_{Gn} \ddot{s}_n + 2(\vec{\Omega}^\circ \times \vec{\lambda}_{Gn})\, \dot{s}_n - (\vec{\lambda}_n \times \mathbf{j})\, s_n] = \mathbf{M}_G. \tag{2.137}$$

$$\left(\mathrm{J}_k^*, \dot{\vec{\Omega}}_k + \delta_k (\dot{\vec{\omega}} + \vec{\Omega}^\circ \times \vec{\omega})\right) + \sum_{l=1}^{2N} (\mathrm{D}_{kl}, \vec{\Omega}_l) - \sum_{n=1}^{\infty} \vec{\alpha}_{kn} \dot{s}_n +$$

$$+ \sum_{l=1}^{2N} \left(\mathrm{B}_{kl}, \int_{-\infty}^{t} \frac{\dot{\vec{\Omega}}_l(\tau)\, d\tau}{\sqrt{t-\tau}}\right) - \sum_{n=1}^{\infty} \vec{\gamma}_{kn} \int_{-\infty}^{t} \frac{\ddot{s}_n(\tau)\, d\tau}{\sqrt{t-\tau}} = 0,$$

$$\mu_n (\ddot{s}_n + \omega_n^2 s_n) + \mu_n \sum_{m=1}^{\infty} \alpha_{nm} \dot{s}_m + (\vec{\lambda}_n, \dot{\mathbf{V}} + \vec{\Omega}^\circ \times \mathbf{V}) +$$

$$+ \left(\vec{\lambda}_{Gn}, \dot{\vec{\omega}} + \vec{\Omega}^\circ \times \vec{\omega}\right) - \left(\vec{\lambda}_n, \mathbf{j}\right) + \sum_{k=1}^{2N} \left(\vec{\alpha}_{kn}, \vec{\Omega}_k\right) +$$

$$+ \mu_n \sum_{m=1}^{\infty} \beta_{nm} \int_{-\infty}^{t} \frac{\ddot{s}_m(\tau)\, d\tau}{\sqrt{t-\tau}} - \sum_{k=1}^{2N} \left(\vec{\gamma}_{kn}, \int_{-\infty}^{t} \frac{\dot{\vec{\Omega}}_k(\tau)\, d\tau}{\sqrt{t-\tau}}\right) = 0,$$

$$k = 1, 2, \ldots, 2N, \quad n = 1, 2, \ldots, \quad \delta_1 = 1, \quad \delta_k = 0 \quad \text{for} \quad k \neq 1.$$

Consider in conclusion the case of an axisymmetric SC, as in Article 2.8. We use the simplifications formulated in Article 2.8, particularly, confine to the first form of liquid wave motions in each plane of stabilization and to the first form of vortex motions (liquid vorticity uniform all over the volume). The planes of stabilization are the same, $O^\circ x^\circ y^\circ$ (the pitch plane) and $O^\circ x^\circ z^\circ$ (the yaw plane). These planes coinciding with Oxy and Oxz in unperturbed motion rotate about $O^\circ X^\circ$ axis at a constant angular velocity $\vec{\Omega}^\circ$. The perturbed motion of the SC in the roll plane, i.e. the motion in Ox axis does not depend on this rotation of the SC for the introduced assumptions, and so the set of Equations (2.122) is valid.

Therefore we further consider perturbed motion in the pitch and yaw planes only, which does not split in this case into independent motions in each of those planes. Because of axial symmetry the following relations hold true

$$J_{11}^\circ + J_{11}^{(0)} = I_{11}^\circ + I_{11}^{(0)} = A, \qquad J_{22}^\circ + J_{22}^{(0)} = J_{33}^\circ + J_{33}^{(0)} = J^\circ + J^{(0)} = C,$$

$$J_{22}^{11} = J_{33}^{11} = J^*, \qquad \mu_2 = \mu_1 = \mu, \qquad \omega_2^2 = \omega_1^2 = \omega^2,$$

$$\lambda_2 = \lambda_1 = \lambda, \qquad \lambda_{G12} = -\lambda_{G13} = \lambda_G,$$

$$\alpha_{12} = -\alpha_{21} = \alpha, \quad \alpha_{112} = \alpha_{113} = \alpha^{\circ}, \quad \delta_{23}^{11} = -\delta_{32}^{11} = \delta,$$

$$\beta_{22}^{11} = \beta_{33}^{11} = \beta^{\circ}, \quad \beta_{11} = \beta_{22} = \beta, \quad \gamma_{112} = -\gamma_{113} = \gamma^{*}. \tag{2.138}$$

Additional subscript n corresponds now, like in (2.119), to the tank number, and N to the quantity of tanks.

Let us introduce the following complex-valued functions of real variable

$$V = \dot{\eta} + i\dot{\zeta}, \quad \omega = \dot{\psi} + i\dot{\theta}, \quad \varepsilon = \psi + i\theta, \quad r_n = p_n + i s_n, \tag{2.139}$$

$$\Omega_n = \Omega_{n2} + i\Omega_{n3} = \dot{\vartheta}_n + i\dot{\Delta}_n, \quad P = P_y + iP_z, \quad M_G = M_{Gy} + iM_{Gz},$$

where p_n is the generalized coordinate for motion in the pitch plane, which is an analogue of s_n for motion in the yaw plane. Using the general Equations (2.137) we can write down the equations of perturbed motion of the SC in the stabilization planes, similar to (2.119). Add equations for generalized coordinates η, ζ and p_n, s_n in pairs, the second equation of each pair multiplied by i. We perform a similar operation with equations for generalized velocities $\dot{\psi}$, $\dot{\theta}$ and $\dot{\vartheta}_n$, $\dot{\Delta}_n$ each second one of the equations multiplied by i. Resulting is the following set of equations of perturbed motion of the SC in $O^{\circ}x^{\circ}y^{\circ}$ and $O^{\circ}x^{\circ}z^{\circ}$ planes, written in complex form:

$$(m^{\circ} + m)(\dot{V} + i\Omega^{\circ}V) + \sum_{n=1}^{N} \lambda_n(\ddot{r}_n + 2i\Omega^{\circ}\dot{r}_n) = P,$$

$$C\dot{\omega} - i(A - C)\Omega^{\circ}\omega + \sum_{n=1}^{N} [J_n^{*}(\dot{\Omega}_n + i\Omega^{\circ}\Omega_n) -$$

$$- i\lambda_{Gn}(\ddot{r}_n + 2i\Omega^{\circ}\dot{r}_n) - ij\lambda_n r_n] = M_G,$$

$$J_n^{*}\dot{\Omega}_n + i\delta_n\Omega_n + J_n^{*}(\dot{\omega} + i\Omega^{\circ}\omega) - \alpha^{\circ}_n\dot{r}_n +$$

$$+ \beta^{\circ}_n \int_{-\infty}^{t} \frac{\dot{\Omega}_n(\tau)\,d\tau}{\sqrt{t-\tau}} + i\gamma_n^{*} \int_{-\infty}^{t} \frac{\ddot{r}_n(\tau)\,d\tau}{\sqrt{t-\tau}} = 0, \tag{2.140}$$

$$\mu_n(\ddot{r}_n - i\alpha_n\dot{r}_n + \omega_n^2 r_n) + \lambda_n(\dot{V} + i\Omega^{\circ}V) + i\lambda_{Gn}(\dot{\omega} + i\Omega^{\circ}\omega) + \alpha^{\circ}_n\Omega_n + ij\lambda_n\varepsilon_n +$$

$$+ \mu_n\beta_n \int_{-\infty}^{t} \frac{\ddot{r}_n(\tau)\,d\tau}{\sqrt{t-\tau}} - i\gamma_n^{*} \int_{-\infty}^{t} \frac{\dot{\Omega}_n(\tau)\,d\tau}{\sqrt{t-\tau}} = 0, \quad n = 1, 2, \ldots, N.$$

For $\Omega^{\circ} = 0$ the set of Equations (2.140) splits after separation of real and imaginary parts into two independent sets of equations of motion – in the pitch and yaw planes. The last of those sets coincides with (2.119).

For $r_n \equiv 0$ we obtain from (2.140) a set of equations corresponding to a rotating SC with tanks completely filled with liquid:

$$(m^{\circ} + m)(\dot{V} + i\Omega^{\circ}V) = P,$$

$$C\dot{\omega} - i(A - C)\Omega^{\circ}\omega + \sum_{n=1}^{N} [J_n^* (\dot{\Omega}_n + i\,\Omega^{\circ}\Omega_n)] = M_G, \tag{2.141}$$

$$J_n^* \dot{\Omega}_n + i\,\delta_n\Omega_n + J_n^*(\dot{\omega} + i\Omega^{\circ}\omega) + \beta^{\circ}{}_n \int_{-\infty}^{t} \frac{\dot{\Omega}_n(\tau)\,d\tau}{\sqrt{t-\tau}} = 0, \qquad n = 1, 2, ..., N.$$

If we assume additionally $\beta^{\circ}{}_n \equiv 0$, $V \equiv 0$, $N = 1$, Equations (2.141) coincide in structure with Equations (3.76) of [19] corresponding to consideration of only the first form of liquid vortex motions in a cavity with smooth walls.

2.10. Determination of Hydrodynamic Coefficients

2.10.1. General remarks. Up to this point the question of determination of hydrodynamic coefficients involved in Equations (2.120), (2.122) has been open. Formulae (2.89) of those coefficients transformation when changing from the 'floating cover' scheme to the 'fixed cover' scheme were derived above. We are going to present the hydrodynamic coefficients in the form suitable for existing numerical values published in literature for cavities of various configurations.

Begin with coefficients associated with potential motion of liquid $\omega_n^2, \lambda_n, \lambda_{On}, \mu_n, J_n, I_n$. Represent them in dimensionless form:

$$\omega_n^2 = \frac{j}{l_n}\,\tilde{\omega}_n^2, \qquad \lambda_n = \rho_n\, l_n^3\, \tilde{\lambda}_n, \qquad \lambda_{On} = \rho_n\, l_n^4\, \tilde{\lambda}_{On}, \tag{2.142}$$

$$\mu_n = \rho_n\, l_n^3\, \tilde{\mu}_n, \qquad J_n = \rho_n\, l_n^5\, \tilde{J}_n, \qquad I_n = \rho_n\, l_n^5\, \tilde{I}_n,$$

where l_n is the characteristic size (e.g. maximum radius of the n -th cavity wetted surface cross section); ρ_n is liquid mass density in the n -th cavity; the tilde sign denotes dimensionless quantity.

Numerous materials on dimensionless coefficients $\tilde{\omega}_n^2, \tilde{\lambda}_n, \tilde{\lambda}_{On}, \tilde{\mu}_n, \tilde{J}$ numerical values are now available for cavities of various configurations with smooth walls (coefficient $\tilde{I}_n$ is identically zero in this case). Most results were obtained by computer numerical solutions of adjacent boundary-value problems (see, e.g. [98, 10, 21, 76]). For cavities of shapes most commonly used in practice complete experimental results for dimensionless coefficients mentioned above demonstrate quite satisfactory agreement, as a rule (see,e.g. [62,63,64]). This makes a good foundation for extensive usage of calculated values of those coefficients for synthesizing mathematical models of real objects. Somewhat more complex is the situation with cavities having inner ribs, which are just the ones corresponding to the mathematical models obtained above.

Solution of internal boundary-value problems of liquid dynamics in moving cavities runs across serious computational difficulties even with modern computers. This caused development of various techniques specially 'adjusted' for one or another cavity configuration.

A general method of calculating coefficients for cavities with narrow inner ribs, according to the scheme of non-stalled flow of ideal liquid, corresponding coefficients for cavities with smooth walls assumed known (perturbations method), is described in [64, 76]. Also presented there are numerical results for some cavities of revolution with radial and ring-shape ribs. This method allows essential simplification of the coefficients computation algorithms by practically avoiding singularities caused by the presence of ribs.

The main problem lies deeper however, and is not mathematical, but physical by nature. The thing is that the scheme itself of nonstalled liquid flow around ribs is, speaking rigorously, correct only for the case of liquid motions with small Strouhal numbers and, at the same time not small Reynolds numbers. That means that there is no noticeable influence of viscosity on the flow picture. There is a certain contradiction here, although such regimes are within the range of parameters values usually occurring in practice. But the most interesting are the values of Strouhal number corresponding to presence of local vortex zones in the ribs' edges vicinity, which cause extensive stalling vortex formation with further growth of this number (see Article 2.2).

The entire phenomenologic description of vibrating rib and liquid interaction forces, used above, is 'adjusted' to this very picture. Therefore it seems reasonable to make the following step in description of the flow around ribs, taking into consideration existence of vortex zones, outside which absolute motion of liquid remains potential. This exactly is made below for the simplest case of a two-dimensional problem corresponding to the experiment discussed in Chapter 4.

Turn now to the coefficients $\beta_n, \beta^{\circ}{}_n, \beta^*_n, \gamma_n$ associated with vortex motion of liquid. The picture here is even more complicated, as to represent them in the form suitable for investigation of closed-loop system 'object – liquid – controller' it is necessary to extract a narrow class of perturbed motions of liquid, in which the relative velocity of liquid in the rib edges vicinity is expressed by a one-term formulae. In the case of motion in the roll plane this problem is solved automatically without any additional assumptions, for the relative motion of liquid is completely determined by velocity field vortex component $\Gamma_n = \dot{\chi}_n$.

The problem is not trivial for motion in the yaw (or pitch) plane, but still it can be solved if we use some heuristic considerations. The thing is that when investigating dynamics of the considered objects, the most interesting is the class of perturbed motions characterized by dominating role of wave motions. This means that in the expression for relative motion of liquid flowing around ribs edges the main is the term proportional to $\dot{s}_n$, while the term proportional to $\Omega_n = \dot{\vartheta}_n$ is of minor importance, and so it can be ignored in the first-order approximation.

Further on we confine to the class of perturbed motions of object and liquid, for which this assumption in correct. Then all the coefficients associated with vortex motion of liquid can be expressed with the help of their dimensionless counterparts denoted by a tilde sign:

$$\beta_n = \widetilde{\beta}_n \left(| \dot{\tilde{s}}_n | \right)^{1/2}, \qquad \beta^{\circ}_n = \rho_n l^4_n \widetilde{\beta}^{\circ}_n \left(| \dot{\tilde{s}}_n | \right)^{1/2}, \qquad \dot{\tilde{s}}_n = \frac{\dot{s}_n}{l_n},$$

$$\gamma_n = \rho_n l_n^5 \tilde{\gamma}_n (|\dot{\tilde{s}}_n|)^{1/2}, \quad \beta_n^* = \rho_n l_n^5 \tilde{\beta}_n^* (|\dot{\chi}_n|)^{1/2}. \tag{2.143}$$

Dimensionless coefficients $\tilde{\beta}_n$, $\tilde{\beta}_n^*$, $\tilde{\beta}^0{}_n$, $\tilde{\gamma}_n$ can be calculated according to Formulae (2.53) using empiric values of coefficients k_s . Their values for some cavities of revolution of the simplest configurations with radial and coaxial ribs are presented in [64, 62, 76].

Sometimes it is expedient instead of $\tilde{\beta}_n$ and $\tilde{\gamma}_n$ to introduce equivalent parameters δ_n and δ^* connected with the former ones by the formulae

$$\tilde{\beta}_n = \frac{\tilde{\delta}_n}{\pi}, \quad \tilde{\gamma}_n = \frac{\tilde{\delta}_n^* \tilde{\mu}_n}{\pi}. \tag{2.144}$$

These very parameters are presented in [64, 62, 76].

2.10.2. Calculation of associated moments of inertia of liquid for the presence of local vortex zones in the ribs edges vicinity. Let us consider the problem determining associated moment of inertia of liquid filling a circular cylindric cavity with inner ribs, which is rotating about its longitudinal axis. It is the cavity for which the most representative experimental information is available. This information is used to identify parameters of the adjacent mathematical model (see [54] and Chapter 4).

The first step in making the mathematical model of flow around ribs more exact, is a change from nonstalled flow scheme to local vortex zones in ribs' edges vicinity, associated with existence of free vortices (intermediate regime, when there is no stalling of those vortices causing vorticity of the liquid kernel). In the considered model liquid flow in the kernel is potential in the absolute (nonrotating) coordinate system. Models of this type are widely known from literature on hydrodynamics (see e.g., N.E.Kochin et al., M.A.Lavrentiev and B.V.Shabat, M.Van Dyke [40, 48, 103]).

Let us introduce complex variable z plane containing S domain shaped as a circle of radius r_0 with inner radial cuts, corresponding to the cross section of a cylindric cavity with liquid. We assume the number of ribs K even. Relative width of the ribs is $\tilde{b} = \frac{b}{r_0}$. Further on we assume $r_0 = 1$.

Construct a conformal mapping of domain S to the inner area of the unit circle S^* of a new complex variable w plane. Denote the mapping function by $z = Z(w)$, complex coordinates of ribs' edges by z_j^*, and write z_j ($j = 1, 2, ..., K$) for the coordinates of vortex points situated at equal distances $r_j = r_1$ from the circle centre.

The w-plane counterparts of points z_j^*, z_j and their inverse z_j' are the following ones

$$w_j^* = e^{i\theta_j}, \quad w_j = \rho e^{i\vartheta_j}, \quad w_j' = \frac{1}{\rho} e^{i\vartheta_j},$$

$$j = 1, 2, ..., K, \quad \rho < 1.$$

For the point $z_0 = 0$ the counterpart is $w_0 = 0$.

Let the cylinder rotate with angular velocity Ω°. We shall determine circulation Γ_j of free vortices from the condition of zero total circulation of the relative motion of liquid

$$\Gamma + \sum_{j=1}^{K} \Gamma_j = \Gamma + K\Gamma_j = 0, \quad \Gamma = -\,2\pi\Omega^\circ. \tag{2.145}$$

Liquid absolute motion complex potential in S^* domain can be written in the form (see [40]):

$$f_\Omega = \Omega^\circ f_\omega(w), \quad f_\omega(w) = f(w) + \frac{1}{Ki}\sum_{j=1}^{K} \ln \frac{w - w_j}{w - \dfrac{1}{\overline{w}_j}}, \tag{2.146}$$

where $f(w)$ is the absolute motion complex potential for absence of vortex zones and $\Omega^\circ = 1$. Dimensionless Joukowski moment of inertia $\widetilde{I}$ corresponding to flow around ribs without vortex zones is determined by the expression

$$\widetilde{I} = \oint_{\Gamma^\circ} \varphi \, d\psi, \tag{2.147}$$

where Γ° is the unit radius circumference, φ and ψ are the real part and the imaginary part of the appropriate complex potential $f(w)$. Associated moment of inertia $\widetilde{I}_\Omega$ taking into account local vortex zones is described by the following expression:

$$\widetilde{I}_\Omega = \oint_{\Gamma^\circ} \operatorname{Re} f_\omega(w)\, d\psi = \widetilde{I} - \oint_{\Gamma^\circ} \operatorname{Re}\left[\frac{i}{K}\sum_{j=1}^{K} \ln \frac{w - w_j}{w - \dfrac{1}{\overline{w}_j}}\right] d\psi. \tag{2.148}$$

Let us write a_s, b_s, and d_s^*, $-\,c_s^*$ for function $\psi\,|_{\rho=1}$ and the integrand function Fourier series expansion coefficients in (2.148), respectively. Expressions for $\widetilde{I}$ and $\widetilde{I}_\Omega$ can be transformed to the following form:

$$\widetilde{I} = \pi \sum_{s=1}^{\infty} s\,(a_s^2 + b_s^2),$$

$$\widetilde{I}_\Omega = \pi \sum_{s=1}^{\infty} s\,[a_s(a_s - c_s^*) + b_s(b_s - d_s^*)]. \tag{2.149}$$

It left only to determine the coordinates of stationary points in which the free vortices are to be situated. Let us introduce an auxiliary complex potential $f_\Gamma(w)$ corresponding to liquid flow in a motionless cylinder with the central vortex having circulation Γ:

$$f_\Gamma(w) = \Gamma f_\gamma(w),$$

$$f_\gamma(w) = \frac{1}{2\pi i}\left[\ln w - \frac{1}{K}\sum_{j=1}^{K} \ln \frac{w - w_j}{w - \dfrac{1}{\overline{w}_j}}\right]. \tag{2.150}$$

Assume coordinates of stationary points corresponding to complex Potential (2.150) for approximate values of w-plane stationary points coordinates corresponding to Potential (2.146). Determine the former ones from the following conditions.

1) Zero velocity of the flow in the points $w = w_j$ if there are no vortices in these points:

$$\left[\frac{1}{w} - \frac{1}{K} \sum_{i=1}^{K}{}' \left(\frac{1}{w - w_i} - \frac{1}{w - \frac{1}{\bar{w}_i}} \right) \right]_{w = w_j} = 0. \tag{2.151}$$

(Prime denotes omitting of the subscript value $i = j$).

2) Zero velocity in the points $w = w_j^*$ which are the images of ribs edges z_j^* (Joukowski – Chaplygin condition):

$$\left[\frac{1}{w} - \frac{1}{K} \sum_{i=1}^{K}{}' \left(\frac{1}{w - w_i} - \frac{1}{w - \frac{1}{\bar{w}_i}} \right) \right]_{w = w_j^*} = 0. \tag{2.152}$$

Table 2.1 presents stationary values of the vortex centre polar radius vector $\rho = \rho^{\circ}$ and angle $\vartheta = \vartheta^{\circ}$ counted off the rib. They are obtained from Conditions (2.151), (2.152) for the cases of 6, 4, and 2 ribs.

Fig. 2.2a,b presents the grid conformally equivalent to the polar coordinate grid for conformal mapping of S domain in z-plane onto S^* domain in w-plane (the case of 2 ribs). Fig. 2.2c presents the picture of streamlines of liquid relative motion with the central vortex and nonstalled flow around the ribs.

Table 2.1

Free Vortex Distribution Stationary Points Polar Coordinates

K	6	4	2
ρ°	0.9352	0.8932	0.8091
ϑ°	7.43	12.62	24.55

Mapping function $z = Z(w)$ was constructed, as in Chapter 1, with the help of numerical RT-algorithm of conformal mapping [82, 84].

Fig.2.3 shows the circulation flow of liquid in w -plane, corresponding to the central vortex and free vortices ($K = 6, 4, 2$), which circulations satisfy Condition (2.145) and coordinates correspond to Table 2.1.

Fig.2.4 shows circulation flow in the physical plane z for the case of two ribs and hence two free vortices.

Fig.2.4*a,b* are the cases, when only Conditions (2.145) and (2.151) are satisfied, and Fig.2.4*c* is the case when all the three Conditions (2.145), (2.151), (2.152) are satisfied. The last case is the closest to the one physically realized.

Fig.2.4 presents a qualitative picture of stalled flow with local vortex zones generation.

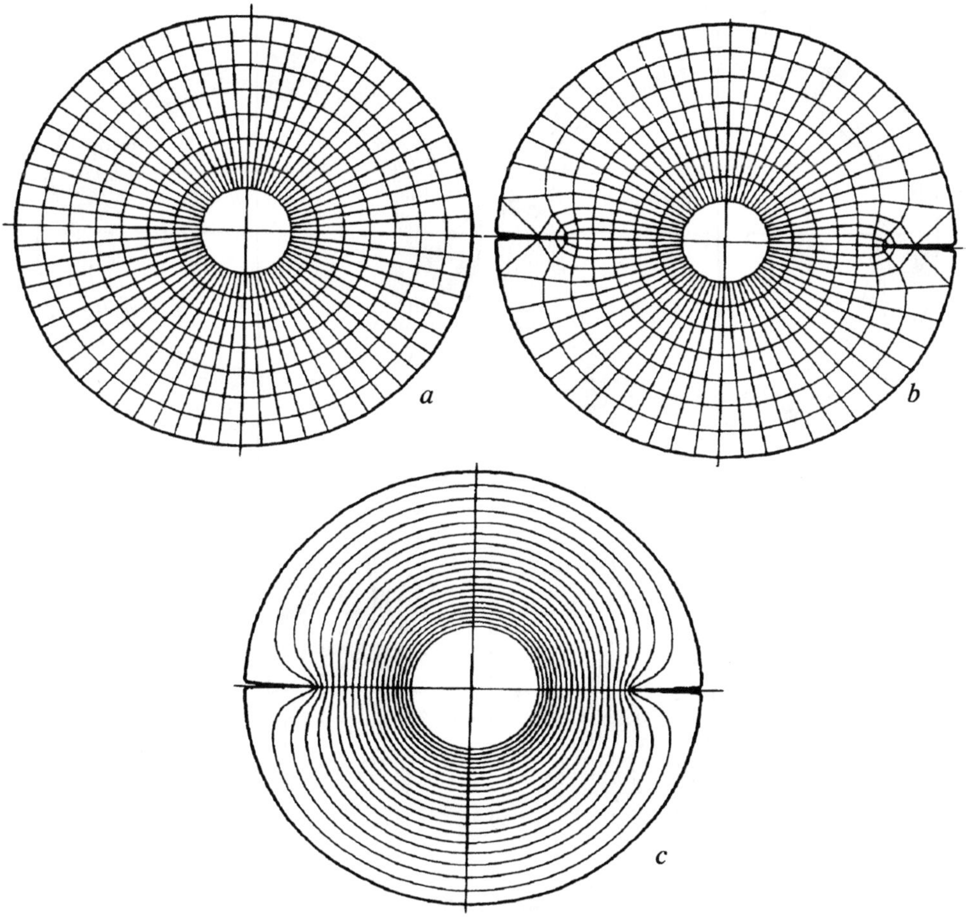

Fig.2.2. Conformal mapping of the domain with two ribs filled with liquid onto the unit circle; *a, b* – construction of a grid conformally equivalent to polar coordinate grid; *c* – streamlines of the circulation slow for the case of the central vortex.

The real picture of vortex flow around ribs is obviously more complicated. This is proved by the fact that the values of associated moment of inertia $\widetilde{I}_\Omega$ of liquid calculated according to Formula (2.149) for $K = 2$ and $\tilde{b} \leq 0.3$ are negative. For $K =$ $= 4; 6$ and $\tilde{b} \geq 0.3$ we obtain somewhat more acceptable from the physical point of view values of the associated moment of inertia of liquid.

Table 2.2 shows the values of associated moment of inertia of liquid $\widetilde{I}_\Gamma$ calculated according to the Formula (2.149) using complex Potential (2.150) and Equations (2.151) and (2.152). Also presented there are values $\widetilde{I}_\Omega$ obtained by the same formula, but using the field of relative velocities, corresponding to the complex potential

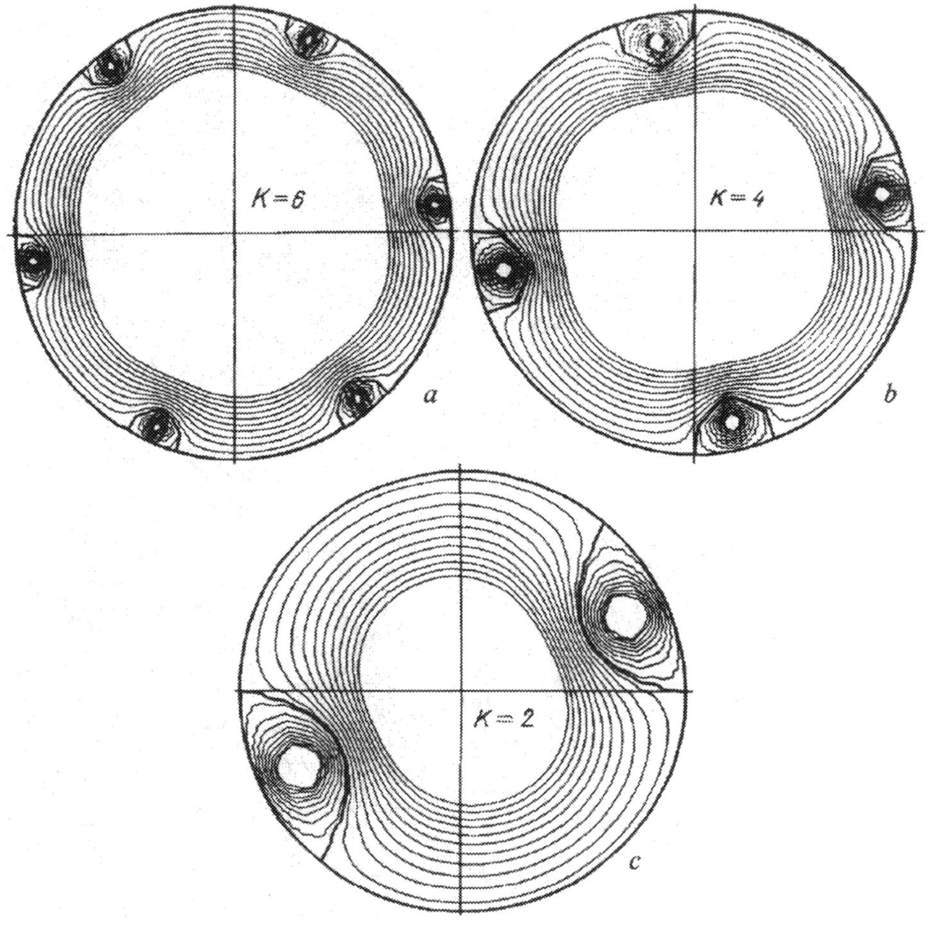

Fig.2.3. Streamlines of the circulation flow of liquid within the circle to which the domain containing liquid is mapped for the case of K inner ribs and free vortices.

(2.146), and equations analogous to (2.151) and (2.152) (the first of them was substituted by the condition of minimum of tangential velocity in the vortex location point, for there are no stationary points for this flow model).

For comparison, Table 2.2 also shows the values of Joukowski moment of inertia $\widetilde{I}$ of (2.149) corresponding to nonstalled flow around ribs, close to those obtained by L.V.Dokuchaev. Dash in the table means that the calculation results in $\widetilde{I}_\Omega < 0$*.

Now we conclude the preliminary discussion of mathematical models of perturbed motion of a solid body with cavities containing vortex LV liquid. This chapter did not

* Results presented in Table 2.2 and pictures of the flow, generated by computer graphics methods, shown in Figs. 2.2,...,2.4, were obtained by Dr. A.A.Livshits, to whom the authors express their sincere acknowledgement.

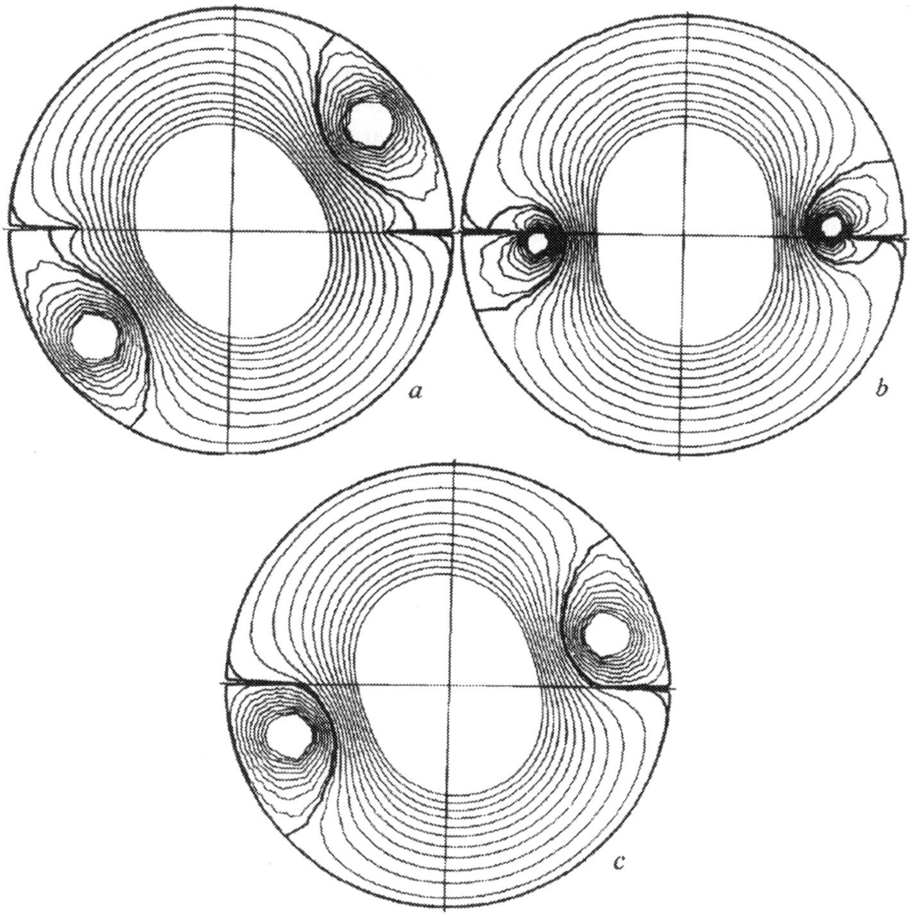

Fig.2.4. Streamlines of circulation flow of liquid in a circular domain for the case of two ribs and two free vortices; a – the angular distance from the vortices to the ribs is greater than it is required according to Joukowski – Chaplygin condition; b – the distance is less than according to the mentioned condition; c – the distance is exactly what is required by the condition.

Table 2.2

Dimensionless Associated Moments of Inertia of Liquid

$\tilde{b}$	K					
	4			6		
	$\tilde{I}$	$\tilde{I}_\Gamma$	$\tilde{I}_\Omega$	$\tilde{I}$	$\tilde{I}_\Gamma$	$\tilde{I}_\Omega$
0.3	0.417	0.012	-	0.572	0.331	0.288
0.4	0.645	0.598	0.180	0.845	0.544	0.622

touch purely mathematical problems associated with integrodifferential equations having singular kernels, to which all the mathematical models considered above, are reduced. Chapter 3 is devoted to this aspect of the problem.

On the other hand, so far open is the question of the correspondence of models of nonvortex absolute motion of liquid outside sharp ribs' edges vicinity expressed by (2.68), models of uniform volume vorticity of liquid expressed by (2.67), and nonuniform vorticity (2.64). Those questions and the corresponding eddy-current problems are discussed based on experimental information in Chapter 4.

Chapter 3

Analytical and Numerical Methods of Dynamics Investigation of Vehicles Described by Vortex Models

There is no certainty in sciences, where none of the mathematical sciences can be applied or which are not in relation with these mathematics.

Leonardo da Vinci *

In the first two chapters we discussed two classes of stabilized objects whose mathematical models are close to each other in a certain sense. These models are used throughout the book as a tool for the analysis of dynamic properties of vehicles described by them, those properties being determined from both experimental research and mathematical simulation of perturbed motion. But before proceeding to direct usage of those models, it is necessary to consider some purely mathematical aspects and introduce some additional constraints. This problem as well as analytical and numerical methods of appropriate equations solution are the contents of this chapter.

The main feature of mathematical models of objects with cavities partially filled with LV liquid and of electromagnetic actuators based on HECF elements with pronounced skin effect is that they are described by integrodifferential equations with singular kernels of $(t-\tau)^{-1/2}$ type. In the general case it is a set of nonlinear integrodifferential equations of Volterra type. In the case of integral operators of particular form these are equations of Abel type.

Similar equations occur in various engineering fields, e.g. the theory of hereditary elasticity (see N.Kh.Arutiunian and V.B.Kolmanovsky, Yu.N.Rabotnov [5, 85]); when investigating controlled systems with aftereffect (see V.B.Kolmanovsky and V.R.Nosov [41]), and in hydromechanics, electrical engineering acoustics, etc. (A.I.Lurie [58]). A developed universal mathematical apparatus is now available for solving integral equations (A.F.Verlan and V.S. Sizikov [101]) and for investigation of processes described by those equations (see [41]). We, however, have rather a narrow problem, namely integral equations with kernels of special kind. That allows to develop problem-oriented analytical and numerical methods specially adjusted for solving such problems, but not intended to be universal.

* The literary works of Leonardo da Vinci compiled and edited from the oroginal manuscripts by Jean Paul Richter and Irma Richter. Oxford University Press. London – New York – Toronto, 1939, Vol. 2, p. 241.

3.1. Preliminary Remarks

Generally, equations describing stabilized vehicles dynamics, considering physical phenomena with aftereffect, are referred to the equations of neutral type, that following V.B.Kolmanovsky and V.R.Nosov [41] can be written as follows:

$$\dot{\mathbf{x}}(t) = \int_0^\infty [d\mathsf{K}(s)]\, \mathbf{x}(t-s) + \int_0^\infty [d\mathsf{K}_1(s)]\, \dot{\mathbf{x}}(t-s) + \mathsf{C}\mathbf{x}(t) + \mathbf{f}(t), \quad (3.1)$$

where $\mathbf{x}(t)$ is n-dimensional vector of system state; $n \times n$ matrices $\mathsf{K}(s)$ and $\mathsf{K}_1(s)$ are to have limited variation in the interval $[0, \infty]$; $\mathbf{f}(t)$ is n-dimensional perturbations vector; C is a square $n \times n$ matrix which is generally a nonlinear function of $\mathbf{x}$ vector. Introduce a new variable

$$\tau = t - s \quad (3.2)$$

and perform change of variables in Equation (3.1). We obtain

$$\dot{\mathbf{x}}(t) + \int_{-\infty}^{t} [d\mathsf{K}(t-\tau)]\, \mathbf{x}(\tau) + \int_{-\infty}^{t} [d\mathsf{K}_1(t-\tau)]\, \dot{\mathbf{x}}(\tau) - \mathsf{C}\mathbf{x}(t) = \mathbf{f}(t). \quad (3.3)$$

In this form Equations (3.3) are exactly the same as those obtained in the first two chapters, and

$$\mathsf{K}(t-\tau) = \mathsf{A}(\mathbf{x})\sqrt{t-\tau}, \quad \mathsf{K}_1(t-\tau) = \mathsf{B}(\mathbf{x})\sqrt{t-\tau}, \quad (3.4)$$

where A and B are $n \times n$ square matrices whose elements are nonlinear functions of $\mathbf{x}$ vector in the general case.

Integration limits 0, ∞ in Equations (3.1) and $-\infty$, t in (3:2) mean that the entire 'infinite' pre-history of the system is considered, that is impossible from the physical point of view, for the investigated real process is always known at a bounded time interval. Therefore we are further investigating processes which have no pre-history before the initial moment.

We can assume $t = 0$ for the initial moment without loss of generality. In this case Equations (3.1) (3.3) with the kernel of (3.4) type can be written in two modifications that are mathematically equivalent.

The first modification:

$$\dot{\mathbf{x}}(t) = \int_0^\infty [d\mathsf{K}(s)]\, \mathbf{x}(t-s) + \int_0^\infty [d\mathsf{K}_1(s)]\, \dot{\mathbf{x}}(t-s) + \mathsf{C}\mathbf{x}(t) + \mathbf{f}(t),$$

or

$$\dot{\mathbf{x}}(t) + \int_{-\infty}^{t} [d\mathsf{K}(t-\tau)]\, \mathbf{x}(\tau) + \int_{-\infty}^{t} [d\mathsf{K}_1(t-\tau)]\, \dot{\mathbf{x}}(\tau) - \mathsf{C}\mathbf{x}(t) = \mathbf{f}(t), \quad (3.5)$$

$$\mathsf{K}(s) = \mathsf{A}\sqrt{s}, \quad \mathsf{K}_1 = \mathsf{B}\sqrt{s}, \quad 0 < s < t, \quad \mathsf{K}(s) = \mathsf{K}_1(s) \equiv 0, \quad s \geq t.$$

The second modification:

$$\dot{\mathbf{x}}(t) = \int_0^t [d\mathsf{K}(s)]\,\mathbf{x}(t-s) + \int_0^t [d\mathsf{K}_1(s)]\,\dot{\mathbf{x}}(t-s) + \mathsf{C}\mathbf{x}(t) + \mathbf{f}(t)\,,$$

or

$$\dot{\mathbf{x}}(t) + \int_0^t [d\mathsf{K}(t-\tau)]\,\mathbf{x}(\tau) + \int_0^t [d\mathsf{K}_1(t-\tau)]\,\dot{\mathbf{x}}(\tau) - \mathsf{C}\mathbf{x}(t) = \mathbf{f}(t)\,, \qquad (3.6)$$

$$\mathsf{K} = \mathsf{A}\sqrt{s}\,, \quad \mathsf{K}_1 = \mathsf{B}\sqrt{s}\,, \quad s > 0\,.$$

This modification is further assumed the main.

Abel's kernel for Modification (3.5) is specified similarly in the theory of hereditary elasticity in [85].

The requirement of limited variation of kernels $\mathsf{K}(s)$ and $\mathsf{K}_1(s)$ in the interval $[0, \infty]$ means that the investigated systems meet the condition of decaying memory (see [85]). Speaking rigorously, Kernel (3.4) does not meet this condition, but Modification (3.5) meets it and for Modification (3.6) it is irrelevant.

The following keen item is worth a special consideration. The mathematical models of vortex phenomena, presented in Chapter 1, 2, 8 adequately describe dynamic processes when vortex motions of the system are already developed. They are, generally speaking, useless for vortex formation and collapse processes description. Elimination of the pre-history, that would otherwise describe vortex formation, means that the constraint introduced results in an inadequate mathematical model at the initial time interval. A compensation for this model imperfection is carried out further in each particular case based on specific character of the considered problem with the help of some heuristic assumptions. The same refers to the process of vortex motions collapse description.

In this chapter we keep this question open as the one associated with the physical correctness of the model. We focus on general methods of investigating dynamics of controlled systems with aftereffect within the scope of a mathematically correct model.

3.2. Analytical Methods of Investigation. Harmonic Balance Method

Let us formulate Cauchy problem for the equations without the right side, written in the form of (3.6) with constant $n \times n$ matrices A, B, and C (the linear case):

$$\dot{\mathbf{x}}(t) + \frac{1}{2}\mathsf{A}\int_0^t \frac{\mathbf{x}(\tau)\,d\tau}{\sqrt{t-\tau}} + \frac{1}{2}\mathsf{B}\int_0^t \frac{\dot{\mathbf{x}}(\tau)\,d\tau}{\sqrt{t-\tau}} - \mathsf{C}\,\mathbf{x}(t) = 0\,, \qquad (3.7)$$

$$\mathbf{x}(0) = \mathbf{x}_0\,.$$

Vector function $\mathbf{x}(t)$ is the solution of this problem. To determine it we can apply the well-developed apparatus of unilateral Laplace transform (see A.I.Lurie [58]).

Let us first consider stability of the trivial solution $\mathbf{x}(t) = 0$ Equation (3.7). Following V.B.Kolmanovsky and V.R.Nosov [41], we describe the approach to stability investigation for an equation of a more general form (3.1). Let us seek particular solutions of homogeneous Equation (3.1) according to Euler in the form

$$\mathbf{x}(t) = e^{pt}\vec{\alpha}, \tag{3.8}$$

where $\vec{\alpha} \neq 0$ is some constant vector determined by the initial conditions. Substituting (3.8) into (3.1) we obtain

$$\left[\int_0^\infty e^{-ps}\, d\,\mathsf{K}(s) + p\int_0^\infty e^{-ps}\, d\,\mathsf{K}_1(s) + \mathsf{C}\right]\vec{\alpha} = p\,\vec{\alpha}. \tag{3.9}$$

It is obvious that for System (3.1) to have nontrivial solutions, p and $\vec{\alpha}$ are to be an eigenvalue and eigenvector respectively of matrix

$$\left[\int_0^\infty e^{-ps}\, d\,\mathsf{K}(s) + p\int_0^\infty e^{-ps}\, d\,\mathsf{K}_1(s) + \mathsf{C}\right],$$

that yields a criterion of asymptotic stability of the trivial solution $\mathbf{x}(t) \equiv 0$ in the form of

$$\operatorname{Re} p_k < 0, \tag{3.10}$$

where p_k are the roots of the characteristic equation

$$\det\left[p\,\mathsf{E} - \int_0^\infty e^{-ps}\, d\,\mathsf{K}(s) - p\int_0^\infty e^{-ps}\, d\,\mathsf{K}_1(s) - \mathsf{C}\right] = 0, \tag{3.11}$$

p is Laplace variable, and E is the unity matrix.

Stability Condition (3.10) means that zeroes of the characteristic quasi-polynomial are to be situated in the left half-plane of complex variable, same as for the case of the systems described by ordinary linear differential equations.

Let us now proceed directly to the investigation of Equation (3.7), confining to the case $\mathsf{A} \equiv 0$ without loss of generality, since in the considered mathematical models (see Chapters 1, 2, 8) dynamic influence of vortex effects is described in terms of generalized velocities only. In this case we obtain the equation

$$\dot{\mathbf{x}}(t) + \frac{1}{2}\mathsf{B}\int_0^t \frac{\dot{\mathbf{x}}(\tau)\,d\tau}{\sqrt{t-\tau}} - \mathsf{C}\,\mathbf{x}(t) = 0, \quad \mathbf{x}(0) = \mathbf{x}_0, \tag{3.12}$$

which is a particular case of (3.7). Let us perform Laplace transform of the left side of Equation (3.12). Resulting is the following equation in the transform domain

$$p\,\mathbf{X}(p) - \mathbf{x}_0 + \frac{1}{2}\mathsf{B}\,\mathbf{J}(p) - \mathsf{C}\,\mathbf{X}(p) = 0, \tag{3.13}$$

where $\mathbf{J}(p)$ is a vector whose components are Laplace transform of the integral

$$j(t) = \int_0^t \frac{\dot{x}(\tau)\,d\tau}{\sqrt{t-\tau}}, \tag{3.14}$$

$x(\tau)$ in (3.14) is a scalar function.

Expression in the right side of (3.14) is the convolution of function $f_1(\tau)$ with $f_2(\tau)$. These functions and Laplace transforms $F_1(p)$ and $F_2(p)$ have the form

$$f_1(\tau) = \dot{x}(\tau), \quad f_2(\tau) = \frac{1}{\sqrt{\tau}}, \tag{3.15}$$

$$F_1(p) = p\,X(p) - x_0, \quad F_2(p) = \left(\frac{\pi}{p}\right)^{1/2}.$$

According to the theorem of convolution (see A.I. Lurie [58]),

$$J(p) = \sqrt{\pi\,p}\;X(p) - \frac{\sqrt{\pi}\;x_0}{\sqrt{p}}. \tag{3.16}$$

Equation (3.13) now gains the form

$$p\,\mathbf{X}(p) - \mathbf{x}_0 + \frac{1}{2}\,\mathsf{B}\,\sqrt{\pi p}\;\mathbf{X}(p) - \frac{\mathsf{B}\;\mathbf{x}_0}{2}\left(\frac{\pi}{p}\right)^{1/2} - \mathsf{C}\,\mathbf{X}(p) = 0\,. \tag{3.17}$$

Its solution written in scalar form is as follows:

$$X_i = \frac{\Delta_i(p)}{\Delta_0(p)}, \quad i = 1, 2, \dots, n, \tag{3.18}$$

where

$$\Delta_0(p) = \det\left(p\,\mathsf{E} + \frac{1}{2}\,\mathsf{B}\,\sqrt{\pi p} - \mathsf{C}\right) = \det\left(q^2\,\mathsf{E} + \frac{1}{2}\,\mathsf{B}\,\sqrt{\pi}\,q - \mathsf{C}\right) = D_0(q) \tag{3.19}$$

is a polynomial including integer powers of variable $q = \sqrt{p}$, the polynomial being a particular representation (for the considered kernel) of the quasi-polynomial in the left side of (3.11); $\Delta_i(p)$ is the determinant obtained from (3.19) by substituting column vector

$$\mathbf{Y} = \mathbf{x}_0 + \frac{\mathsf{B}\mathbf{x}_0}{2}\left(\frac{\pi}{p}\right)^{1/2} \tag{3.19a}$$

for i-th column of (3.19).

A similar solution can be obtained for a more general case $\mathsf{A} \neq 0$. Therefore all the further discussion refers to the general case of Equation (3.6).

The universal stability Criterion (3.10) for this particular case is written in the following form:

$$\frac{\pi}{4} < \operatorname{Arg} q_s < \frac{7\pi}{4}, \tag{3.20}$$

where q_s are roots of equation

$$D_0(q) = 0\,, \tag{3.21}$$

some of which roots can be zero in a particular case.

Stability Domain (3.20) in q plane is presented in Fig. 3.1 Stability Criterion (3.20) can be obtained with the help of Equation (3.6) solution asymptotic representation for $t \to \infty$, according to the method described in [58].

Let us write the right side of Expression (3.18) taking into account (3.19a), in the form of the following meromorphic function:

$$Z_i(q) = \frac{D_i(q)}{q^m D(q)}, \tag{3.21a}$$

where $D_i(q)$ and $D(q)$ are polynomials including integer powers of variable q and having no zero roots; m is the maximum order of multiplicity of zero poles appearing, on one hand, because of zero roots of Equation (3.21), and on the other hand, engendered by nonzero initial conditions with respect to variables under the integral operators. Meromorphic Function (3.21a) can be represented by the following expansion in elementary fractions:

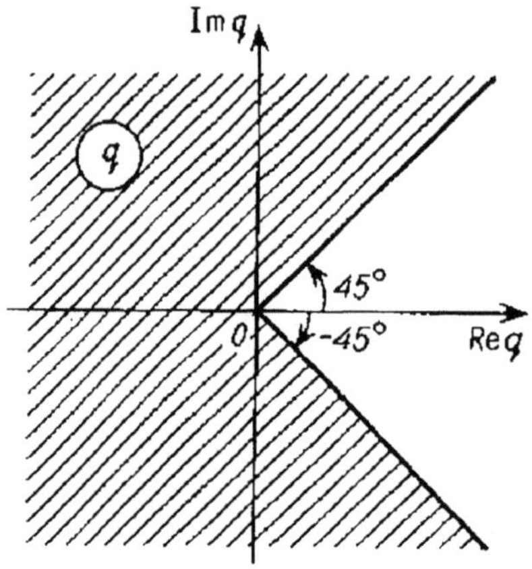

Fig. 3.1. Stability domain in q plane. Stability is at the hatched side.

$$Z_i(q) = \frac{C_i}{q} + \sum_{s=2}^{m} \frac{G_{is}}{q_s} + \sum_{s=1}^{n} \frac{D_{is}}{q - q_s}, \tag{3.21b}$$

where q_s ($s = 1, 2, ..., n$) are simple roots of polynomial $D(q)$.

The components of Expression (3.21b) corresponding to zero poles of q of multiplicity higher than unity cause non-decaying in time transient components in the solution. These components are caused by incomplete adequacy of the considered mathematical models to real physical processes and are primarily due to impossibility to describe vortex motions formation processes. Therefore such particular solutions are to be excluded from further consideration.

Taking into account those circumstances let us group separately terms corresponding to multiple zero roots. Besides that, join the terms $\frac{C_i}{q}$ and $\sum_{s=1}^{n} \frac{D_{is}}{(q - q_s)}$, that brings the right side of (3.21b) to the form

$$Z_i(q) = \sum_{s=1}^{n} \frac{B_{is}}{q(q - q_s)} + \Phi(q), \tag{3.21c}$$

where $\Phi(q)$ is a function that can be represented by an expansion in terms of $1/q$ of the order higher than unity. Such representation form allows to recover the appropriate original function from each transform component (see [58]).

It is necessary to make a very important remark here, that should be held in mind further on. It consists in that the characteristic equation corresponding to Expansion (3.21c) allows existence of a simple zero root with respect to variable $q = \sqrt{p}$ (even in one-dimensional case when instead of vector function $\mathbf{x}(t)$ a scalar function $x(t)$ is included in (3.12)). The component of $1/\sqrt{t}$ type corresponds to this root. (An analogue of this situation for system described by ordinary differential equations is a simple zero root with respect to p corresponding to a system neutral relative to coordinate). At the same time stability analysis should be carried out using nonzero roots $q_s \neq 0$ $(s = 1, 2, \ldots, n)$. That is done below when considering particular examples.

Recovering the original function from transform (3.21a), i.e. inverse Laplace transform, is thus reduced to inverting functions of $1/q(q - q_s)$ type (see [58]) that is carried out by the formula

$$\frac{1}{q(q - q_s)} \to e^{q_s^2 t}\left(1 - \frac{2}{\sqrt{\pi}} \int_0^{-q_s\sqrt{t}} e^{-u^2} du\right). \tag{3.22}$$

The integral in (3.22) cannot be solved by quadratures. For real roots q it is known as *probability integral* (see [58]) and can be evaluated with the help of appropriate tables. For complex roots, which are specific for controlled systems, the problem is much more complicated.

Consider expressions

$$\Psi_s = e^{q_s^2 t}(1 - \Phi_s), \quad \Phi_s = \frac{2}{\sqrt{\pi}} \int_0^{-q_s\sqrt{t}} e^{-u^2} du, \tag{3.23}$$

included in (3.22). It is clear that the system stability is determined by function Φ_s behaviour. Introduce function

$$\Phi(\xi) = \frac{2}{\sqrt{\pi}} \int_0^{\xi} e^{-u^2} du, \tag{3.24}$$

corresponding to Φ_s. Consider the function for great and small values of ξ. The following asymptotic expansion is known for Integral (3.24) (see [58]):

$$\Phi(\xi) = \frac{2}{\sqrt{\pi}} \int_0^{\xi} e^{-u^2} du \sim 1 - \frac{e^{-\xi^2}}{\xi\sqrt{\pi}}\left[1 - \frac{1}{2\xi^2} + \frac{1 \cdot 3}{\left(2\xi^2\right)^2} - \ldots\right]. \tag{3.25}$$

This asymptotic expansion is obtained for the branch of two-valued function $q = \sqrt{p}$, which is positive for real positive p, i.e. it corresponds to function $q(p)$ Riemann surface (p-plane) mapping onto the right half-plane of q. Therefore the integration variable in the right-side integral of (3.25) should be selected so, that the upper limit should have a positive real part:

$$\xi = q\sqrt{t}, \quad \operatorname{Re} q > 0, \quad \operatorname{Re} \xi > 0.$$

When considering q values located in the left half-plane $\operatorname{Re} q < 0$, a new variable $v = -u$ should be introduced. Then

$$\xi = -q\sqrt{t}\,, \quad \operatorname{Re} q < 0\,, \quad \operatorname{Re}\xi > 0,$$

and hence

$$\Phi(q\sqrt{t}) = \Phi(-\xi) = \frac{2}{\sqrt{\pi}}\int_0^{-\xi} e^{-u^2}du = -\frac{2}{\sqrt{\pi}}\int_0^{\xi} e^{-v^2}dv = -\Phi(\xi)\,. \tag{3.26}$$

This yields

$$\Phi(-q\sqrt{t}) = \begin{cases} -\Phi(\xi) \sim 1 + \dfrac{e^{-\xi^2}}{\xi\sqrt{\pi}}\left[1 - \dfrac{1}{2\xi^2} + \dfrac{1\cdot 3}{(2\xi^2)^2} - \dots\right] \\ \qquad \text{for } \operatorname{Re} q > 0\,,\ \xi = q\sqrt{t}\,, \\ \Phi(\xi) \sim 1 - \dfrac{e^{-\xi^2}}{\xi\sqrt{\pi}}\left[1 - \dfrac{1}{2\xi^2} + \dfrac{1\cdot 3}{(2\xi^2)^2} - \dots\right] \\ \qquad \text{for } \operatorname{Re} q < 0\,,\ \xi = -q\sqrt{t}\,. \end{cases} \tag{3.27}$$

From (3.23) and (3.27) we obtain

$$\Psi_s = e^{q_s^2 t}(1 - \Phi_s) =$$

$$= \begin{cases} e^{q_s^2 t} - \dfrac{1}{q_s\sqrt{\pi t}}\left(1 - \dfrac{1}{2q_s^2 t} + \dots\right) & \text{for } \operatorname{Re} q_s > 0\,, \\ -\dfrac{1}{q_s\sqrt{\pi t}}\left(1 - \dfrac{1}{2q_s^2 t} + \dots\right) & \text{for } \operatorname{Re} q_s < 0\,. \end{cases} \tag{3.28}$$

In the limit, for $t \to \infty$, (3.28) yields quadrature formulae known from the theory of probability integral and Fresnel integrals:

$$\frac{2}{\sqrt{\pi}}\int_0^{\infty} e^{-u^2}du = -\frac{2}{\sqrt{\pi}}\int_0^{-\infty} e^{-v^2}dv = 1\,. \tag{3.29}$$

$$2\left(\frac{2}{\pi}\right)^{1/2}\int_0^{\infty}\sin u^2 du = 2\left(\frac{2}{\pi}\right)^{1/2}\int_0^{\infty}\cos u^2 du = 1. \tag{3.30}$$

For the case of small values of $|\xi|$ and $\operatorname{Re}\xi \gtrless 0$ the following expansion is correct:

$$\Phi(\xi) = -\Phi(-\xi) = \frac{2}{\sqrt{\pi}}\int_0^{\xi} e^{-u^2}du = \frac{2\xi}{\sqrt{\pi}}\left(1 - \frac{\xi^2}{1!3} + \frac{\xi^4}{2!5} - \dots\right), \tag{3.31}$$

which results in

$$\Phi_s = \frac{2q_s\sqrt{t}}{\sqrt{\pi}}\left(1 - \frac{q_s^2 t}{1!3} + \frac{q_s^4 t^2}{2!5} - \dots\right),$$

$$\Psi_s = e^{q_s^2 t}\left[1 - \frac{2q_s\sqrt{t}}{\sqrt{\pi}}\left(1 - \frac{q_s^2 t}{1!3} + \frac{q_s^4 t^2}{2!5} - \dots\right)\right]. \tag{3.32}$$

Expressions (3.28) and (3.32) can be made the foundation for analyzing the considered systems stability and transient processes for various q_s values.

Let us consider q-plane stability and instability domains structure using Expressions (3.29). We assume

$$q_s = \rho_s e^{i\varphi_s}, \quad 0 \le \varphi_s < \frac{\pi}{2}, \quad \frac{\pi}{2} < \varphi_s \le \pi, \tag{3.33}$$

$$\pi \le \varphi_s < \frac{3\pi}{2}, \quad \frac{3\pi}{2} < \varphi_s \le 2\pi,$$

and introduce into Φ_s variable v, defined this time as follows:

$$v = ue^{-i(\varphi_s - \pi)}; \quad du = e^{i(\varphi_s - \pi)} dv, \quad v = \rho_s \sqrt{t} \quad \text{for } u = -\rho_s e^{i\varphi_s \sqrt{t}}. \tag{3.34}$$

Inequalities (3.33) exclude the case $q_s = \pm i q^\circ$, where q° is a real number. In this case solution includes the following component:

$$\frac{1}{q^2 + q^{\circ 2}} = \frac{1}{p + q^{\circ 2}}, \tag{3.35}$$

the original function from which is

$$\frac{1}{p + q^{\circ 2}} \to e^{-q^{\circ 2} t}, \tag{3.36}$$

determining the stable component of the transient process.

Let us rewrite Integral (3.23) taking into account (3.33), as follows:

$$\Phi_s = \frac{2}{\sqrt{\pi}} \int_0^{\rho_s\sqrt{t}} \exp\left\{-v^2 \left[\cos 2(\varphi_s - \pi) + i \sin 2(\varphi_s - \pi)\right] + i(\varphi_s - \pi)\right\} dv, \tag{3.37}$$

and consider its asymptotics for $t \to \infty$ for some specific values of q_s root's argument φ_s. It is obvious from (3.37) that the change of variables performed results in the positive upper integration limit, that enables direct usage of asymptotic Expansion (3.25) and quadrature Formulae (3.29) and (3.30).

Consider function Φ_s for $\varphi_s = 0; \pi/4; 3\pi/4; \pi$.

Case 1. $\varphi_s = 0$; $\varphi_s - \pi = -\pi$; $2(\varphi_s - \pi) = -2\pi$:

$$\Phi_s = -\frac{2}{\sqrt{\pi}} \int_0^{\rho_s\sqrt{t}} e^{-v^2} dv = -\Phi(\rho_s\sqrt{t}) \to -1 \quad \text{for } t \to \infty.$$

Case 2. $\varphi_s = \pi/4$; $\varphi_s - \pi = -3\pi/4$; $2(\varphi_s - \pi) = -3\pi/2$:

$$\Phi_s = \frac{2}{\sqrt{\pi}} \int_0^{\rho_s\sqrt{t}} e^{-iv^2} \left(\cos\frac{3\pi}{4} - i \sin\frac{3\pi}{4}\right) dv =$$

$$= -\left(\frac{2}{\pi}\right)^{1/2} \left[(1 + i) \int_0^{\rho_s\sqrt{t}} \cos v^2 dv + (1 - i) \int_0^{\rho_s\sqrt{t}} \sin v^2 dv\right] \to$$

$$\to -\frac{1}{2}\left[(1+i)+(1-i)\right] = -1 \quad \text{for} \quad t\to\infty .$$

Case 3. $\varphi_s = 3\pi/4$; $\varphi_s - \pi = -\pi/4$; $2(\varphi_s - \pi) = -\pi/2$:

$$\Phi_s = \frac{2}{\sqrt{\pi}} \int_0^{\rho_s\sqrt{t}} e^{iv^2}\left(\cos\frac{\pi}{4} - i\sin\frac{\pi}{4}\right) dv =$$

$$= \left(\frac{2}{\pi}\right)^{1/2} \left[(1-i)\int_0^{\rho_s\sqrt{t}} \cos v^2\, dv + (1+i)\int_0^{\rho_s\sqrt{t}} \sin v^2\, dv\right] \to$$

$$\to \frac{1}{2}\left[(1-i)+(1+i)\right] = 1 \quad \text{for} \quad t\to\infty .$$

Case 4. $\varphi_s = \pi$; $\varphi_s - \pi = 0$:

$$\Phi_s = \frac{2}{\sqrt{\pi}} \int_0^{\rho_s\sqrt{t}} e^{-v^2} dv = \Phi_s(\rho_s\sqrt{t}) \to 1 \quad \text{for } t\to\infty .$$

For $\varphi_s = 5\pi/4$; $7\pi/4$ we obtain similarly:

Case 5. $\varphi_s = 5\pi/4$; $\Phi_s \to 1$ for $t\to\infty$.

Case 6. $\varphi_s = 7\pi/4$; $\Phi_s \to -1$ for $t\to\infty$.

Now we can consider stability and instability domains in Lyapunov sense over the entire variable q plane.

1. The right half-plane with excluded imaginary axis and real positive axis.

a) $-\pi/4 < \varphi_s < \pi/4$, $\varphi_s \neq 0$:

$$p_s = q_s^2, \quad -\frac{\pi}{2} < \operatorname{Arg} p_s < \frac{\pi}{2}, \quad \operatorname{Re} p_s > 0,$$

$$\Psi_s \to 2e^{p_s t} \to \infty \quad \text{for} \quad t\to\infty .$$

b) $\varphi_s = \pm\pi/4$:

$$p_s = q_s^2 = \pm iq^{\circ 2}, \quad \operatorname{Arg} p_s \pm \frac{\pi}{2}, \quad \operatorname{Re} p_s = 0,$$

$$\Psi_s \to 2e^{\pm iq^{\circ 2}t} \quad \text{for} \quad t\to\infty . \tag{3.38}$$

c) $\pi/4 < \varphi_s < \pi/2$ or $-\pi/2 < \varphi_s < -\pi/4$:

$$p_s = q_s^2, \quad \frac{\pi}{2} < \operatorname{Arg} p_s < \frac{3\pi}{2}, \quad \operatorname{Re} p_s < 0,$$

$$\Psi_s \to 2e^{p_s t} - \frac{1}{q_s\sqrt{\pi t}} \to 0 \quad \text{for} \quad t\to\infty . \tag{3.39}$$

2. The left half-plane with excluded imaginary axis and real negative axis.

$$\frac{\pi}{2} < \varphi_s < \frac{3\pi}{2}, \quad \varphi_s \neq \pi:$$

$$\Psi_s \to -\frac{1}{q_s\sqrt{\pi t}} \to 0 \quad \text{for} \quad t \to \infty . \tag{3.40}$$

3. Real and imaginary axis.

a) $\varphi_s = 0 , \ s = 1, 2 :$

$$\operatorname{Im} q_s = 0 , \ q_s > 0 , \ \operatorname{Im} p_s = 0 , \ p_s > 0 ,$$

$$\Psi_s \to 2e^{p_s t} \to \infty \quad \text{for} \quad t \to \infty .$$

b) $\varphi_s = \pi , \ s = 1, 2 :$

$$\operatorname{Im} q_s = 0 , \ q_s < 0 , \ \operatorname{Im} p_s = 0 , \ p_s < 0 ,$$

$$\Psi_s \to -\frac{1}{q_s\sqrt{\pi t}} \to 0 \quad \text{for} \quad t \to \infty . \tag{3.41}$$

c) $\varphi_1 = \pi/2 , \ \varphi_2 = 3\pi/2 , \ \rho_1 = \rho_2 = q^\circ :$

$$\operatorname{Im} p_s = 0 , \ p_1 = p_2 = -q^{\circ 2} < 0 , \ e^{-q^{\circ 2} t} \to 0 \quad \text{for} \quad t \to \infty . \tag{3.42}$$

Thus, based on asymptotic representation of the initial equations solution, we have obtained the condition of trivial solution's asymptotic stability in the form of (3.20), that follows from the general theorems of stability of controlled systems with aftereffect (see V.B.Kolmanovsky and V.R. Nosov [41]). The possibilities of this method are much wider, however, for using it we can judge on qualitative behaviour of the considered systems in transient processes based on location of the characteristic equation roots in q plane.

In the instability domain $-\pi/4 < \varphi_s < \pi/4$, for instance, functions describing the transient process grow exponentially, the instability being of aperiodic type for $\varphi_s = 0$ and oscillatory for $\varphi_s \neq 0$. At the stability limit, as it follows from (3.38), the transient process tends to steady-state harmonic oscillations.

For the stability domain $\pi/4 < \varphi_s < 7\pi/4$ the problem is more complicated. Particularly, in the left half-plane $\pi/2 < \varphi_s < 3\pi/2$ transient processes decay as $1/q_s\sqrt{\pi t}$ for $t \to \infty$, and in the part of the right half-plane, where $\pi/4 < \varphi_s < \pi/2$ or $-\pi/4 > \varphi_s > -\pi/2$, they decay as the sum $2e^{p_s t} - (1/q_s\sqrt{\pi t}\,)$.

The starting moment of the transient process can be described with asymptotic Expansion (3.31), (3.32).

Let us now consider the question of the possibility to reduce the set of integrodifferential equations to ordinary differential equations. A physical reason for the introduction of the appropriate simplified mathematical model is that for the considered controlled systems the principal process for investigation is in the vicinity of some characteristic frequency which we denote by ω_0 . For vehicles with compartments partially filled with liquid for oscillations in the pitch and yawing planes it is the partial frequency of the principal tone of liquid wave motions in the tank playing the dominant role in the closed-loop 'object – controller' system dynamics.

When investigating rotation dynamics of axisymmetric objects rotation about their longitudinal axis and dynamics of electromagnetic actuators of stabilization and orientation systems, this characteristic frequency is the closed-loop system frequency, generated by the regulator.

Let us return now to the interval $(-\infty, t)$ involved in the initial mathematical models of Chapters 1 and 2, applying special constraints formulated in this chapter neither to the kernel, nor to function $x(t)$ in that interval. Suppose the process is steady-state harmonic oscillations with ω frequency, and hence function $x(t)$ has the form $x(t) = x_0 e^{i\omega t}$. Consider the expression

$$G(t) = g \int_{-\infty}^{t} \frac{\dot{x}(\tau)\, d\tau}{\sqrt{t-\tau}} = g x_0\, i\omega \int_{-\infty}^{t} \frac{e^{i\omega t}\, d\tau}{\sqrt{t-\tau}}. \tag{3.43}$$

Introduce a new variable $u^2 = t - \tau$ so that

$$\tau = t - u^2, \quad \sqrt{t-\tau} = u, \quad d\tau = -2u\,du. \tag{3.44}$$

and apply Quadratures (3.30) known from Fresnel integrals theory. This yields:

$$G(t) = 2igx_0\, \omega\, e^{i\omega t} \int_0^{\infty} \exp(-i\,\omega u^2)\, du =$$

$$= 2igx_0\, \omega\, e^{i\omega t} \int_0^{\infty} (\cos \omega u^2 - i \sin \omega u^2)\, du = g(i+1)\left(\frac{\pi\omega}{2}\right)^{1/2} x_0\, e^{i\omega t}. \tag{3.45}$$

Expression (3.45) can be written in the form

$$G(t) = g_L\, \dot{x}(t) + g_D\, x(t), \tag{3.46}$$

where

$$g_L = g\left(\frac{\pi}{2\omega}\right)^{1/2}, \quad g_D = g\left(\frac{\pi\omega}{2}\right)^{1/2}. \tag{3.47}$$

Coefficients (3.47) can be given the following physical interpretation. For objects with liquid g_L is an additional associated mass (moment of inertia) characterizing the inertial effect of the integral term, and g_D is the equivalent damping coefficient. For electromagnetic actuators g_L is an additional inductance, and g_D is the equivalent ohmic resistance characterizing energy dissipation due to heating up of the core material by eddy currents.

3.3. An Algorithm for Numerical Solution of a Set of Integrodifferential Equations with a Singular Kernel of the Type $(t - \tau)^{-1/2}$

Analytical methods of investigation discussed above, based on application of operational calculus methods, allow analyzing asymptotic stability of linear systems and approximate calculation of transient processes for great and small t values and a limited class of perturbations. When we have to deal with pronounced nonlinear effects and perturbations of arbitrary composition, particularly specified by a table of values, as well as for coefficients of perturbed motion equations, variable in time, the only possible way to solve the problem is direct numerical integration of the initial equations.

Consider an algorithm for numerical integration of a set of integrodifferential equations with a singular kernel of the type $(t - \tau)^{-1/2}$, which method is based on application of quadrature formulae for calculation of a definite integral with consequent stepwise transformation of the initial set of equations to a set of differential equations of Lyapunov–Cauchy form and application of standard numerical methods to integration of that set of equations.

An important element of this algorithm is the solution of the integral equation

$$\int_0^t \frac{\sigma(\tau)\, d\tau}{\sqrt{t - \tau}} = f(t), \tag{3.48}$$

where $\sigma(\tau)$ is the unknown function (vectors $\mathbf{x}$ or $\dot{\mathbf{x}}$ components) and $f(t)$ is the function considered known when solving Equation (3.48). A specific feature of the integral Equation (3.48) is the singular kernel which becomes infinite for $\tau = t$. Therefore the necessary condition of the integral Equation (3.48) solution existence

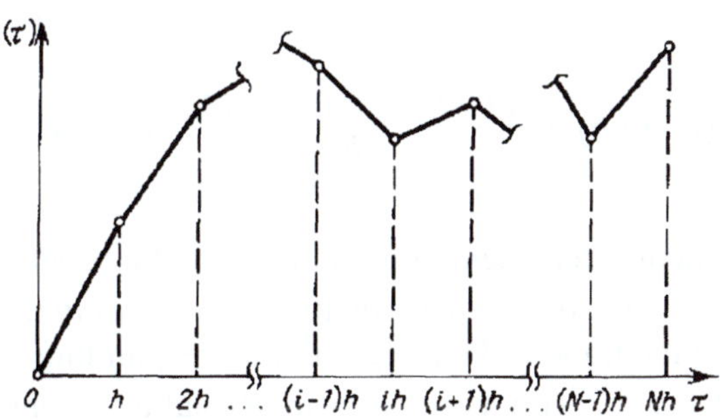

Fig. 3.2. General shape of function $\sigma(\tau)$

is meeting the following conditions (see A.F. Verlan and V.S. Sizikov [101]):

$$\sigma(0) = 0. \tag{3.49}$$

$$f(0) = 0. \tag{3.50}$$

Condition (3.49) is met automatically, since we consider only processes without pre-history. As to Condition (3.50), it can be violated for the considered class of problems. In that case so-called 'splice condition' is not met, and this circumstance is one of the problems emerging when solving equations describing dynamics of controlled systems with aftereffect (see V.B.Kolmanovsky and V.R.Nosov [41]). Hence when simulating the processes of real systems one has either to introduce some additional constraints for function $f(t)$, as it is done below in the investigation of electromagnetic levitation system dynamics, or reconcile oneself to a discontinuity of the solution at the start.

It should be noted that when applying so-called 'step-by-step methods', a modification of which is the method described herein, we obtain 'solution smoothing' (according to [41]).

Let us return to the integral Equation (3.48). We shall calculate the values of the integral function $\sigma(\tau)$ in discrete moments of time $0, h, 2h, \ldots, ih, \ldots, Nh,$ where h is the step of integration of the initial set of integrodifferential equations, and $Nh = t$. Within the intervals $0-h, h-2h, \ldots, (i-1)h-ih, \ldots, (N-1)h-Nh$ we consider function $\sigma(\tau)$ linear (Fig. 3.2). As a result, Equation (3.48) changes to

$$f(t) = \int_0^h \frac{\frac{\sigma(h)}{h}\tau}{\sqrt{t-\tau}}\,d\tau + \int_0^{2h} \frac{\sigma(h) + \frac{\sigma(2h)-\sigma(h)}{h}(\tau - h)}{\sqrt{t-\tau}}\,d\tau + \ldots$$

$$\ldots + \int_{(i-1)h}^{ih} \frac{\sigma((i-1)h) + \frac{\sigma(ih)-\sigma((i-1)h)}{h}(\tau-(i-1)h)}{\sqrt{t-\tau}}\,d\tau +$$

$$+ \int_{ih}^{(i+1)h} \frac{\sigma(ih) + \frac{\sigma((i+1)h)-\sigma(ih)}{h}(\tau - ih)}{\sqrt{t-\tau}}\,d\tau + \ldots$$

$$\ldots + \int_{(N-1)h}^{Nh} \frac{\sigma((N-1)h) + \frac{\sigma(Nh)-\sigma((N-1)h)}{h}(\tau-(N-1)h)}{\sqrt{t-\tau}}\,d\tau . \qquad (3.51)$$

The integral function representation in the form of a product of a linear function by the kernel $(t-\tau)^{-1/2}$ enables to perform the quadratures in the right side of (3.51) more accurately than the trapezoidal integration corresponding to linear interpolation of function $\sigma(\tau)(t-\tau)^{-1/2}$ (for $\tau = t$ such interpolation is impossible at all). This technique lets us avoid difficulties caused by the kernel singularity and make the accuracy of the integral Equation (3.48) numerical solution approach the accuracy of numerical integration of the initial set of integrodifferential equations by one of standard methods. If necessary it is possible to apply a higher order interpolation of function $\sigma(\tau)$.

For a more compact form of the right side of Equation (3.51) let us consider the following integral

$$\int_{t_0}^{t_1} \frac{\sigma(t_0) + \dfrac{\sigma(t_1) - \sigma(t_0)}{h}(\tau - t_0)}{\sqrt{t-\tau}}\, d\tau =$$

$$= \int_{t_0}^{t_1} \frac{\left[\sigma(t_0)\left(1 + \dfrac{t_0}{h}\right) - \dfrac{\sigma(t_1)\, t_0}{h}\right] + \dfrac{\sigma(t_1) - \sigma(t_0)}{h}\tau}{\sqrt{t-\tau}}\, d\tau =$$

$$= 2\left[\sigma(t_0)\left(1 + \frac{t_0}{h}\right) - \frac{\sigma(t_1)\, t_0}{h}\right][\sqrt{t-t_0} - \sqrt{t-t_1}] +$$

$$+ 2\frac{\sigma(t_1) - \sigma(t_0)}{h}\left[t_0\sqrt{t-t_0} + \tfrac{2}{3}(t-t_0)^{3/2} - t_1\sqrt{t-t_1} - \tfrac{2}{3}(t-t_1)^{3/2}\right]. \quad (3.52)$$

As it is clear from (3.52), the integral function singularity for $\tau = t$ is integrable, i.e. the integral has a finite value.

Let us substitute Expression (3.52) with the appropriate integration limits into the right side of (3.51). After simple but clumsy calculations we obtain:

$$f(Nh) = \sqrt{h}\left\{\tfrac{4}{3}\sigma(h)\left[N\sqrt{N} - \sqrt{N-1}\left(N + \tfrac{1}{2}\right)\right] +\right.$$

$$+ 2\sum_{i=2}^{N-1}\left\{[\sigma((i-1)h)[1 + (i-1)] - \sigma(ih)(i-1)] \times\right.$$

$$\times\left[\sqrt{N-(i-1)} - \sqrt{N-i}\right] + \tfrac{2}{3}\left[\sigma(ih) - \sigma((i-1)h)\right] \times$$

$$\times\left.\left[\sqrt{N-(i-1)}\left(N + \frac{i-1}{2}\right) - \sqrt{N-i}\left(N + \frac{i}{2}\right)\right]\right\} +$$

$$\left. + \tfrac{2}{3}\sigma((N-1)h) + \tfrac{4}{3}\sigma(Nh)\right\}. \quad (3.53)$$

Assigning successive integer values to N we obtain the following resolving set of equations:

$$\int_0^t \frac{\sigma(\tau)\, d\tau}{\sqrt{t-\tau}} \cong \sqrt{h}\left\{G_N + \tfrac{4}{3}\sigma(Nh)\right\} = f(Nh), \quad N = 1, 2, 3, \ldots, \quad (3.54)$$

where

$$G_0 = G_1 = 0, \quad G_2 = \tfrac{4}{3}\sigma(h)\left[2(\sqrt{2} - 1)\right],$$

$$G_N = \tfrac{4}{3}\sigma(h)\left[N\sqrt{N} - \sqrt{N-1}\left(N + \tfrac{1}{2}\right)\right] +$$

$$+ \tfrac{4}{3}\sum_{i=2}^{N-1}\left\{\sigma((i-1)h)\left[\sqrt{N-i}\,(N-i) - \sqrt{N-(i-1)}\left(N - \frac{2i+1}{2}\right)\right] +\right.$$

$$+ \ \sigma(ih)\left[\sqrt{N-(i-1)}\,(N-(i-1)) - \sqrt{N-i}\left(N-\frac{2i-3}{2}\right)\right]\Bigg\} +$$

$$+ \ \frac{2}{3}\sigma((N-1)h), \quad N > 2. \qquad (3.55)$$

Denoting $G_N = G(t)$ and $\sigma(Nh) = \sigma(t)$, that is correct for discrete moments of time, we bring (3.55) to the form

$$\sqrt{h}\left\{G(t) + \frac{4}{3}\sigma(t)\right\} = f(t). \qquad (3.56)$$

Further, assuming it possible to factor nonlinear matrices A and B outside the integral sign in Equations (3.6), let us make scalar functions $\sigma(t)$ correspond to vector $\mathbf{x}$ components and introduce vector $\mathbf{G}(t)$, which components are determined according to Formulae (3.55), and a similar vector $\mathbf{G}_1(t)$ for $\dot{\mathbf{x}}(t)$. In that case the set of Equation (3.7) changes to a set of ordinary differential equations, if the right side is added to it.

$$\dot{\mathbf{x}}(t)\left[\mathsf{E} + \frac{2}{3}\mathsf{B}\sqrt{h}\right] + \mathbf{x}(t)\left[\frac{2}{3}\mathsf{A}\sqrt{h} - \mathsf{C}\right] + \frac{1}{2}\sqrt{h}\left[\mathsf{A}\mathbf{G}(t) + \mathsf{B}\mathbf{G}_1(t)\right] = \mathbf{f}(t). \qquad (3.57)$$

Equations (3.57) are integrated numerically with one of standard methods (Runge – Kutta, Adams – Stormer, etc.). The following procedures are performed at each step:

a) Vectors $\mathbf{G}(t)$ and $\mathbf{G}_1(t)$ are calculated according to quadrature Formulae (3.55);

b) Equations (3.57) are reduced to the normalized form of Lyapunov – Cauchy by resolving for vector $\dot{\mathbf{x}}$ components;

c) One step of integration is performed.

3.4. Methodologic Example

Consider the following Cauchy problem solution for a methodologic example:

$$\dot{x}(t) + b\int_0^t \frac{\dot{x}(\tau)\,d\tau}{\sqrt{t-\tau}} + c\,x(t) = 0, \quad x(0) = x_0, \qquad (3.58)$$

where $x(t)$ is an unknown scalar function. With this simple example we can, as it is shown below, study all the variety of processes specific for the considered class of systems, and illustrate the possibilities of the presented analytical and numerical methods.

Let us perform Laplace transform of the left side of Equation (3.58) for zero initial conditions

$$X(p)\,(p + b\sqrt{\pi p} + c) = 0, \qquad (3.59)$$

where $X(p)$ is Laplace transform of function $x(t)$ This yields the characteristic equation in terms of the new complex variable q:

$$q^2 + b\sqrt{\pi}\, q + c = 0\,. \tag{3.60}$$

Roots of the 2nd-order Equation (3.60) have the form

$$q_{1,2} = -\frac{b\sqrt{\pi}}{2} \pm \left(\frac{\pi b^2}{2} - c\right)^{1/2}. \tag{3.61}$$

Coefficients b, c unambiguously determine locations of roots in variable q plane. Analyzing Expression (3.61) we can resolve 6 specific domains which characteristics are united in Table 3.1. The last row of the table contains information on stability of the trivial solution $x(t) = 0$ of Equation (3.58), according to Criterion (3.20). Axis $b = 0$ corresponds to the singular case, namely to 1-order ordinary differential equation

$$\dot{x} + cx = 0\,. \tag{3.62}$$

Table 3.1

Specific domains for roots locations

Domain #	b	c	Roots characteristics	Solution character
1	$b > 0$	$c < \frac{\pi b^2}{4}$	Negative real roots	Stability
2	$b > 0$	$c > \frac{\pi b^2}{4}$	Complex conjugated roots in the left half-plane	Stability
3	$b < 0$	$c > \frac{\pi b^2}{2}$	Complex conjugated roots in the right half-plane	Stability
4	$b < 0$	$\frac{\pi b^2}{4} < c < \frac{\pi b^2}{2}$	Complex conjugated roots in the right half-plane	Instability
5	$b < 0$	$0 < c < \frac{\pi b^2}{2}$	Two positive real roots	Instability
6	$b < 0$	$c < 0$	Two real roots: a positive, and a negative one	Instability

The total of domains *1*, *2*, and *3* makes the stability domain; domains *1* and *2* make the left half-plane; domains *1*, *5*, and *6* the real roots domain; domains *2*, *3*, and *4* the image roots domain. The domains are shown in Fig. 3.3, where stability domain is cross-hatched. Domain numbers are encircled.

Consider the following modifications of design cases.

1. $c = 1\ \mathrm{s}^{-1}$ **– a statically stable system.**

1) $b = 1.5\ \mathrm{s}^{-1/2}$ – two real postive roots: $q_1 = 2.205\ \mathrm{s}^{-1/2}$; $q_2 = 0.453\ \mathrm{s}^{-1/2}$.

2) $b = -1\ \mathrm{s}^{-1/2}$ – two complex conjugated roots with a positive real part, greater than the imaginary one:

$$q_{1,2} = 0.886 \pm i\,0.463\ \mathrm{s}^{-1/2}.$$

3) $b = -0.798\ \mathrm{s}^{-1/2}$ – complex conjugated roots with coinciding real and

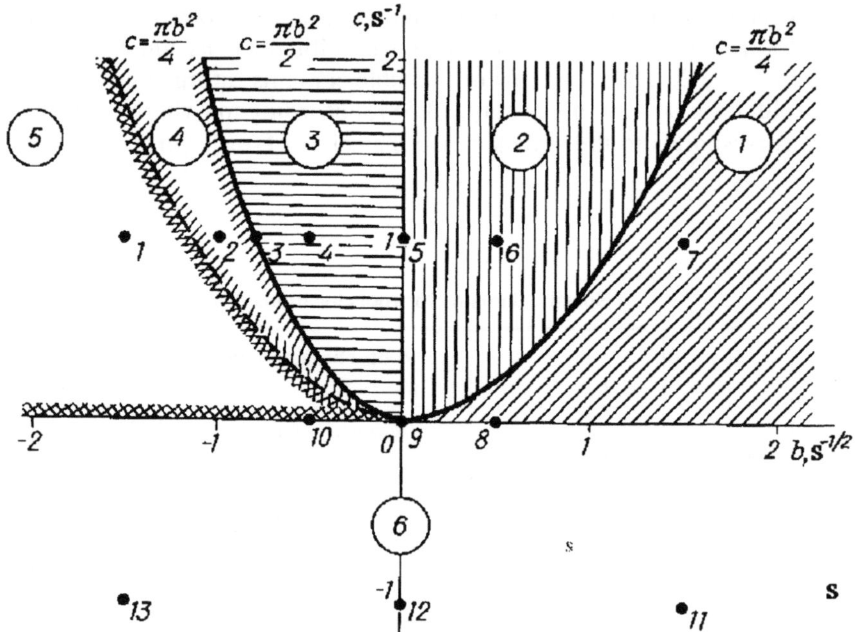

Fig. 3.3. Specific domains in the plane of Equation (3.58) coefficients b and c.

● denotes the calculated parameters values

imaginary parts: $q_{1,2} = 0.707 \pm i\,0.707\ \mathrm{s}^{-1/2}$, corresponding to nondecaying harmonic oscillations with angular frequency

$$\omega_c = 2\cdot(0.707)^2 = 1\ \mathrm{s}^{-1}.$$

4) $b = -0.5\ \mathrm{s}^{-1/2}$ – two complex conjugated roots with a positive real part, less than the imaginary one:

$$q_{1,2} = 0.443 \pm i\,0.896\ \mathrm{s}^{-1/2}.$$

5) $b = 0\ \mathrm{s}^{-1/2}$ – two conjugated pure imaginary roots $q_{1,2} = \pm i\ \mathrm{s}^{-1/2}$, corresponding to the singular case for which the transient process is a decaying component

$$x(t) = x_0 e^{-t}. \tag{3.63}$$

6) $b = 0.5\ s^{-1/2}$ – two complex conjugated roots with a negative real part, less than the imaginary one:

$$q_{1,2} = -0.443 \pm i\,0.896\ s^{-1/2}.$$

7) $b = 1.5\ s^{-1/2}$ – two real negative roots:

$$q_1 = -2.205\ s^{-1/2}, \quad q_2 = -0.453\ s^{-1/2}.$$

It is easy to notice that for the first six cases the roots are situated in the circumference of unit radius.

2. $c = 0\ s^{-1}$ – a statically neutral system.

8) $b = 0.5\ s^{-1/2}$ – two real roots, a negative and a zero one:

$$q_1 = -0.5\ s^{-1/2}, \quad q_2 = 0\ s^{-1/2}.$$

Root q_1 corresponds to the stable component of the transient process.

9) $b = 0\ s^{-1/2}$ – two zero roots $q_{1,2} = 0\ s^{-1/2}$. Cases 8 and 9 correspond to solution $x(t) = \text{const}$.

10) $b = 0.5\ s^{-1/2}$ – two real roots, a positive one and a zero one.

$$q_1 = 0.5\ s^{-1/2}, \quad q_2 = 0\ s^{-1/2}.$$

Root q_1 corresponds to the unstable component of the transient process.

Cases 8-10, rigorously speaking correspond not to integrodifferential equations, but to an integral equation of Volterra type of the 2nd kind with respect to function $\dot{x}(t)$.

3. $c = -1\ s^{-1}$ – a statically unstable system.

11) $b = 1.5\ s^{-1/2}$ – two real roots, a negative and a positive one:

$$q_1 = -2.922\ s^{-1/2}, \quad q_2 = 0.334\ s^{-1/2}.$$

12) $b = 0\ s^{-1/2}$ – two identical real roots of opposite signs; $q_{1,2} = \pm 1\ s^{-1/2}$ – the singular case corresponding to an unsteady exponential process.

$$x(t) = x_0 e^{t}. \tag{3.64}$$

13) $b = -1.5\ s^{-1/2}$ – two real roots, a positive and a negative one:

$$q_1 = 2.922\ s^{-1/2}, \quad q_2 = -0.334\ s^{-1/2}.$$

The points corresponding to the mentioned 13 cases in the plane of parameters b and c are denoted by black circles in Fig. 3.3, a one-valued correspondence being set between the circles and case numbers. In q variable plane the 13 cases of roots are presented in Fig.3.4, where the unit radius circumference is shown with a dashed line.

For all the 13 cases of parameters b and c combinations (or for the respective combinations of the characteristic Equation (3.60) roots) Equation (3.58) with the initial condition $x_0 = 1$ was numerically integrated. To do this, according to the algorithm described above, the integral in the left side of (3.58) was presented in the form

$$\int_0^t \frac{\dot{x}(\tau)\, d\tau}{\sqrt{t-\tau}} \cong \sqrt{h}\left\{G(t) + \frac{4}{3}\dot{x}(t)\right\}. \qquad (3.65)$$

Substituting (3.65) into (3.58) reduces the integrodifferential equation to an ordinary differential equation that can be written in Lyapunov – Cauchy form:

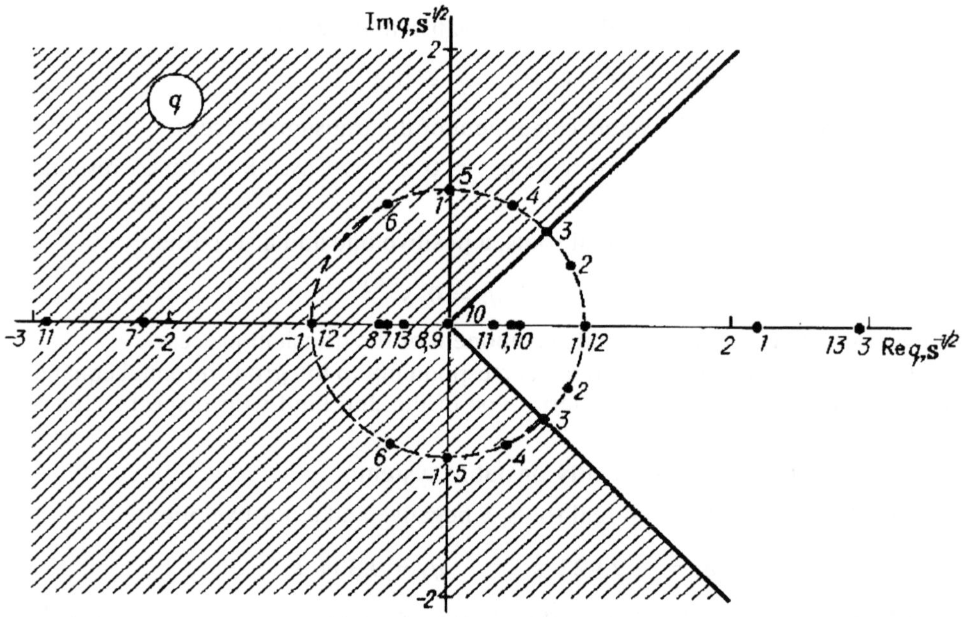

Fig. 3.4. Characteristic Equation (3.60) roots locations in q plane for the calculated cases. Stability domain is hatched

$$\dot{x}(t) = -\frac{b\sqrt{h}\,G(t) + cx(t)}{1 + \frac{4}{3}b\sqrt{h}}. \qquad (3.66)$$

Equation (3.66) was numerically integrated with Runge – Kutta method with 4-point approximation, the right side being calculated according to Formulae (3.55) at each step.

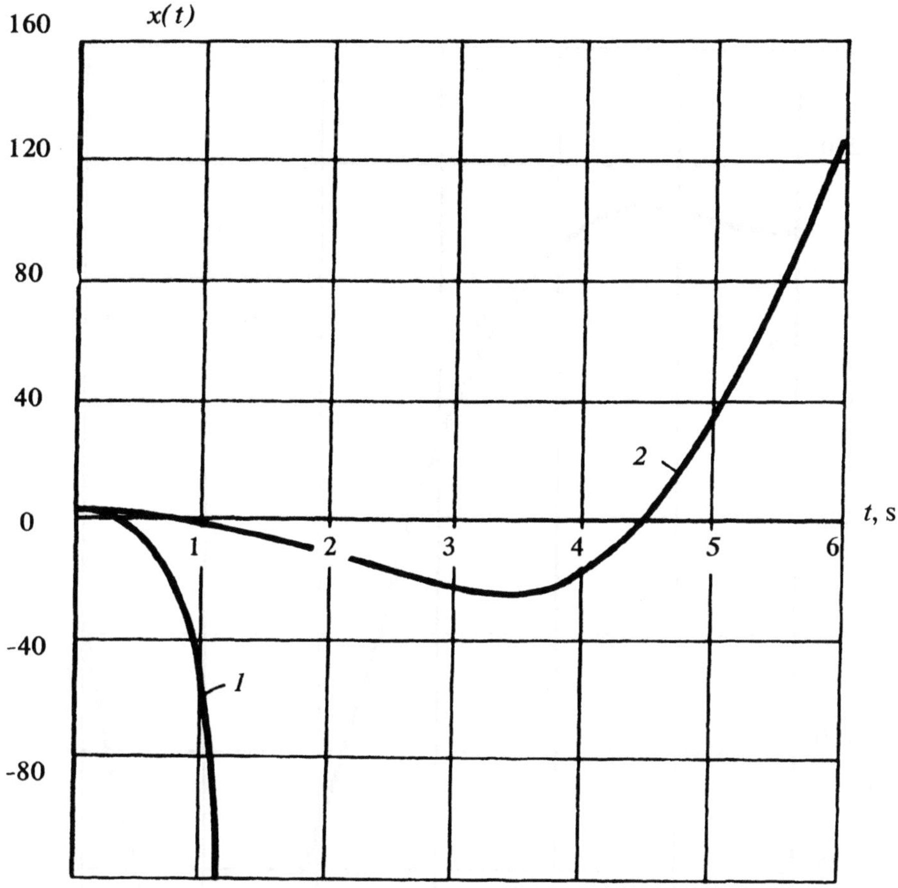

Fig. 3.5. Transient processes $x(t)$ in System (3.58); *1* – the system with two positive real roots; *2* – two complex conjugated roots with positive real part greater than the imaginary one

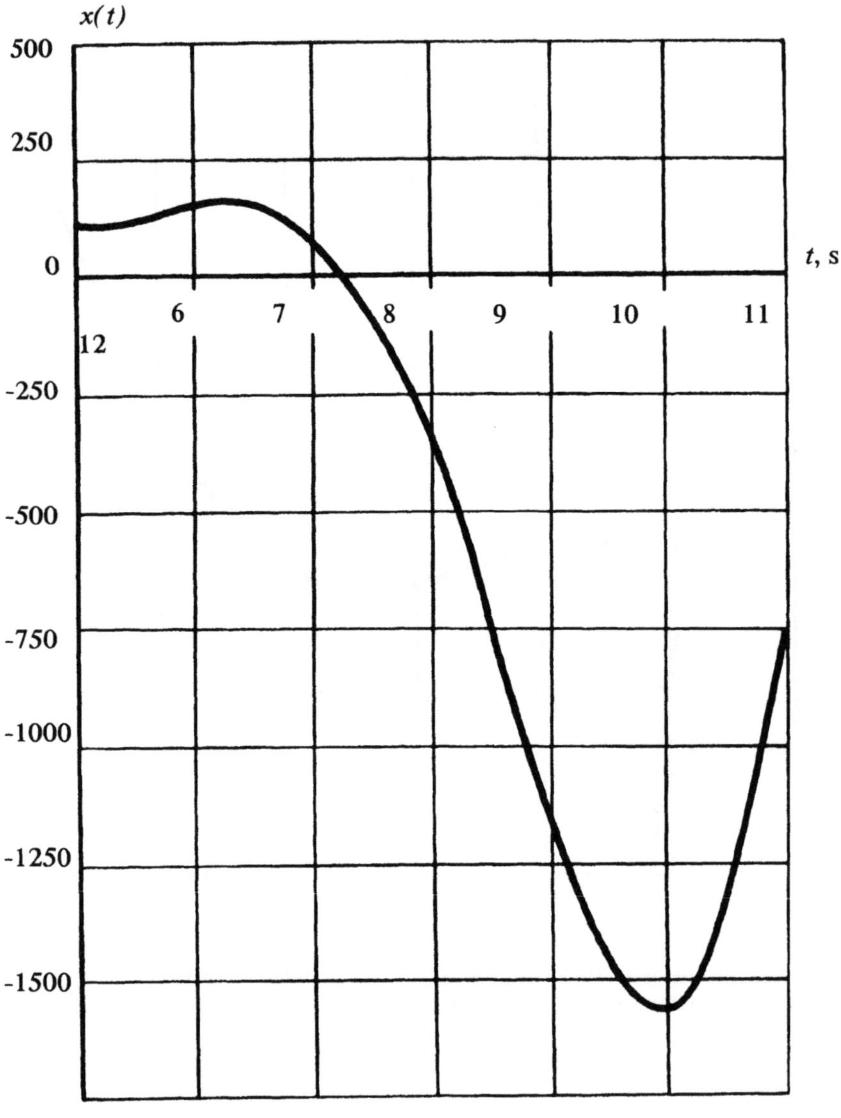

Fig. 3.6. Continuation of transient process $x(t)$ in System (3.58 with two complex conjugated roots having positive real part greater than the imaginary one, starting with the 6th second

It is necessary to outline the following subtle detail. At $t = 0$ the solution has a discontinuity, because for $h \to 0$, $\dot{x}(0) = -cx_0 \neq 0$, i.e. the 'splice condition' $\dot{x}(t) = 0$ (see [41]) is not satisfied. The effect of this discontinuity is soon eliminated, however; thus the process of 'solution smoothing' (see [41]) is practically implemented.

Fig. 3.5 shows the results of integration in cases 1 and 2, and Fig. 3.6 shows the solution for case 2 starting with the 6th second. As it is clear from those figures, the

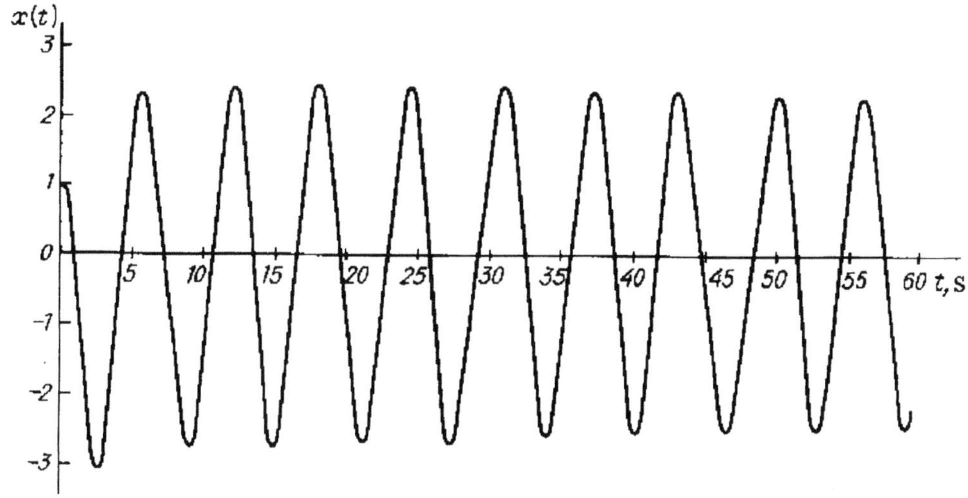

Fig. 3.7. The process of non-decaying harmonic oscillations establishment in System (3.58) with complex conjugated roots having the same real and imaginary parts

numerical integration corroborates the instability of Equation (3.58) solution for those combinations of parameters. For case 1 (real roots) instability develops aperiodically, and for case 2 (complex conjugated roots) it is oscillatory that tends to establishment of steady harmonic oscillations with circular frequency $\omega_c = 1\ s^{-1}$, when coefficient b tends to stability limit ($b = -0.798$). This is completely corroborated by the results of integration in case 3 shown in Fig. 3.7.

The results of numerical integration in cases 4, 5, 6, 7 corresponding to parameters lying within the stability domain are shown in Fig. 3.8. In case 4 (complex conjugated roots in the right half-plane) the solution is an oscillatory one, and for cases 6 (complex conjugated roots in the left half-plane) and 7 (two real roots in the left half-plane) it is aperiodic. For the singular case 5 the solution is the decaying exponent of (3.63).

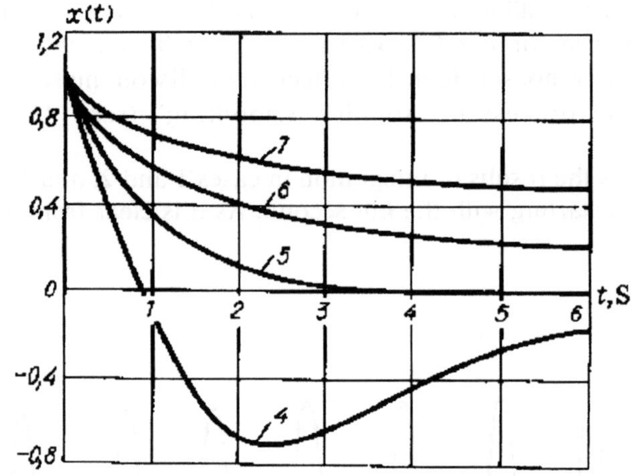

Fig. 3.8. Transient processes *x(t)* in System (3.58); *4* – two complex conjugated roots with a positive real partless than the imaginary one; *5* – two conjugated imaginary roots; *6* – two complex conjugated roots with a negative real part less than the imaginary one; *7* – two negative real roots

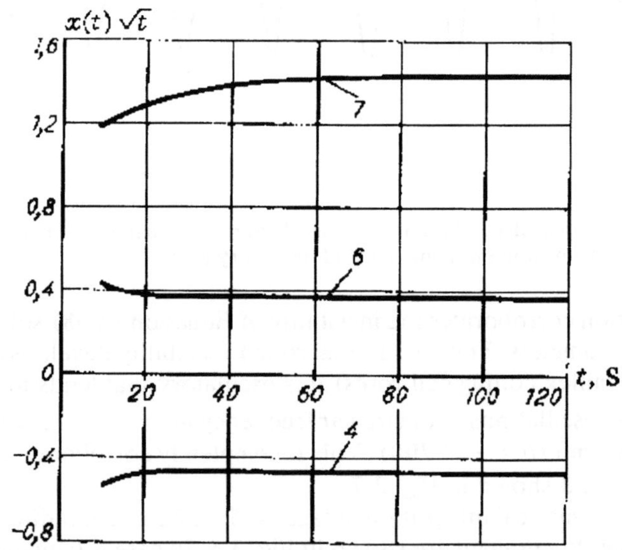

Fig. 3.9. Function $x(t)\ \sqrt{t}$ behaviour in System (3.58); *4* – two complex conjugated roots with a positive real partless than the imaginary one; *6* – two complex conjugated roots with a negative real part less than the imaginary one; *7* – two negative real roots

As it can be seen, in cases 4, 6, and 7 the transient processes are considerably 'prolonged' in time. To perceive the cause of this phenomenon let us write Equation (3.58) solution in Laplace transform domain:

$$X(p) = \frac{x_0(\sqrt{p} + b\sqrt{\pi})}{\sqrt{p}\,(p + b\sqrt{\pi p} + c)}\,. \tag{3.67}$$

Let us consider the behaviour of Solution (3.67) in the time domain for $t \to \infty$. It is necessary to pass to the limit $p \to 0$ in (3.67). We obtain as a result

$$X(p) = \frac{x_0 b}{c}\left(\frac{\pi}{p}\right)^{1/2} \to x(t)\big|_{t\to\infty} = \frac{x_0 b}{c\sqrt{t}}\,. \tag{3.68}$$

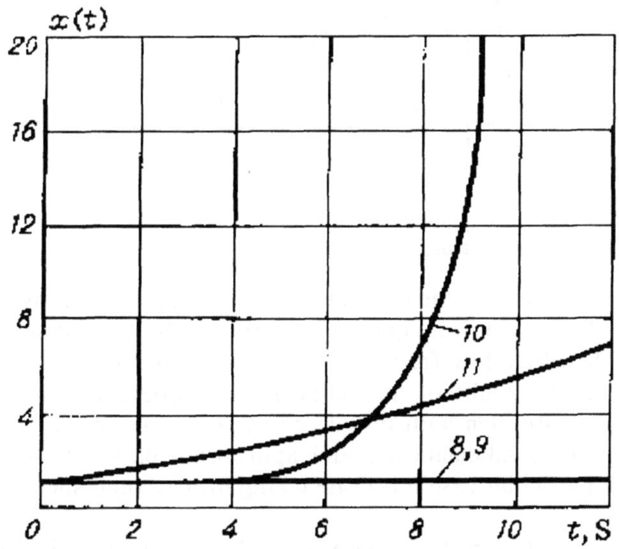

Fig. 3.10. Transient processes $x(t)$ in System (3.58); *8* – two real roots, a negative and a zero one; *9* – two zero roots; *10* – two real roots, a positive and a zero one; *11* – two real roots, a negative and a positive one

For easy understanding Fig. 3.9 shows function $x\sqrt{t}$ for cases 4, 6, and 7. It is obvious that results of integration corroborate the character of asymptotic behaviour of Equation (3.58) solution in the form of (3.68).

Thus, Equation (3.58) solution (it is specific for any equations of similar type) for nonzero initial conditions $x_0 \neq 0$ will always include the slowly decaying component of (3.68) independent of characteristic frequencies of the system. On the other side, mathematical models of LV liquid vortex motions and eddy currents in HECF elements obtained in Chapters 1 and 2 assume that magnetic (hydrodynamic)

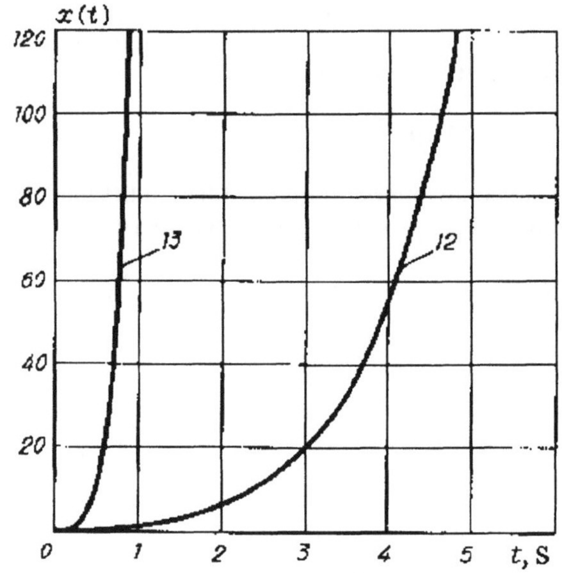

Fig. 3.11. Transient processes *x(t)* in System (3.58); *12* – two equal real roots with opposite signs; *13* – two real roots, a positive and a negative one

Reynolds number is great, i.e. the characteristic frequency of the investigated process is always greater than some finite quantity determined by the physical essence of the initial problem. Therefore low-frequency components of the solution of (3.68) type lie beyond applicability limits of the mathematical models discussed. Hence, when solving the physical problem with initial conditions on coordinates included in the integral operator of the mathematical model, appropriate particular solutions should be excluded. In each particular case it is done taking into account specific features of the investigated object.

Fig. 3.10 shows solutions for cases 8, 9, 10 (statically neutral system, $c = 0$) and case 11 (statically unstable system, $c < 0$) . For cases 8, 9 there is a solution $x(t) = x_0$, and in case 10 the existence of a positive real root causes severe instability that is more pronounced than in case 11 of a statically unstable system.

Fig. 3.11 shows the results of integration of equations corresponding to cases 12, 13 of a statically unstable system. For the singular case 12 instability develops exponentially, (3.64).

The example considered provides a good picture of the possibilities, computational stability, and accuracy of the described analytical and numerical methods. Questions concerning application of these methods to solution of particular applied problems will be discussed in Chapters 5, 6, 7, and 8.

Chapter 4

Experimental Verification of Mathematical Models for Eddy Currents and Vortex Motions of Liquid.

Experience never errs; it is only your judgements that err by promising themselves effects such as are not caused by your experiments.

Leonardo da Vinci*

In the previous chapters we discussed equations of perturbed motion for objects, dynamics of which is determined by vortex fields. We also discussed some mathematical research aspects of those equation solutions. Mathematical models of eddy currents in HECF elements and vortex motions of LV liquid lying in the foundation of the appropriate integrodifferential equations were obtained based on a number of heuristic considerations and simplifying assumptions (neglecting nonlinear dependence of magnetic field induction in the core on the field intensity, neglecting hysteresis; neglecting liquid viscosity all over the volume except ribs edges vicinity, approximate representation of hydrodynamic forces associated with vortex formation at the ribs' edges, and so on). Truthfulness of the assumptions lying in the foundation of the mathematical models introduced can be verified experimentally only. This chapter is devoted to presentation of some results obtained in this direction.

4.1. Mathematical Models Used for Experimental Data Processing

4.1.1 Analogy of eddy currents and vortex motions of liquid. Before proceeding directly to experimental investigation of eddy current in HECF elements and vortex motions of LV liquid for the purpose of the appropriate mathematical models verification, let us discuss the analogy of those physical phenomena that seem so different. As a matter of fact we have already used in Chapter 3 the fact that fields associated with eddy currents in HECF elements and with vortex motions of LV liquid are described by integral operators with the same singular kernels. This very circumstance lets us consider integrodifferential equations representing the corresponding mathematical models with a unified approach. Proceeding to the

---

* The literary works of Leonardo da Vinci. Comp. and ed. from the original manuscripts by Jean Paul Richter and Irma Richter. Oxford Univ. Press. London – N.Y. – Toronto, 1939, Vol. 2, p. 240

analysis of experimental investigations, let us try to outline a narrower class of investigated objects by a certain limitation of generality. For those objects the analogy is the most complete, i.e. the mathematical models coincide identically in the linear approximation. Let us consider the mathematical models of Chapter 1 and 2. Since in equations of Chapter 2, describing the motion of a solid body with liquid, all the external forces, and moments are moved to the right side, the same is to be done with equations of Chapter 1. For a cavity with impenetrable walls, containing LV liquid, with zero solid body mass an analogue is a HECF element shaped as a torus $G^{(i)}$ made of a material with infinite magnetic permeability ($\mu \to \infty$). Complete similarity is achieved at small fixed relative displacement of the torus halves and complete filling of the torus with liquid (absence of wave motions). Thus, comparison should be begun with the sets of Equations (1.65) and (2.66).

Let us write down the last two equations of each of those sets, assuming $\mathbf{U} = \mathbf{V} - (\mathsf{R}, \mathbf{I})$:

$$
\begin{aligned}
&(\mathsf{L}, \dot{\mathbf{I}}) + (\mathsf{K}, \dot{\mathbf{J}}) = \mathbf{U},\\
&(\mathsf{M}, \dot{\mathbf{J}}) + (\bar{\mathsf{K}}, \dot{\mathbf{I}}) + \left(\mathsf{T}, \int_{-\infty}^{t} \frac{\dot{\mathbf{J}}(\tau)\, d\tau}{\sqrt{t-\tau}}\right) = 0.
\end{aligned}
\tag{4.1}
$$

$$
\begin{aligned}
&(\mathsf{J}^{(0)}, \dot{\vec{\omega}}) + (\mathsf{J}^{*}, \dot{\vec{\Omega}}^{*}) = \mathbf{M}_{G},\\
&(\mathsf{J}^{**}, \dot{\vec{\Omega}}^{*}) + (\bar{\mathsf{J}}^{*}, \dot{\vec{\omega}}) + \left(\mathsf{B}^{*}, \int_{-\infty}^{t} \frac{\dot{\vec{\Omega}}^{*}(\tau)\, d\tau}{\sqrt{t-\tau}}\right) = 0.
\end{aligned}
\tag{4.2}
$$

Comparing (4.1) and (4.2) one can see that current **I** variation vector in (4.1) is analogous to angular velocity $\vec{\omega}$ of the body (that is considered infinitesimal of the first order); and eddy currents **J** vector is analogous to generalized relative angular velocity $\vec{\Omega}^{*}$ of liquid vortex motion (also considered a first-order infinitesimal); generalized voltage **U** vector is an analogue to the resulting moment $\mathbf{M}_G$ vector of external exciting forces with respect to the body – liquid system mass centre *G*. It is obvious that mathematical Models (4.1) and (4.2) coincide completely (up to denominations) with the only difference that tensor B^{*} elements are nonlinear functions of argument $|\vec{\Omega}^{*}|$.

It is clear from above, that the most interesting from the methodological point of view is the experimental investigation of objects described by models being particular cases of (4.1) and (4.2). The following objects were used in reality: an electromagnet with elongated HECF core of U-shape cross section, that is separated with an air gap from the ferromagnetic rail with C-shaped cross section, or from a flat ferromagnetic rail, and a tank of circular cylinder shape with flat heads and 6 radial ribs, completely filled with LV liquid (water), that can rotate along its longitudinal axis.

4.1.2. Mathematical models of eddy currents, used for experimental data processing

a) M a t h e m a t i c a l m o d e l s o f t h e e l e c t r o m a g n e t , c o n s i d e r i n g e d d y c u r r e n t s i n t h e H E C F c o r e . Let us use the simplest of the models obtained in Chapter 1, namely (1.92) that is nevertheless essentially different from conventional ones. Since the air gap in the electromagnet core was kept constant in the experiment, let us rewrite the set of Equations (1.92) ignoring the mechanical degree of freedom ($\eta \equiv 0$) :

$$c_l'(\dot{I}+\dot{J}) + RI = v_1(t), \quad c_l'(\dot{I}+\dot{J}) + c_\gamma \int_{-\infty}^{t} \frac{\dot{J}(\tau)\, d\tau}{\sqrt{t-\tau}} = 0. \tag{4.3}$$

$v_1(t)$ here is the voltage applied to the electromagnet coil.

b) M a t h e m a t i c a l m o d e l s i g n o r i n g e d d y c u r r e n t s i n H E C F c o r e :

$$L_1\dot{I} + R_1 I = v_1(t), \tag{4.4}$$

where L_1 is the coil's inductance, R_1 is its active resistance.

c) M a t h e m a t i c a l m o d e l s d e s c r i b i n g t h e i n f l u e n c e o f e d d y c u r r e n t s i n t h e H E C F c o r e w i t h t h e h e l p o f a s h o r t - c i r c u i t l o o p i n d u c t i v e l y c o u p l e d w i t h t h e c o i l .

$$L_2(\dot{I}+\dot{I}_1) + RI = v_1(t), \quad L_2(\dot{I}+\dot{I}_1) + rI_1 = 0. \tag{4.5}$$

I_1 here denotes the equivalent current in the short-circuit loop, r is the ohmic resistance of the loop.

4.1.3. Mathematical models of LV liquid vortex motions, used for experimental data processing

Let us consider vortex motions of liquid, caused by the rotation of the cylindric tank with radial ribs about its longitudinal axis. In order to identify the structure and parameters of the mathematical models for those motions the following three modifications of the body – liquid system mathematical models were used.

a) The mathematical model corresponding to linear damping:

$$I^1\ddot{\varphi} + \varepsilon\dot{\varphi} = M(t). \tag{4.6}$$

b) The mathematical model corresponding to nonvortex absolute motion of LV liquid outside sharp edges of the ribs (a particular case of (2.68)):

$$(I^\circ + I)\ddot{\varphi} + \beta_1 \int_{-\infty}^{t} \frac{\ddot{\varphi}(\tau)\, d\tau}{\sqrt{t-\tau}} = M(t), \tag{4.7}$$

where I is Joukowski's associated moment of inertia for liquid; $M(t)$ is the resulting moment of the external forces system with respect to Ox axis; $\beta_1 = \tilde{\beta}\,|\,\dot{\varphi}\,|^{1/2}$; $\tilde{\beta}$ is the parameter for reconciliation with experimental data.

c) The mathematical model corresponding to developed stalled flow around the ribs (a particular case of (2.64)).

Let us consider the general equations corresponding to rotation of the object with an axially symmetric cavity about its longitudinal axis, (2.117). Assume $N = 1$ and substitute $I_1^{*'}$ symbol for I_1^*:

$$(I^{\circ} + I^{(0)})\ddot{\varphi} + I_1^{*'}\ddot{\chi}_1 = M(t),$$

$$I_1^{**}\ddot{\chi}_1 + I_1^{*'}\ddot{\varphi} + \beta_{11}\int_{-\infty}^{t}\frac{\ddot{\chi}_1(\tau)\,d\tau}{\sqrt{t-\tau}} = 0, \tag{4.8}$$

$$\beta_{11} = \tilde{\beta}_{11}\,|\,\dot{\chi}_1\,|^{1/2}.$$

Note, that relation $\kappa = I_1^{*'}/I_1^{**}$ characterizes in the integral sense the difference of velocities space distribution from the distribution corresponding to vorticity uniform all over volume, $\tilde{\beta}_{11}$ is an integral characteristic for generalized forces of LV liquid interaction with a rib for a developed stalled flow.

Let us perform some transformations with Equations (4.8), in order to reduce them to simpler form (2.122). Multiply the left side of the second equation by κ and introduce a new generalized velocity $\dot{\chi}$ and parameters I^*, β^*, defined by the formulae

$$\dot{\chi}_1 = \kappa\,\dot{\chi}, \quad I^* = \frac{I_1^{*'2}}{I_1^{**}}, \quad \beta^* = \tilde{\beta}^*\;|\dot{\chi}\,|^{1/2}, \quad \tilde{\beta}^* = \tilde{\beta}_{11}\,\kappa^{5/2}. \tag{4.9}$$

In terms of new variable φ, χ mathematical Model (4.8) gains the form of (2.122) for $N = 1$ (one cavity):

$$(I^{\circ} + I^{(0)})\ddot{\varphi} + I^*\ddot{\chi} = M(t), \quad I^*(\ddot{\varphi} + \ddot{\chi}) + \beta^*\int_{-\infty}^{t}\frac{\ddot{\chi}(\tau)\,d\tau}{\sqrt{t-\tau}} = 0. \tag{4.10}$$

Here, unlike (4.8), only two parameters for reconciliation with experimental data are left, namely I^* and $\tilde{\beta}^*$.

4.2. Determination of Mathematical Model Parameters Based on Experimental Results

Transfer function W(p) for the investigated object, calculated from the input action to the measured quantity, can be written in the form

$$W(p) = \frac{a_0 + \sum_{k=1}^{K1} a_k f_k(p)}{1 + \sum_{k=1}^{K2} b_k g_k(p)}, \tag{4.11}$$

where a_0, a_k, b_k are parameters to be determined; $f_k(p)$, $g_k(p)$ are known functions of Laplace variable, in the simplest case they are integer powers of p. Resulting from the experiment, the values of experimental amplitude $A_E(\omega_j)$, and phase $\varphi_E(\omega_j)$ response were determined for a discrete set of the input action frequency values ω_j ($j = 1, 2, ..., N$). Thus

$$W_E(i\omega_j) = A_E(\omega_j)\, e^{i\varphi_E(\omega_j)} = \mathrm{Re}_E(\omega_j) + i\,\mathrm{Im}_E(\omega_j). \tag{4.12}$$

The experiment was planned so, that it ensured satisfaction of condition $N > K_1 + K_2 + 1$ providing redundant information to determine the mathematical model parameters. This redundancy is necessary to increase reliability of the obtained results.

There are various ways to take advantage of the obtained redundancy, e.g. to seek the mathematical model parameters from the minimizing condition for functionals of the type (ξ^2 is a weight coefficient)

$$J_1 = \sum_{j=1}^{N} \left\{ [\mathrm{Re}_W - \mathrm{Re}_E(\omega_j)]^2 + [\mathrm{Im}_W - \mathrm{Im}_E(\omega_j)]^2 \right\}, \tag{4.13}$$

or

$$J_2 = \sum_{j=1}^{N} \left\{ [A_W - A_E(\omega_j)]^2 + \xi^2\, [\varphi_W - \varphi_E(\omega_j)]^2 \right\}. \tag{4.14}$$

In (4.13) and (4.14) Re_W, Im_W, A_W, φ_W are the respective real part, imaginary part, amplitude, and phase response of transfer Function (4.11), which are obtained from the latter by substituting $p = i\omega_j$. At the same time, to find the stationary points of Functionals (4.13) and (4.14) in the space of parameters a_0, a_k, b_k it is necessary to solve nonlinear equations, that are not so often solvable.

To avoid this difficulty let us try the following technique. Assume $p = i\omega_j$ in transfer Function (4.11) and equate it to the experimentally obtained value of (4.12):

$$\frac{a_0 + \sum_{k=1}^{K1} a_k \,\mathrm{Re}\, f_k(i\omega_j) + i \sum_{k=1}^{K1} a_k \,\mathrm{Im}\, f_k(i\omega_j)}{1 + \sum_{k=1}^{K2} b_k \,\mathrm{Re}\, g_k(i\omega_j) + i \sum_{k=1}^{K2} b_k \,\mathrm{Im}\, g_k(i\omega_j)} = \mathrm{Re}_E(\omega_j) + i\,\mathrm{Im}_E(\omega_j). \tag{4.15}$$

Multiplying the left and right sides of Equation (4.15) by the left side denominator and separating the real and imaginary parts we obtain a set of $2N$ equations:

$$a_0 + \sum_{k=1}^{K1} a_k \operatorname{Re} f_{kj} + \sum_{k=1}^{K2} b_k (\operatorname{Im}_{Ej} \operatorname{Im} g_{kj} - \operatorname{Re}_{Ej} \operatorname{Re} g_{kj}) = \operatorname{Re}_{Ej}, \tag{4.16}$$

$$\sum_{k=1}^{K1} a_k \operatorname{Im} f_{kj} - \sum_{k=1}^{K2} b_k (\operatorname{Im}_{Ej} \operatorname{Re} g_{kj} + \operatorname{Re}_{Ej} \operatorname{Im} g_{kj}) = \operatorname{Im}_{Ej} \quad (j = 1, 2, ..., N).$$

Subscript j here symbolizes the value of the appropriate function for $\omega = \omega_j$. We thus have $2N$ linear algebraic Equations (4.16) to determine $K_1 + K_2 + 1$ parameters of the mathematical model.

The set of Equations (4.16) is redundant, so let us solve it with the least squares method. To construct the solving set of $(K_1 + K_2 + 1)$ order it is necessary to multiply each equation of (4.16) by the coefficient at a certain variable and add all the $2N$ equations. Repeating this procedure for each variable we obtain a set of $K_1 + K_2 + 1$ equations. Solution to this set of equations minimized the functional equal to sum of error squares characterizing mismatch of the left and right sides of Equations (4.16).

Let us write down the equation corresponding to variable a_0:

$$Na_0 + \sum_{k=1}^{K1} a_k \sum_{j=1}^{N} \operatorname{Re} f_{kj} +$$

$$+ \sum_{k=1}^{K2} b_k \sum_{j=1}^{N} (\operatorname{Im}_{Ej} \operatorname{Im} g_{kj} - \operatorname{Re}_{Ej} \operatorname{Re} g_{kj}) = \sum_{j=1}^{N} \operatorname{Re}_{Ej}. \tag{4.17}$$

The other $K_1 + K_2$ equations have the form

$$a_0 c_l + \sum_{k=1}^{K1} a_k c_{lk}^{(1)} + \sum_{k=1}^{K2} b_k c_{lk}^{(2)} = d_l^{(1)},$$

$$a_0 m_n + \sum_{k=1}^{K1} a_k m_{nk}^{(1)} + \sum_{k=1}^{K2} b_k m_{nk}^{(2)} = d_n^{(2)}, \tag{4.18}$$

$$l = 1, 2, ..., K_1, \quad n = 1, 2, ..., K_2.$$

The following symbols are introduced in (4.18):

$$c_l = \sum_{j=1}^{N} \operatorname{Re} f_{lj}, \qquad c_{lk}^{(1)} = \sum_{j=1}^{N} (\operatorname{Re} f_{lj} \operatorname{Re} f_{kj} + \operatorname{Im} f_{lj} \operatorname{Im} f_{kj}),$$

$$c_{lk}^{(2)} = \sum_{j=1}^{N} [\operatorname{Re} f_{lj} (\operatorname{Im}_{Ej} \operatorname{Im} g_{kj} - \operatorname{Re}_{Ej} \operatorname{Re} g_{kj}) -$$

$$- \operatorname{Im} f_{lj} (\operatorname{Im}_{Ej} \operatorname{Re} g_{kj} + \operatorname{Re}_{Ej} \operatorname{Im} g_{kj})],$$

$$d_l^{(1)} = \sum_{j=1}^{N} (\operatorname{Re}_{Ej} \operatorname{Re} f_{lj} + \operatorname{Im}_{Ej} \operatorname{Im} f_{lj}),$$

$$m_n = \sum_{j=1}^{N} (Im_{Ej} \operatorname{Im} g_{nj} - \operatorname{Re}_{Ej} \operatorname{Re} g_{nj}),$$

$$m_{nk}^{(1)} = \sum_{j=1}^{N} [\operatorname{Re} f_{kj} (\operatorname{Im}_{Ej} \operatorname{Im} g_{nj} - \operatorname{Re}_{Ej} \operatorname{Re} g_{nj}) - \tag{4.19}$$

$$- \operatorname{Im} f_{kj} (\operatorname{Im}_{Ej} \operatorname{Re} g_{nj} + \operatorname{Re}_{Ej} \operatorname{Im} g_{nj})],$$

$$m_{nk}^{(2)} = \sum_{j=1}^{N} [(\operatorname{Im}_{Ej} \operatorname{Im} g_{kj} - \operatorname{Re}_{Ej} \operatorname{Re} g_{kj}) (\operatorname{Im}_{Ej} \operatorname{Im} g_{nj} - \operatorname{Re}_{Ej} \operatorname{Re} g_{nj}) -$$

$$- (\operatorname{Im}_{Ej} \operatorname{Re} g_{kj} + \operatorname{Re}_{Ej} \operatorname{Im} g_{kj}) (\operatorname{Im}_{Ej} \operatorname{Re} g_{nj} + \operatorname{Re}_{Ej} \operatorname{Im} g_{nj})],$$

$$d_n^{(2)} = \sum_{j=1}^{N} [\operatorname{Re}_{Ej} (\operatorname{Im}_{Ej} \operatorname{Im} g_{nj} - \operatorname{Re}_{Ej} \operatorname{Re} g_{nj}) -$$

$$- \operatorname{Im}_{Ej} (\operatorname{Im}_{Ej} \operatorname{Re} g_{nj} + \operatorname{Re}_{Ej} \operatorname{Im} g_{nj})].$$

Finally setting $d_0 = \sum_{j=1}^{N} \operatorname{Re}_{Ej}$ we obtain

$$a_0 N + \sum_{k=1}^{K1} a_k c_k + \sum_{k=1}^{K2} b_k m_k = d_0,$$

$$a_0 c_l + \sum_{k=1}^{K1} a_k c_{lk}^{(1)} + \sum_{k=1}^{K2} b_k c_{lk}^{(2)} = d_l^{(1)}, \tag{4.20}$$

$$a_0 m_n + \sum_{k=1}^{K1} a_k m_{nk}^{(1)} + \sum_{k=1}^{K2} b_k m_{nk}^{(2)} = d_n^{(2)},$$

$$l = 1, 2, ..., K_1, \quad n = 1, 2, ..., K_2.$$

It is clear from (4.19) that System (4.20) has a symmetric matrix of coefficients. System (4.20) solution determines parameters of transfer Function (4.11) of the investigated object from the experimental response. The choice of the form for functions $f_k(p)$, $g_k(p)$ and values K_1 and K_2 depends on the choice of structure and dimension for the mathematical model involved in the investigated object description.

To determine parameters of the mathematical model based on transfer Function (4.11) parameters it is necessary in each certain case to solve an additional set of equations connecting those parameters. This set of equations is essentially nonlinear, as a rule, therefore, when planning the experiment and choosing point of application for the input action and the observed coordinate, one should take as a guide the possibility to simplify the following processing of results.

4.3. Scheme of the Experiment and Primary Processing of Results

The mathematical models allowing for eddy currents HECF elements and for vortex motions of LV liquid were obtained with a number of restrictions (1.5), (2.35) imposed on the dimensionless parameters equivalent to Strouhal number and Reynolds number. Analysis of the restrictions (1.5) shows, that the introduced models with real parameters of the levitation electromagnets remain valid in the frequency band of tenth fractions of Hertz to several kilohertz. Extremely important is the frequency band lower bound, which indicates, that the introduced models do not describe the finish of the process of reaching the stationary mode, i.e. the process of vortex motions vanishing.

The purpose of the experiments carried out was, on one side, to verify the adequacy of the introduced mathematical models, and on the other side, to determine the levitation electromagnets' parameters in order to synthesize the electromagnetic levitation system control laws.

The mathematical model parameters identification was based on the experimental frequency response functions (amplitude-phase-frequency characteristics) of the electromagnet coil terminal voltage to the current in that coil. When investigating liquid vortex motions frequency response functions were determined of the external moment acting about longitudinal axis of the cylindrical tank with radial ribs to the tank angular displacement.

Fig. 4.1 shows the scheme of the experimental plant for electromagnet frequency response determination. The experimental plant includes the investigated electromagnet (EM) with a cut core, the controlled amplifier-converter (CAC) with pulse-duration modulation (PDM) with the carrier frequency of 2 kHz, and the control block consisting of a summer (Σ), a constant-voltage source (CVS), an audio-frequency

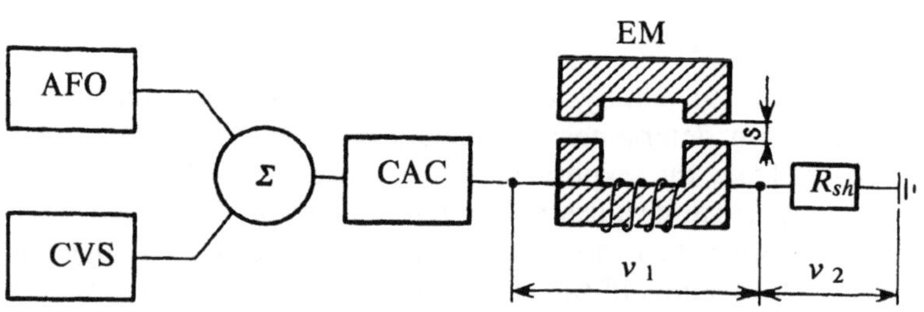

Fig. 4.1. The scheme of the experimental plant for electromagnet testing

oscillator (AFO), and a calibrated thermostabilized shunt (R_{sh}) connected in a series with the electromagnet coil (see V.G.Lebedev *et al.* [53]).

The air gap s of the core is fixed with calibrated spacers made of non-magnetic material.

Voltage drop on the electromagnet coil terminals (v_1) and on the shunt (v_2) were measured through identical *RC*-filter circuits with the time constant of 0.005 s. The measured values v_1 and v_2 were converted to the digital form (via an analog-to-digital converter having the conversion time of 20 ms) recorded to a magnetic disk storage of the minicomputer Eclipse C/300 (Data General Corp., U.S.A.) The *RC*-circuits were introduced in order to smooth out the carrier frequency.

The control voltage generated by the constant-voltage source determines the value of the direct bias current in the electromagnet coil. The frequency and amplitude of the sinusoidal component of voltage and current in the coil are set by the audio-frequency oscillator, the control signal amplitude being such, that the alternating component of the current in the electromagnet coil constitutes about 10% of the direct component, independent of frequency.

The possibility of further usage of linearized models was tested by repeating the experiment with double control signal amplitude.

The experiment was carried out in the following way. A number of fixed base frequencies ω_j was selected from the operating frequency band of the electromagnetic levitation stabilization system. For those base frequencies measurements were made for certain values of air gap s and of mean bias current in the electromagnetic coil. Then the bias current was changed, in order to cover the operating range of current; after that the air gap value was changed. Thus frequency response matrix of the electromagnet was constructed for the operating range of the air gap and electromagnet coil current.

For each fixed value of ω_j values v_1 and v_2 were measured with the interval of $T_{\omega j}$, to make about 50 measurements during the period of $2\pi/\omega_j$. Total computer disk record length exceeded 10 periods for steady-state oscillations.

The values v_1 and v_2, as functions of time, can be written in the form

$$v_1 = U^\circ + A_U \sin(\omega_j t + \varphi_U), \quad v_2 = R_{sh}\,[I^\circ + A_I \sin(\omega_j t + \varphi_I)\,]. \tag{4.21}$$

Since accuracy of the frequency values ω_j setting did not exceed 5%, especially in the low-frequency area, an estimate of the instant spectral density was made in the interval of ± 10% of the base value set on the audio-frequency oscillator.

$$S(\omega) = \left(S_1^2(\omega) + S_2^2(\omega)\right)^{1/2},$$

$$S_1(\omega) = \int_0^T [v_1(t) - \tilde{v}_1] \sin \omega t \, dt, \quad S_2(\omega) = \int_0^T [v_1(t) - \tilde{v}_1] \cos \omega t \, dt. \tag{4.22}$$

Here $\tilde{v}_1$ is the mean value of $v_1(t)$ over the interval of processing $[0, T\,]$; T is equal to 10 base periods.

Function $S(\omega)$ values were calculated with the step of 0.5% of the base value $\omega_{j\,bas}$; then max $S(\omega)$ and the corresponding frequency value were obtained, the latter being assumed for ω_j. Further, in order to evaluate U°, $R_{sh}\, I^\circ$, A_U, A_I, φ_U, φ_I

functions $v_1(t)$ and $v_2(t)$ were approximated in the interval $[0,T]$ with least-squares technique by functions of the form

$$C + A\sin\omega_j t + B\cos\omega_j t\,. \tag{4.23}$$

Unknown C, A, and B are evaluated from the set of linear algebraic equations

$$\begin{bmatrix} N+1 & \sum_{k=0}^{N}\sin\omega_j t_k & \sum_{k=0}^{N}\cos\omega_j t_k \\ \sum_{k=0}^{N}\sin\omega_j t_k & \sum_{k=0}^{N}\sin^2\omega_j t_k & \frac{1}{2}\sum_{k=0}^{N}\sin 2\omega_j t_k \\ \sum_{k=0}^{N}\cos\omega_j t_k & \frac{1}{2}\sum_{k=0}^{N}\sin 2\omega_j t_k & \sum_{k=0}^{N}\cos^2\omega_j t_k \end{bmatrix} \begin{bmatrix} C \\ A \\ B \end{bmatrix} = \begin{bmatrix} \sum_{k=0}^{N} v(t_k) \\ \sum_{k=0}^{N} v(t_k)\sin\omega_j t_k \\ \sum_{k=0}^{N} v(t_k)\cos\omega_j t_k \end{bmatrix}, \tag{4.24}$$

where $N = \dfrac{T}{T_{\omega j}}$, $t_k = kT_{\omega j}$, $k = 1, 2, \ldots, N$.

Comparing (4.23) to (4.21) we obtain

$$U^\circ = C_1,\ R_{sh}I^\circ = C_2,\ A_U = \left(A_1^2 + B_1^2\right)^{1/2},\ A_I = \left(A_2^2 + B_2^2\right)^{1/2},$$
$$\cos\varphi_U = \frac{A_1}{A_U},\quad \cos\varphi_I = \frac{A_2}{A_I},\quad \sin\varphi_U = \frac{B_1}{A_U},\quad \sin\varphi_I = \frac{B_2}{A_I}\,. \tag{4.25}$$

Here subscript 1 corresponds to quantities determined by v_1 and 2 corresponds to v_2. The amplitude $A(\omega)$ and phase $\varphi(\omega)$ frequency characteristics of the electromagnet are calculated from Expressions (4.25):

$$A(\omega_j) = \frac{A_I(\omega_j)}{A_U(\omega_j)},\quad \varphi(\omega_j) = \varphi_I(\omega_j) - \varphi_U(\omega_j)\,. \tag{4.26}$$

When checking the adequacy of the mathematical model of LV liquid vortex motions, the experimental results by G.N.Mikishev and G.A.Churilov were used.

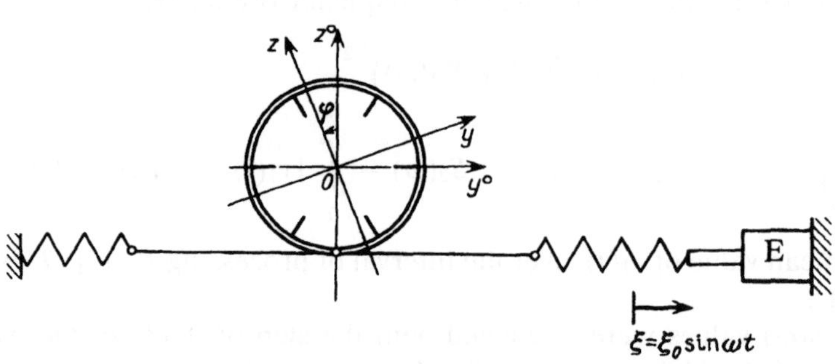

Fig. 4.2. The kinematic scheme of the experiment to study liquid vortex motions

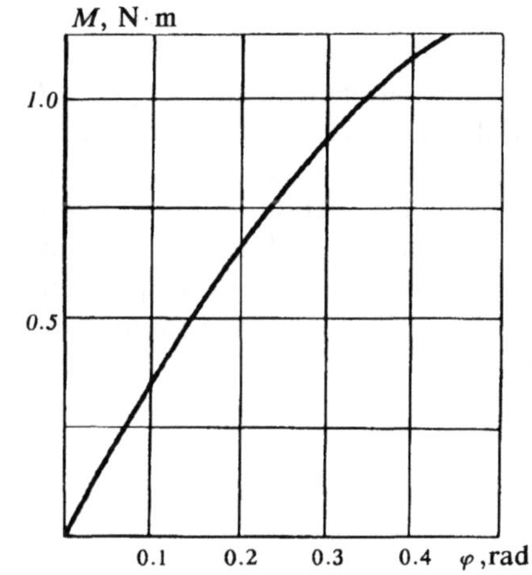

Fig. 4.3. The spring rigidity characteristic

Descriptions of the experimental plant, of the method of carrying out the experiment, and of the primary data processing are presented in [2,54].

Determined from the experiments were frequency response functions of the cylindrical tank partially filled with water having the temperature of 20° C, the tank rotating about its longitudinal axis.

The kinematic scheme of the experiment is shown in Fig. 4.2. The radius of the tank is 0.175 m, the rib width is 0.3 of the radius, the liquid depth is 0.34 m, the lateral moment of inertia of the dry tank is $I^{\circ} = 0.324$ kg·m^2.

The spring rigidity characteristic is presented in Fig. 4.3.

The exciting moment is created by sinusoidal displacement of the exciters's rod E. Measured quantities: amplitude of angle φ oscillations about the equilibrium position and phase shift between the angle φ and exciter's rod oscillations, with the amplitude of rod oscillations set to ξ_0. The viscous friction coefficient in the tank suspension and excitation systems ε_0 is equal to 0.02 N·m·s /rad.

4.4. Verification of Models for Eddy Currents in HECF Elements

Mathematical models for the object of control are developed in order to create an adequate description of the object in the operating frequency range of the closed-loop system in the form most suitable for synthesizing control algorithms. Quite natural is the tendency to have a possibly linear model with steady parameters, since for such models there are the most developed methods of synthesizing control systems.

By virtue of the mentioned above, for the criterion of the considered mathematical Models (4.3) – (4.5) adequacy we choose the deviation degree of those models from some constant values, when the external voltage frequency varies within the control system operating frequency range. The standard electromagnet was used for the experiment. It has the following parameters: the length is 0.75 m, the core has an U-shaped cross-section, and the ferromagnetic rail has a C-shaped cross-section (see Fig. 4.4), the core width is 0.02 m, the average length of the cross-section is 0.7 m, the number of turns is 400, the nominal induction in the air gap is 0.7 T.

Fig. 4.4 shows typical amplitude and phase response characteristics of the electromagnet from voltage v_1 to the coil current. Experimental values are shown by circles. Note, that phase response, after reaching its minimum at the frequency about 5 Hz, starts to increase slowly.

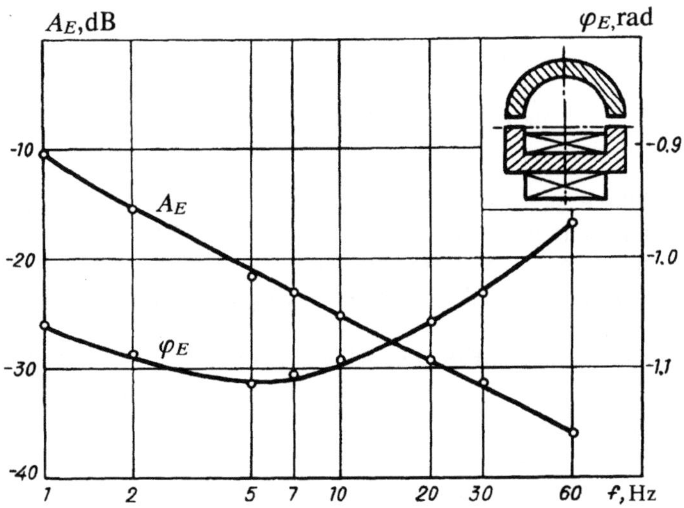

Fig. 4.4. Typical amplitude and phase response of an electromagnet

Let us write down transfer functions from voltage to current for mathematical models (4.3) – (4.5) respectively, which yield

$$W_1(p) = \frac{c'_I\sqrt{p} + c_\gamma\sqrt{\pi}}{c'_I c_\gamma \sqrt{\pi}\, p + Rc'_I\sqrt{p} + Rc_\gamma\sqrt{\pi}} . \tag{4.27}$$

$$W_2(p) = \frac{1}{L_1\, p + R_1} . \tag{4.28}$$

$$W_3(p) = \frac{L_2\, p + r}{(R + r)\; L_2\, p + Rr} , \tag{4.29}$$

where p is Laplace variable. Divide the numerator and denominator for the right sides of (4.27), (4.28), (4.29) by $Rc_\gamma\sqrt{\pi}$, R_1 and Rr respectively:

$$W_1(p) = \frac{1}{R}\frac{\tau\sqrt{p}+1}{Tp+\tau\sqrt{p}+1}. \tag{4.30}$$

$$W_2(p) = \frac{1}{R_1}\frac{1}{T_1 p+1}. \tag{4.31}$$

$$W_3(p) = \frac{1}{R}\frac{\tau_2 p+1}{(T_2+\tau_2)p+1}. \tag{4.32}$$

The following symbols are introduced herein:

$$\tau = \frac{c'_I}{c_\gamma\sqrt{\pi}}, \quad T = \frac{c'_I}{R}, \quad T_1 = \frac{L_1}{R_1}, \quad \tau_2 = \frac{L_2}{r}, \quad T_2 = \frac{L_2}{R}. \tag{4.33}$$

Assuming $p = i\omega$ in (4.30) – (4.32), we obtain frequency response characteristics corresponding to Models (4.3) – (4.5):

$$W_1(i\omega) = \frac{1}{R}\frac{\tau\sqrt{\omega/2}+1+i\tau\sqrt{\omega/2}}{\tau\sqrt{\omega/2}+1+i(T\omega+\tau\sqrt{\omega/2})}. \tag{4.34}$$

$$W_2(i\omega) = \frac{1}{R_1}\frac{1}{1+iT_1\omega}. \tag{4.35}$$

$$W_3(i\omega) = \frac{1}{R}\frac{1+i\tau_2\omega}{1+i(T_2+\tau_2)\omega}. \tag{4.36}$$

Frequency response determined from (4.34) – (4.36) tends to $-\pi/4$, $-\pi/2$, 0, respectively, with $\omega \to \infty$. Separating real and imaginary parts in the right sides of (4.34) – (4.36) and equating them to $A_E(\omega_j)\cos\varphi_E(\omega_j)$ and $A_E(\omega_j)\sin\varphi_E(\omega_j)$ determined from the experiment, we obtain three independent sets of equations, two in each, to determine the parameters of the appropriate mathematical model. Hence we can determine two parameters of the transfer function for each frequency ω_j. As those, we choose T and τ for Function (4.30), T_2 and τ_2 for (4.32), and T_1 and R_1 for (4.31).

The active resistance R of the electromagnet coil is known and is practically independent of frequency in the considered frequency range. In order to reconcile with experimental data it is however necessary to assume, that R_1 in Model (4.4) is a function of frequency. This function deviation from the known value is the measure of Model (4.4) adequacy to the real electromagnet.

To determine parameters for reconciliation with the experiment let us perform some simple transformations with Equations (4.34) – (4.36) using Expressions (4.12); we obtain:

$$\tau(\omega_j) = \frac{\sqrt{2}\,[\cos\varphi_E(\omega_j) - RA_E(\omega_j)]}{\sqrt{\omega_j}\,[RA_E(\omega_j) - \sqrt{2}\sin(\varphi_E(\omega_j)+\pi/4)]},$$

$$T(\omega_j) = \frac{1}{\omega_j RA_E(\omega_j)}\frac{1+R^2A_E^2(\omega_j) - 2RA_E(\omega_j)\cos\varphi_E(\omega_j)}{RA_E(\omega_j) - \sqrt{2}\sin(\varphi_E(\omega_j)+\pi/4)}, \tag{4.37}$$

$$R_1(\omega_j) = \frac{\cos \varphi_E(\omega_j)}{A_E(\omega_j)}, \quad T_1(\omega_j) = -\frac{\operatorname{tg} \varphi_E(\omega_j)}{\omega_j}. \tag{4.38}$$

$$\tau_2(\omega_j) = \frac{RA_E(\omega_j) - \cos \varphi_E(\omega_j)}{\omega_j \sin \varphi_E(\omega_j)},$$

$$T_2(\omega_j) = -\frac{R^2 A_E^2(\omega_j) - 2A_E(\omega_j)\cos \varphi_E(\omega_j) + 1}{\omega_j A_E(\omega_j) \sin \varphi_E(\omega_j)}. \tag{4.39}$$

Using (4.37) – (4.39) and Expressions (4.33) it is simple to calculate the parameters of mathematical Models (4.3) – (4.5):

$$c'_I(\omega_j) = RT(\omega_j), \quad c_\gamma(\omega_j) = \frac{c'_I(\omega_j)}{\sqrt{\pi}\,\tau(\omega_j)},$$

$$L_1(\omega_j) = R_1(\omega_j)\,T_1(\omega_j), \quad L_2(\omega_j) = RT_2(\omega_j), \quad r(\omega_j) = \frac{L_2(\omega_j)}{\tau_2(\omega_j)}. \tag{4.40}$$

Fig. 4.5 shows parameters L_1 and R_1 of Model (4.4) as functions of frequency, those parameters being determined based on experimental amplitude and phase response characteristics shown in Fig. 4.4. When frequency varies from 1 to 60 Hz, L_1 value doubles, and R_1 value variation is almost 20-fold.

Fig. 4.6 shows parameters L_2 and r of Model (4.5) as functions of frequency; L_2 variation is almost by a factor of 1.7, and r variation is more than 5-fold. The variation of parameters c'_I and c_γ for Model (4.3) (see Fig. 4.7) does not exceed 10% and 25% respectively.

Models (4.4) and (4.5) are obviously unsuitable for the description of an electromagnet with non-laminated core as an element of control system over the whole system operating frequency range. Model (4.3) presents a satisfactory agreement with experimental results, therefore it is made the foundation for control algorithms synthesis and analysis (see Chapter 7).

To determine Model (4.3) parameters dependence on air gap s and average bias current I^o each frequency response characteristic obtained for fixed s and I^o was processed according to the method stated in Article 4.2. Mismatch of experimental and calculated values of amplitude and phase response obtained according to Formula (4.34) for the whole range of the gap, bias current, and frequency variation did not exceed 10%, it was less than 5% as a rule. The results of calculation are shown in Fig. 4.4 with solid line, and experimental results with circles.

Coefficient c'_I is a linearization coefficient, so in order to determine the electromagnet inductance it is necessary to take into account the inductance dependence on the average bias current.

$$c'_I = L(s, I)\big|_{s=s^o,\, I=I^o} + I^o \frac{\partial L}{\partial I}\Big|_{s=s^o,\, I=I^o}. \tag{4.41}$$

Resulting from the experiment we obtain c'_I as a function of I^o for fixed s in a number of points.

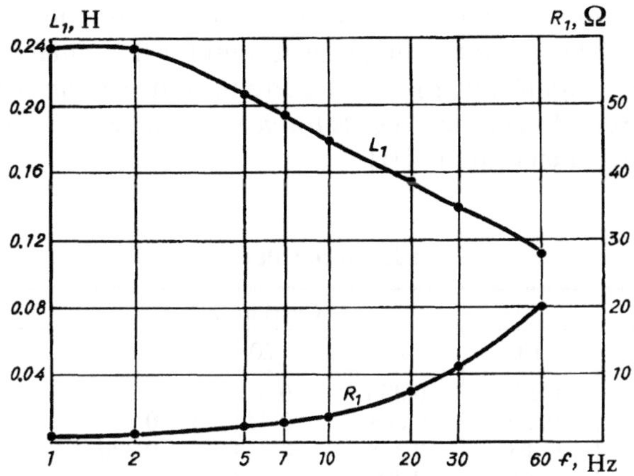

Fig. 4.5. Parameters L_1, R_1 as functions of frequency

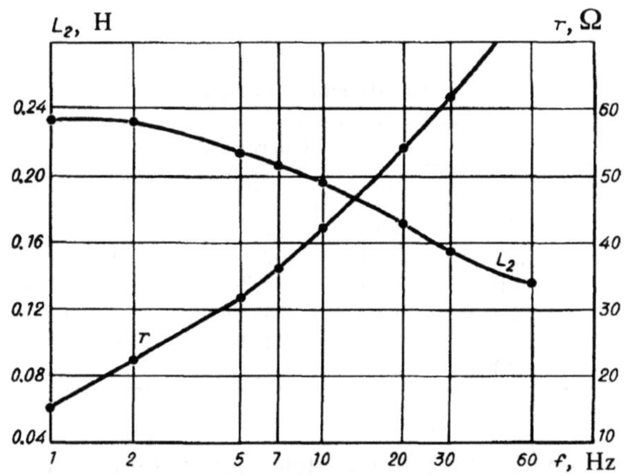

Fig. 4.6. Parameters L_2, r as functions of frequency

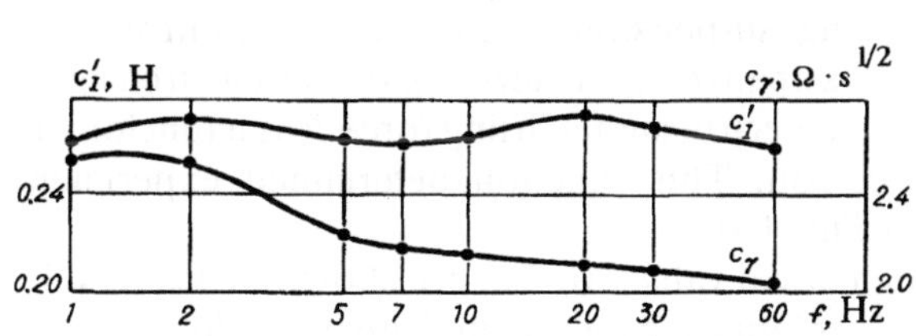

Fig. 4.7. Parameters c'_1, c_γ as functions of frequency

Considering (4.41) as a differential equation with respect to L, let us integrate it by Euler's method. Take c'_l at the minimum current for the initial value, because the saturation effect is insignificant for small current, and we can assume inductance independent of current. Integration results for various air gap values are presented in Table 4.1. Inductance dimension is [H].

Table 4.1

Inductance Values

Air gap, mm	Current, A						
	5	10	15	20	25	40	50
14	0.330	0.330	0.330	0.312	0.314	0.200	0.179
12	0.350	0.374	0.364	0.344	0.318	0.218	0.181
10	0.374	0.380	0.378	0.354	0.320	0.223	0.184
8	0.400	0.398	0.390	0.358	0.324	0.226	0.186
6	0.450	0.440	0.410	0.370	0.330	0.230	0.187
4	0.550	0.520	0.465	0.393	0.341	0.250	0.195

Table 4.2

Coefficient c_γ Values

Air gap, mm	Current, A						
	5	10	15	20	25	40	50
14	3.08	3.26	2.97	2.54	1.78	1.50	1.20
12	3.17	3.32	3.05	2.14	1.81	1.40	1.19
10	3.10	3.00	2.70	1.74	1.84	1.35	1.18
8	3.14	2.65	2.10	1.76	1.86	1.33	1.17
6	2.88	1.81	1.94	1.79	1.88	1.30	1.16
4	2.60	2.23	1.90	1.71	1.41	1.25	1.15

Table 4.2 shows coefficient c_γ values determined at the same points, dimension is $[\Omega \cdot s^{1/2}]$. Dependence of self-induction coefficient L and coefficient c_γ on air gap and current are shown in Fig. 4.8 and 4.9 respectively.

When synthesizing suspension electromagnets control algorithms further on, besides inertial and dissipative characteristics of the electromagnet we shall need its force characteristic, i.e. electromagnet lifting force F as a funcion of air gap and current in the electromagnet coil. This function determined experimentally by Dr. V.G. Gorelov is shown in Fig. 4.10.

Lifting force numerical values (in N) are shown in Table 4.3.

Results of the performed experimental investigations prove that mathematical Model (4.3) describes the electromagnet satisfactorily throughout the operating range of air gap and coil current. Besides that, it enables consideration of effects caused by eddy currents in the ferromagnetic core.

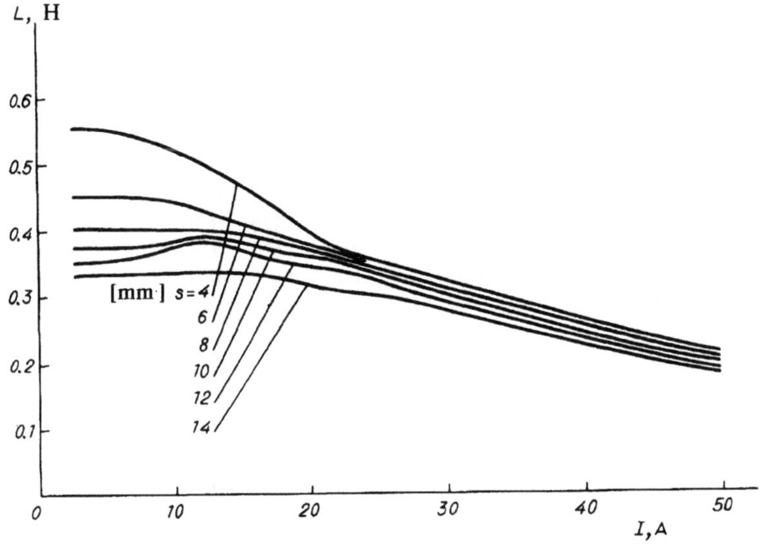

Fig. 4.8. Inductance as a function of air gap and current

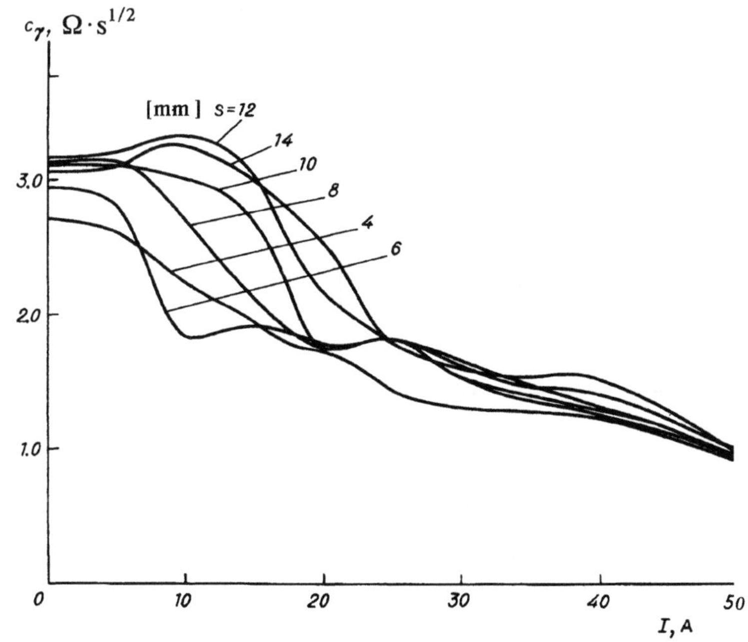

Fig. 4.9. Coefficient c_γ as a function of air gap and current

Table 4.3

Lifting force Values, N

Air gap, mm	Current, A												
	4	8	12	16	20	24	26.4	28	32	36	40	44	48
14	78.2	306	682	1100	1560	1950	2150	2270	2500	2730	2960	3150	3300
12	125	390	842	1410	1950	2390	2610	2770	3010	3240	3480	3680	3800
10	137	527	1116	1840	2500	2970	3190	3320	3670	3920	4160	4350	4500
8	208	747	1591	2550	3300	3840	4050	4220	4600	4960	5310	5610	5860
6	333	1334	2697	3820	4600	5250	5580	5800	6230	6620	6970	7270	7520
4	686	2602	4818	6100	7210	8050	8520	8760	9380	9890	10340	10700	11000

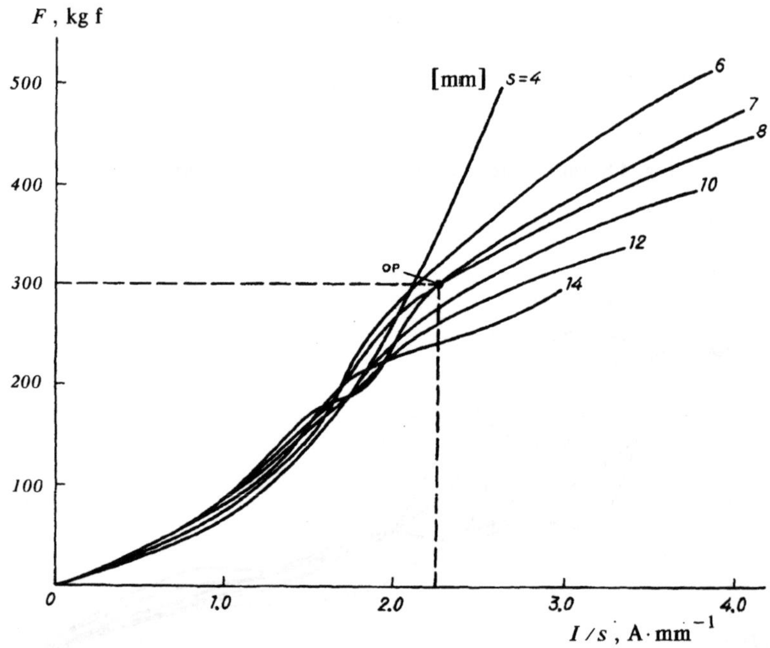

Fig. 4.10. Lifting force as a function of current related to air gap. OP - operating point.

In order to determine applicability limits for the suggested mathematical model of HECF core eddy currents influence on the electromagnet dynamics, it is necessary however to carry out the experiment for a wider frequency variation range. It was impossible to accomplish this with the experimental plant described in Article 4.3 because of limited power of the amplifier converter and insufficient resolution of the electromagnet coil current measurement system.

For the expanded frequency variation range experimental response characteristics were determined for the model electromagnet with U-shaped cross-section and a flat ferromagnetic rail, having the length one sixth of that for the real electromagnet. Average length of the cross-section was 0.3 m. For a fixed air gap the following three quantities were measured: electromagnet coil current I, electromagnet – rail attraction force F, and measurement winding voltage, the winding having 10 turns applied to the electromagnet core close to the pole. Voltage on this winding is proportional to time derivative $\dot{\Phi}$ of magnetic flux Φ through the pole section. This experiment was performed by D.A. Kochetov [39]. The input quantity, like in the previous experiment, was the electromagnet coil voltage. Frequency f of the voltage sinusoidal component varied continuously in the interval $2 - 2000$ Hz. The number of turns applied to the electromagnet and the air gap value were chosen such, that induction in the gap was equal to the induction of the full-scale electromagnet described above.

Fig. 4.11 shows amplitude response of the model electromagnet with solid lines; subscripts denote that the characteristic belongs to the corresponding measured quantity. Cross section of the model electromagnet is shown in the upper right corner of the picture, the magnetic circuit area is hatched.

Zero decibel level corresponds to current amplitude of 2 A, force 100 N, and measurement winding voltage 30 mV. The step of amplitude response $A_F(f)$ and $A_I(f)$ is caused by increase of measurement paths amplification coefficients by 30 and 20 dB respectively.

Fig. 4.12 shows phase response, denominations are the same as in Fig. 4.11.

In the frequency band of 200 to 2000 Hz resonant excitation of the electromagnet elements and its mounting system took place. This affected the character of force amplitude and phase response variation.

To analyze the obtained experimental results we shall use not the simple Model (4.3), but a little more complete one, obtained from (1.90), if we set $\eta \equiv s \equiv 0$ (constant air gap) in it:

$$c_I' \dot{I} + c_J' \dot{J}^* + RI = v_1(t), \quad c_J' (\dot{I} + \dot{J}^*) + c_\gamma' \int_{-\infty}^{t} \frac{\dot{J}^*(\tau)\, d\tau}{\sqrt{t-\tau}} = 0, \tag{4.42}$$

$$F = c_I (I + J^*) .$$

Model (4.42), unlike (4.3), allows to consider the influence of electromagnet coil magnetic flux, closed outside the ferromagnetic rail. Besides that,

$$c'_I = L^\circ + c'_J, \tag{4.43}$$

where L° is the inductance corresponding to leakage flux. The measurement winding voltage is proportional to time derivative of the total flux through the electromagnet face section, induced by the electromagnet coil current and eddy currents in the ferromagnet.

$$u_{\dot{\Phi}} = k(\dot{I} + \dot{J}^*), \tag{4.44}$$

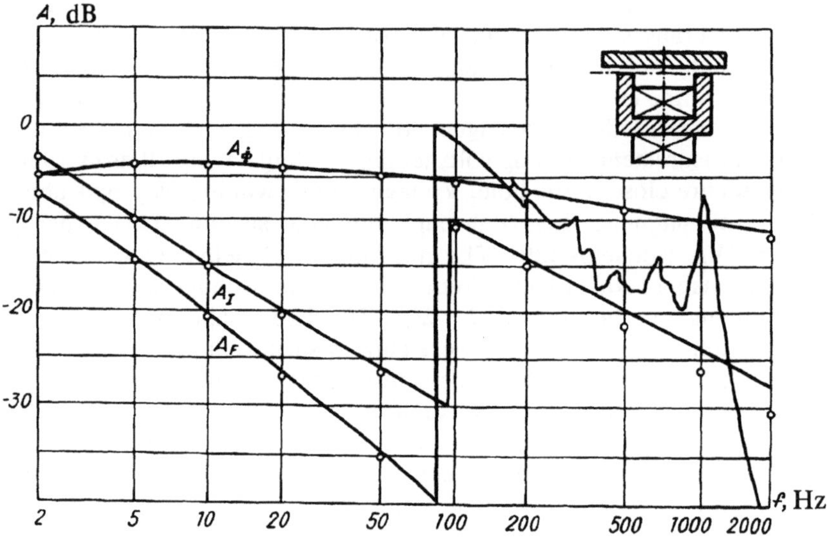

Fig. 4.11. Amplitude response of the model electromagnet

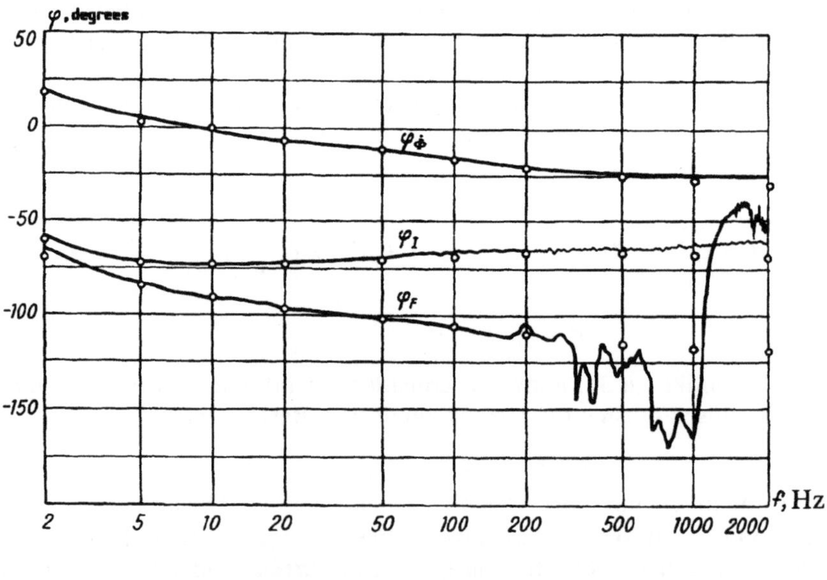

Fig. 4.12. Phase response of the model electromagnet

where k is the proportionality coefficient, $k = 1.25$ mV · s/A .

Let us write down transfer functions from input voltage v_1 to the measured quantities, considering Equality (4.44):

$$W_I(p) = \frac{1}{R}\frac{\tau\sqrt{p}+1}{\varepsilon T\tau p\sqrt{p}+Tp+\tau\sqrt{p}+1}, \tag{4.45}$$

$$W_F(p) = \frac{c_I}{R}\frac{1}{\varepsilon T\tau p\sqrt{p}+Tp+\tau\sqrt{p}+1}, \tag{4.46}$$

$$W_{\dot\Phi}(p) = \frac{k}{R}\frac{p}{\varepsilon T\tau p\sqrt{p}+Tp+\tau\sqrt{p}+1}, \tag{4.47}$$

where R is the ohmic resistance of the electromagnet coil. The following symbols are introduced in (4.45) – (4.47):

$$\tau = \frac{c'_J}{c'_\gamma\sqrt{\pi}}, \quad T = \frac{c'_J}{R}, \quad \varepsilon = \frac{L^\circ}{c'_J}. \tag{4.48}$$

It is clear from comparison of (4.47) and (4.46), that phase responses φ_F and $\varphi_{\dot\Phi}$ are to differ by 90°. This fact is corroborated experimentally up to accuracy of 2 – 3° in the frequency range of 2 – 200 Hz, in which there was no resonant excitation of the plant. In order to determine transfer functions parameters τ, T, ε from experimental amplitude and phase responses, the method described in Article 4.2 was used. The following values were obtained: $\tau = 0.1\ \mathrm{s}^{1/2}$, $T = 0.2$ s, $\varepsilon = 0.17$. Using Expressions (4.48) and taking into account, that R = 0.3 Ω, we obtain $c'_\gamma = 0.34\ \Omega\cdot\mathrm{s}^{1/2}$, $c'_J = 0.06$ H.

Calculation of self-induction coefficient based on magnetostatics problem solution (see V. K. Chudnov and L. E. Lomonosova [16]) for the model electromagnet yields the value of 0.054 H. To estimate the value of c'_γ Expressions (1.88) and (1.76) can be used, assumed $c'_\gamma \approx c_\gamma = \gamma^\circ$

$$c'_\gamma = \frac{\mu_0\, l_1\, N^{\circ 2}}{s_\mu}\left(\frac{\nu_M}{\pi}\right)^{1/2}, \tag{4.49}$$

where l_1 is the magnet length, s_μ is average length of the core cross section, related to double relative magnetic permeability; N° is the number of coil turns; ν_M is magnetic viscosity of (1.2).

Relative magnetic permeability of unsaturated structural steel is in the interval of 700 – 1000 units. Using (4.49) and taking into account $l_1 = 0.13$ m , $N^\circ = 236$, and average core length approximately equal to 0.3 m, we obtain the estimate of coefficient c'_γ value:

$$0.28 < c'_\gamma < 0.35. \tag{4.50}$$

The experimentally obtained value $c'_\gamma = 0.34$ is within the Interval (4.50).

Thus mathematical Model (4.42) parameters values obtained from experimental amplitude and phase response match the calculated ones with accuracy of about 10%.

In Figs. 4.11 and 4.12 the circles represent the calculated amplitude and phase response characteristics corresponding to transfer Functions (4.45) – (4.47) with parameters $\varepsilon = 0.17$, $\tau = 0.1\ s^{1/2}$, $T = 0.2$ s. In the frequency range of 2 to 500 Hz the calculated amplitude response functions match the experimental ones up to accuracy of 1 dB, the difference of phase response does not exceed 5°. In the frequency interval of 500 to 2000 Hz mismatch of the calculated and experimental characteristics increases and reaches the values of 3 dB in amplitude and 10° in phase at the frequency of 2000 Hz.

The above analysis shows that we can use the simplest Model (4.3) for the investigation of processes which characteristic frequencies do not exceed 50 – 100 Hz. If it is necessary to study processes with higher frequencies (up to 200 – 500 Hz) we have to take into consideration leakage fields and apply a more complete Model (4.42). If the characteristic frequencies of the process exceed 1000 Hz, the error of the proposed model is 20 – 40%, and to describe eddy current effect more accurately a model of more accurate structure is required.

4.5. Low-Viscous Liquid Vortex Motions Model Verification

The foundation of the perturbed motion equations for a solid with cavities partially filled with LV liquid is the phenomenologic Formula (2.49) determining the force applied to the rib from the liquid. Those equations take into account the liquid vortex motion caused by vortex formation at sharp edges. When processing experimental data, we will pay principal attention to validity verification for the description, introduced by V.M.Rogovoi [86] of hydrodynamic force applied to the rib. To verify the introduced models we used experimental data obtained when determining frequency response characteristics of a tank with ribs, rotating about its longitudinal axis. This experiment allows the most explicit manifestation of vortex formation effects, that are not shaded by the liquid free surface oscillatory motions.

The kinematic scheme of the experimental plant is shown in Fig. 4.2. The excitation system consists of two prestressed springs connected with a tension member embracing the tank and a force exciter, whose rod performs harmonic vibrations

$$\xi(t) = \xi_0 \sin \omega t \,, \tag{4.51}$$

with specified amplitude ξ_0 and discretely variable frequency ω.

The moment applied to the tank is

$$M = -M_1(\varphi) - M_2\left(\varphi - \frac{1}{r_0}\xi\right) - \varepsilon_0 \dot{\varphi} \,, \tag{4.52}$$

where M_1 is the moment generated by the left spring, M_2 is the moment generated by the right spring, r_0 is the outer radius of tank, ε_0 is the excitation system damping coefficient.

Having approximated the spring rigidity characteristic (Fig. 4.3) with a cubic parabola and having performed harmonic linearization of that characteristic, we obtain from (4.52), taking into account (4.51), the following expression for the moment applied to the tank:

$$M(t) = -(C_0 - C_2 A_\varphi^2) A_\varphi \sin(\omega t + \alpha_\varphi) - \varepsilon_0 \omega A_\varphi \cos(\omega t + \alpha_\varphi) -$$
$$- \left[C_0 + C_2 \left(A_\varphi^2 + \frac{\xi_0^2}{r_0^2} - 2A_\varphi \frac{\xi_0}{r_0} \cos\alpha_\varphi\right)\right] \left[\left(A_\varphi \cos\alpha_\varphi - \frac{\xi_0}{r_0}\right) \sin\omega t + \right.$$
$$\left. + A_\varphi \sin\alpha_\varphi \cos\omega t\right]. \tag{4.53}$$

$A_\varphi, \alpha_\varphi$ in (4.53) are the amplitude and phase shift of angle φ oscillations, determined during the experiment, as functions of frequency. For the springs used in the experiment coefficients C_o and C_2 are equal to 1.74 N·m /rad and 1.85 N·m /rad^3 respectively. Expression (4.53) allows for inequality of deformations for the left and right spring of the experimental plant, caused by the force exciter rod displacement.

Let us rewrite (4.53) in a more compact form

$$M(t) = M_s \sin\omega t + M_c \cos\omega t, \tag{4.54}$$

where

$$M_s = -(C_0 - C_2 A_\varphi^2) A_\varphi \cos\alpha_\varphi + \varepsilon_0 \omega A_\varphi \sin\alpha_\varphi -$$
$$- \left[C_0 + C_2 \left(A_\varphi^2 + \frac{\xi_0^2}{r_0^2} - 2A_\varphi \frac{\xi_0}{r_0} \cos\alpha_\varphi\right)\right] \left(A_\varphi \cos\alpha_\varphi - \frac{\xi_0}{r_0}\right),$$
$$M_c = -(C_0 - C_2 A_\varphi^2) A_\varphi \sin\alpha_\varphi - \varepsilon_0 \omega A_\varphi \cos\alpha_\varphi - \tag{4.55}$$
$$- \left[C_0 + C_2 \left(A_\varphi^2 + \frac{\xi_0^2}{r_0^2} - 2A_\varphi \frac{\xi_0}{r_0} \cos\alpha_\varphi\right)\right] A_\varphi \sin\alpha_\varphi .$$

Let us substitute Expression (4.54) for the moment into Model (4.6); assume $\varphi = A_\varphi \sin(\omega t + \alpha_\varphi)$ and equate coefficients at $\cos\omega t$ and $\sin\omega t$ in the left and right sides of Equation (4.6). With simple transformations we obtain the formula for damping coefficient ε determination from experimental values of amplitude and phase response:

$$\varepsilon = \frac{1}{\omega A_\varphi} (M_c \cos\alpha_\varphi - M_s \sin\alpha_\varphi). \tag{4.56}$$

Formula (4.56) allows independent determination of the damping coefficient for each frequency value, for which the tank angular displacement amplitude A_φ and phase α_φ were determined.

Amplitude and phase response were experimentally determined for 5 values of the force exciter rod vibrations amplitude ξ_0 within the resonance frequency interval of tank vibrations. That frequency is determined by the excitation system springs

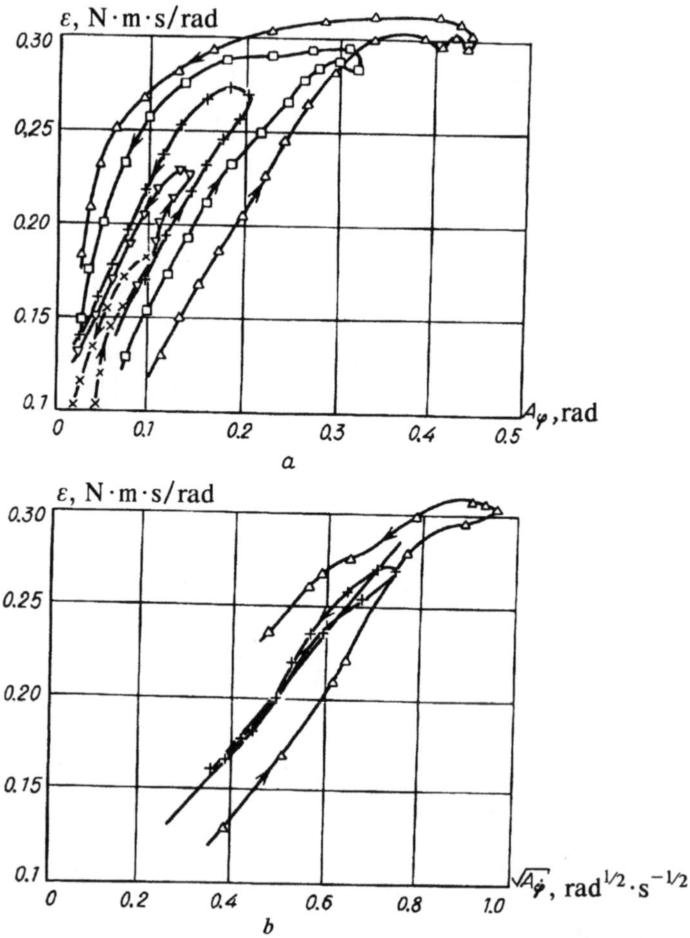

Fig. 4.13. Damping coefficient ε as a function of: *a)* tank oscillation amplitude A_φ ; *b)* variable $\sqrt{A_{\dot\varphi}}$; for various amplitude ξ_0 values: × for ξ_0=0.0068 m, ∇ for ξ_0=0.0102 m, + for ξ_0=0.0169 m, ◇ for ξ_0=0.0254 m, Δ for ξ_0=0.0339 m

rigidity. The values of ξ_0 were chosen in order to cover the range of tank rotation angle of 0 to 0.4 rad, i.e. up to 25°. Fig. 4.13 *a* shows damping coefficient ε dependence upon tank vibrations amplitude, according to Formula (4.56). Curves with arrows in this figure show direction of path-tracing around experimental points for the increase of excitation frequency. Return points of those curves correspond to resonance peaks of amplitude response characteristics.

It should be noted that damping coefficient values at the upgoing branch of the amplitude response are not equal to its values at the downgoing branch for equal amplitudes of tank vibrations (a hysteresis-type effect). That means that the damping coefficient depends not only on vibrations amplitude, but on the frequency of those vibrations too. With growth of vibrations frequency the damping coefficient grows too. Fig. 4.13, *b* shows dependence of damping coefficient ε on variable $|A_{\dot\varphi}|^{1/2}$ for two ξ_0 values, where $A_{\dot\varphi} = \omega A_\varphi$. It is clear that, up to amplitudes of about 0.3 rad damping coefficient ε variation is described quite satisfactorily with function $|\dot\varphi|^{1/2}$ (the curve for $\xi_0 = 0.0169$ m is practically linear and has small hysteresis, and the curves for lesser values ξ_0 , that are not shown, run in the same line with insignificant scatter). This is an experimental corroboration for the nonlinear character of the hydrodynamic force expressed by (2.49), that is suggested by V.M.Rogovoi and S.V.Cheremnykh [86]. For great values of vibrations amplitude an effect occurs, that can be conventionally called 'saturation' in damping, i.e. further increase of vibration amplitude does not result in a more vigorous energy dissipation. Those effects cannot be described in terms of Simplified Model (4.6) suitable only for qualitative analysis of the system behaviour for infinitesimal displacements of the tank and liquid, i.e. for local stability analysis. Therefore we proceed to consideration of more sophisticated models.

Mathematical Models (4.7) and (4.8) are nonlinear, that is why before starting analysis of their adequacy to the considered physical phenomena we are going to perform one-frequency harmonic linearization of nonlinear terms in those equations. Nonlinear elements of Equations (4.7) and (4.8) have identical structure, so let us consider the function of the form

$$f(t) = |\dot y(t)|^{1/2} \int_{-\infty}^{t} \frac{\ddot y(\tau)\, d\tau}{\sqrt{t-\tau}}, \tag{4.57}$$

assuming

$$y(t) = A \sin(\omega t + \alpha) . \tag{4.58}$$

Including (4.58), let us take the convolution integral in (4.57):

$$\int_{-\infty}^{t} \frac{\ddot y(\tau)\, d\tau}{\sqrt{t-\tau}} = -\,\omega^2 A \int_{-\infty}^{t} \frac{\sin(\omega t + \alpha)}{\sqrt{t-\tau}}\, d\tau =$$

$$= -\omega^2 A \left(\frac{\pi}{2\omega}\right)^{1/2} (\sin(\omega t + \alpha) - \cos(\omega t + \alpha)) . \qquad (4.59)$$

Substituting (4.59) into (4.57) and including (4.58), we obtain, up to the first Fourier series term:

$$f(t) = -\sqrt{\pi/2}\, A^{1.5}\, \omega^2 \sqrt{|\cos(\omega t + \alpha)|}\, (\sin(\omega t + \alpha) - \cos(\omega t + \alpha)) \approx$$

$$\approx -\sqrt{\pi/2}\, A^{1.5}\, \omega^2 (a_1 \cos(\omega t + \alpha) + b_1 \sin(\omega t + \alpha)) . \qquad (4.60)$$

Here a_1 and b_1 are Fourier coefficients. Calculate them, supposed $\omega t + \alpha = \Theta$:

$$a_1 = \frac{1}{\pi} \int_0^{2\pi} \sqrt{|\cos\Theta|}\, \sin\Theta \cos\Theta\, d\Theta - \frac{1}{\pi} \int_0^{2\pi} \sqrt{|\cos\Theta|}\, \cos^2\Theta\, d\Theta =$$

$$= -\frac{4}{\pi} \int_0^{\pi/2} \cos^{2.5}\Theta\, d\Theta = -\frac{4}{\pi}\, \frac{\Gamma(0.5)\,\Gamma(1.75)}{2\Gamma(2.25)} \approx -0.915 , \qquad (4.61)$$

$$b_1 = \frac{1}{\pi} \int_0^{2\pi} \sqrt{|\cos\Theta|}\, \sin^2\Theta\, d\Theta - \frac{1}{\pi} \int_0^{2\pi} \sqrt{|\cos\Theta|}\, \cos\Theta \sin\Theta\, d\Theta =$$

$$= \frac{4}{\pi} \int_0^{\pi/2} \sin^2\Theta \cos^{0.5}\Theta\, d\Theta = \frac{4}{\pi}\, \frac{\Gamma(1.5)\,\Gamma(0.75)}{2\Gamma(2.25)} \approx 0.610 .$$

where $\Gamma(x)$ is Gamma function.

Hence, in the harmonic oscillations mode, up to the first term of Fourier series expansion, Function (4.57) can be written in the form

$$f(t) \approx \kappa_\varepsilon \omega \sqrt{A}\, \dot{y} + \kappa_\mu \sqrt{A}\, \ddot{y} , \qquad (4.62)$$

where

$$\kappa_\varepsilon = -\sqrt{\pi/2}\, a_1 \approx 1.147, \qquad \kappa_\mu = \sqrt{\pi/2}\, b_1 \approx 0.765 .$$

Using nonlinear term representation of the form similar to (4.62) and external moment Expression (4.54), let us rewrite Model (4.7) in the following form:

$$(I^\circ + I + \kappa_\mu \tilde{\beta} \sqrt{A_\varphi})\, \ddot{\varphi} + \kappa_\varepsilon \tilde{\beta}\, \omega \sqrt{A_\varphi}\, \dot{\varphi} = M_s \sin\omega t + M_c \cos\omega t . \qquad (4.63)$$

The Joukowski associated moment of inertia for the tank used in the experiment is equal to $0.176\ \mathrm{kg \cdot m^2}$.

Equation (4.63) coincides in structure with Equation (4.6), with the exception, that in Equation (4.63) the equivalent moment of inertia and damping coefficient are functions of the amplitude and frequency of the tank harmonic oscillations. In Model

(4.7) the parameter for reconciliation with experimental data is coefficient $\widetilde{\beta}$ value. Coefficient $\widetilde{\beta}$ apparently can be expressed in terms of experimentally determined parameters with the formula

$$\widetilde{\beta} = \frac{1}{\kappa_\varepsilon\, \omega \sqrt{A_\varphi}}\, \varepsilon\,, \tag{4.64}$$

where $\varepsilon = \varepsilon\,(A_\varphi, \alpha_\varphi, \omega)$ is determined from Expression (4.56). In Model (4.10) parameters for matching experiment are $\widetilde{\beta}^*$ and I^*. Solidfied liquid's moment of inertia $I^{(0)}$ is equal to 0.501 kg·m^2.

Supposed in (4.10)

$$\varphi = A_\varphi \sin\,(\omega t + \alpha_\varphi)\,, \quad \chi = A_\chi \sin\,(\omega t + \alpha_\chi)\,,$$

and using Representation (4.62) and (4.54) we obtain from (4.10)

$$\begin{gathered} -\,(I^\circ + I^{(0)}\,)\,\omega^2 A_\varphi \sin\,(\omega t + \alpha_\varphi) - I^*\omega^2 A_\chi \sin\,(\omega t + \alpha_\chi) = \\ = M_s \sin \omega t + M_c \cos \omega t\,, \\ -\,I^* A_\varphi \sin\,(\omega t + \alpha_\varphi) - I^* A_\chi \sin\,(\omega t + \alpha_\chi) - \\ -\,\kappa_\mu\,\widetilde{\beta}^* A_\chi^{1.5} \sin\,(\omega t + \alpha_\chi) + \kappa_\varepsilon\,\widetilde{\beta}^* A_\chi^{1.5} \cos\,(\omega t + \alpha_\chi) = 0\,. \end{gathered} \tag{4.65}$$

Equating coefficients at sin ωt and cos ωt in the left and right sides of Equations (4.65) we come to a set of four nonlinear equations for determination of parameters $I^*, \widetilde{\beta}^*, A_\chi, \alpha_\chi$ based on experimental data for each frequency value, for which A_φ and α_φ were measured:

$$\begin{aligned} -\,(I^\circ + I^{(0)}\,)\,A_\varphi \cos \alpha_\varphi - I^* A_\chi \cos \alpha_\chi &= \frac{1}{\omega^2} M_s\,, \\ -\,(I^\circ + I^{(0)}\,)\,A_\varphi \sin \alpha_\varphi - I^* A_\chi \sin \alpha_\chi &= \frac{1}{\omega^2} M_c\,, \\ I^* A_\chi \cos \alpha_\chi + I^* A_\varphi \cos \alpha_\varphi + \kappa_\mu \widetilde{\beta}^* A_\chi^{1.5} \cos \alpha_\chi + \kappa_\varepsilon \widetilde{\beta}^* A_\chi^{1.5} \sin \alpha_\chi &= 0, \\ I^* A_\chi \sin \alpha_\chi + I^* A_\varphi \sin \alpha_\varphi + \kappa_\mu \widetilde{\beta}^* A_\chi^{1.5} \sin \alpha_\chi - \kappa_\varepsilon \widetilde{\beta}^* A_\chi \cos \alpha_\chi &= 0. \end{aligned} \tag{4.66}$$

Let us introduce the following denominations:

$$I^* A_\chi \cos \alpha_\chi = X\,, \quad I^* A_\chi \sin \alpha_\chi = Y\,, \quad \frac{1}{I^*} = x\,, \quad \frac{\widetilde{\beta}^*}{I^{*2.5}} = y\,. \tag{4.67}$$

Then the first two equations of (4.66) yield

$$\begin{aligned} X &= -\,(I^\circ + I^{(0)}\,)\,A_\varphi \cos \alpha_\varphi - \frac{1}{\omega^2} M_s\,, \\ Y &= -\,(I^\circ + I^{(0)}\,)\,A_\varphi \sin \alpha_\varphi - \frac{1}{\omega^2} M_c\,. \end{aligned} \tag{4.68}$$

The last two equations of the Set (4.66) result in two linear equations determining x and y:

$$xX + y\left(X^2 + Y^2\right)^{0.25}\left(\kappa_\varepsilon Y + \kappa_\mu X\right) = -A_\varphi \cos\alpha_\varphi ,$$
$$xY + y\left(X^2 + Y^2\right)^{0.25}\left(\kappa_\mu Y - \kappa_\varepsilon X\right) = -A_\varphi \sin\alpha_\varphi . \qquad (4.69)$$

Having obtained X and Y values from (4.68) and having solved System (4.69) for x and y, we determine the values of $\tilde{I}^*$, $\tilde{\beta}^*$, A_χ and α_χ from Relationships (4.67).

Thus, using the sequence of Equations (4.68), (4.69), (4.67) and (4.56), (4.64), we can determine parameters $\tilde{I}^*$ and $\tilde{\beta}^*$ for Model (4.10) and $\tilde{\beta}$ or Model (4.7), respectively, for each frequency value. The calculation for all the available experimental points yields for coefficient $\tilde{\beta}$ the value of 0.2 $N \cdot m \cdot s^2/rad^{1.5}$ with the error of ± 20 %. Parameters $\tilde{I}^*$ and $\tilde{\beta}$ of Model (4.10) are equal to 0.45 $kg \cdot m^2$ and 0.6 $N \cdot m \cdot s^2/rad^{1.5}$, respectively, with the error of ± 40 %.

It should be noted that the scheme of Model (4.10) parameters calculation is more sensitive to initial quantities ω, A_φ, α_φ measurement errors, because generalized coordinate χ is not measured. Besides that, it should be noted, that Model (4.10) usage provides a possibility to describe developed vortex motion of LV liquid with vortex field nonuniform over the volume, retaining only one additional 'vortex' degree of freedom. The difference of the real vortex field from that uniform over the volume is formally included in this model by varying parameters $\tilde{I}^*$ and $\tilde{\beta}^*$, which play the role of parameters for reconciliation with the experiment. A more accurate approximation of the real velocity field can be achieved by increasing the quantity of additional degrees of freedom taken into account when describing the vortex field.

Fig. 4.14 shows angular displacement amplitude of the system calculated according to Model (4.7) (dashed line) and according to Model (4.10) (solid line) for various values of force exciter rod vibrations amplitude. Convention for the experimental points is the same as in Fig. 4.13. Fig. 4.15 shows angular displacement phase according to Models (4.7) and (4.10) for maximum and minimum amplitudes of rod vibrations. Amplitude and phase response characteristics were constructed based on Equations (4.7) and (4.10) integration with experimentally determined values of parameters $\tilde{\beta}$, $\tilde{I}^*$, $\tilde{\beta}^*$ and external moment specified by Expression (4.52). The integration method was presented in Chapter 3. Initial conditions for integration were specified zero, and the integration process continued till establishment of harmonic oscillations in all the generalized coordinates. After that amplitudes and phases of steady oscillations were determined. Accuracy control is accomplished by reconstructing the values of model parameters for reconciliation with the experiment based on calculated values of A_φ, α_φ and comparing the reconstructed values with the initial ones. Mismatch of those values did not exceed fractions of a percent, that, on one side, is an evidence of satisfactory integration accuracy, and on the other side of

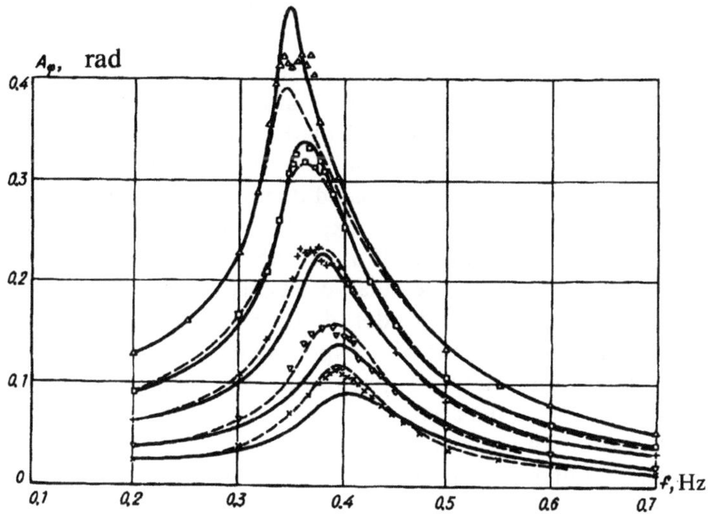

Fig. 4.14. Angular displacement amplitude . Denominations are the same as in Fig. 4.11

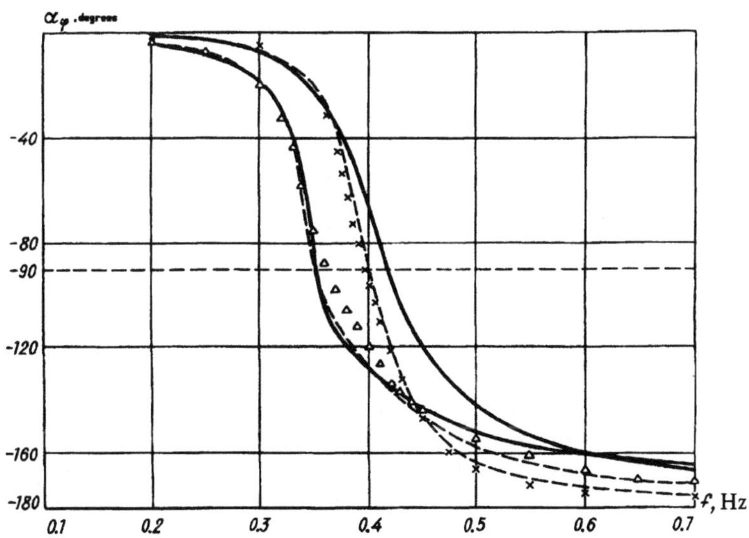

Fig. 4.15. Angular displacement phase. Denominations are the same as in Fig. 4.11

legitimacy of one-frequency harmonic linearization of the nonlinear term in Models (4.7) and (4.10).

Results presented in Fig. 4.14 show that Model (4.7) is in better agreement with experimental data than Model (4.10) in the domain of small oscillation amplitudes (less than 0.15 rad). In the amplitude interval of 0.15 to 0.3 rad both models provide approximately equal agreement with the experiment. For greater amplitudes of oscillations Model (4.7) yields overestimated damping, compared to the experiment, although it is seen from Fig. 4.14 that experimental points in the vicinity of the resonance peak are higher, making as if a 'cut' peak. It is obviously caused by some additional factor ignored by this model, that results in decrease of oscillations amplitude in the resonance frequency vicinity. Model (4.10) in this amplitude interval (more than 0.3 rad) provides better agreement with experiment, and the amplitude response of Model (4.10) has a more natural appearance embracing the 'cut' peak of the experimental characteristic from above.

Consider Equation (4.63) being a harmonically linearized equivalent of Model (4.7). The coefficient at $\dot{\varphi}$ in that equation, which determines energy dissipation in liquid, grows as $\sqrt{A_\varphi}$ with the growth of oscillations amplitude. Unlike Model (4.10),

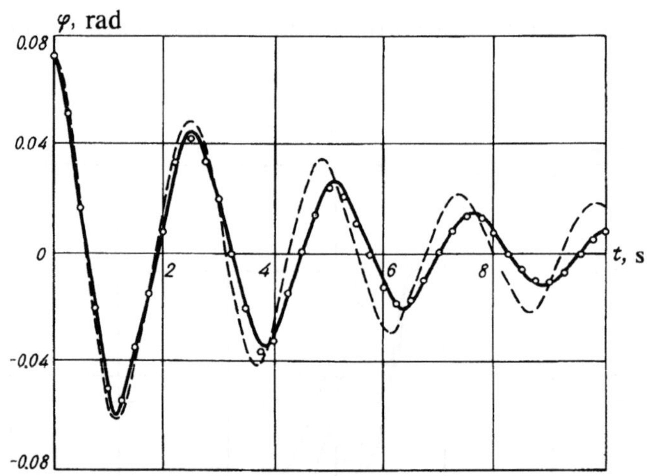

Fig. 4.16. Transient process: ∘ here denotes experimental points, the solid line shows calculation using Model (4.10), the dashed line shows calculation using Model (4.7).

that is essentially unable to cause damping saturation at great oscillation amplitudes, which is observed in the experiment (see Fig. 4.13). This difference of the considered models is a fundamental one and requires an additional analysis based on independent experiments, as far as possible. Such analysis can be performed, particularly, based on comparison of calculated and experimental transient processes. The appropriate results are presented below.

The solid line in Fig. 4.16 shows the transient process calculated according to Model (4.10) for $\Gamma^{*} = 0.4$, $\beta^{*} = 0.8$. The dashed line in the figure shows the transient process calculated according to Model (4.7) for $\tilde{\beta} = 0.24$, the experimental points are shown with circles. Parameters of the models were chosen so, in order to obtain maximum agreement with the experiment, beginning with $t = 0$. As it is seen in Fig. 4.16, it succeeded at the first half-period for both models. Further on, however, the curves begin to disagree noticeably. Model (4.10) provides satisfactory agreement with the experiment for the entire time interval considered, whereas the solution corresponding to Model (4.7) has a lower damping than the experimental transient process and is in worse agreement with it in frequency. An attempt to make this solution

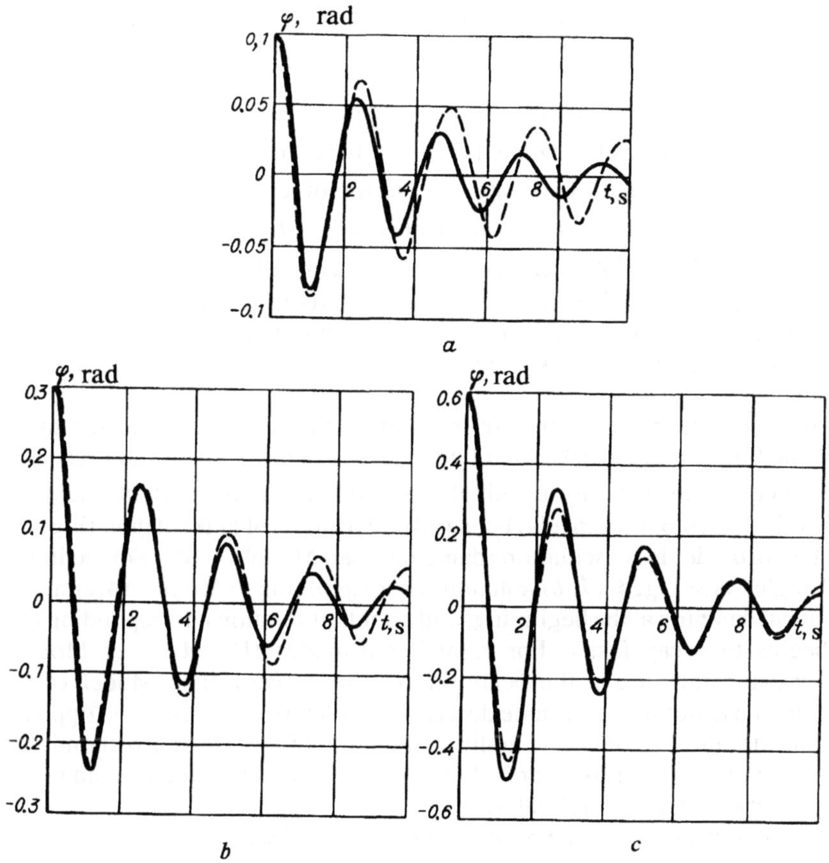

Fig. 4.17. Transient processes: calculation according to Models (4.7) (dashed lines) and (4.10) (solid lines); *a)* initial angle $\varphi_0 = 0.1$ rad; *b)* $\varphi_0 = 0.3$ rad; *c)* $\varphi_0 = 0.6$ rad .

agree with the experiment by increasing $\widetilde{\beta}$ to make the amplitudes at the end of the transient process match is unsuccessful because of noticeable decrease of calculated amplitudes of oscillations at their start.

Thus we should assume for parameter I^* the value 0.4 $\text{kg}\cdot\text{m}^2$ as the most reliable, that coincides (up to I^* values straggling) with that obtained for the harmonic oscillations mode.

It is interesting to note the following circumstance. Suppose that at the very beginning of the transient process accelerated vortices making a vortex kernel of liquid and free vortices at the ribs edges appear. Circulations of the free vortices obey Joukowski – Chaplygin condition not only in the initial moment but in all the successive moments of time. We can use then the results of the model problem of Article 2.10. Convert dimensionless Joukowski moments of inertia $\widetilde{I}_\Gamma$, $\widetilde{I}_\Omega$ for the considered case ($K = 6$; $\widetilde{b} = 0.3$) from Table 2.1 to dimensioned ones, according to the formulae

$$I_\Omega = \rho h r_0^4 \widetilde{I}_\Omega\,, \quad I_\Gamma = \rho h r_0^4 \widetilde{I}_\Gamma\,,$$

where ρ is mass density of liquid, h is its depth, r_0 is the inner radius of the tank. Instead of the above associated moment of inertia for liquid $I = 0.176$ $\text{kg}\cdot\text{m}^2$ we obtain then $I_\Gamma = 0.079$ $\text{kg}\cdot\text{m}^2$; $I_\Omega = 0.105$ $\text{kg}\cdot\text{m}^2$. On the other hand, the moment of inertia for solidified liquid $I^{(0)} = 0.501$ $\text{kg}\cdot\text{m}^2$. Hence $I^{(0)} - I_\Omega = 0.396$ $\text{kg}\cdot\text{m}^2$; $I^{(0)} - I_\Gamma = 0.422$ $\text{kg}\cdot\text{m}^2$; $0.396 \le I^* \le 0.422$ that practically coincides the value $I^* = 0.4$ $\text{kg}\cdot\text{m}^2$ obtained from processing of the transient process. This can be considered an additional argument in favour of existence in the discussed experiment of quasistationary local vortex zones near the ribs' edges and a corroboration of the introduced mathematical models adequacy.

To demonstrate the difference of the transient process characterized by Models (4.7) (dashed line) and (4.10) (solid line) Fig. 4.17 shows integration results for the appropriate equations and various initial conditions for angle φ. Parameters $\widetilde{\beta}$, I^*, $\widetilde{\beta}^*$ values are 0.2, 0.45, 0.6 respectively. For small initial angles of tank rotation (Fig. 4.17, *a*) Model (4.7) provides less oscillations damping, than Model (4.10). For medium values of initial angles (see Fig. 4.17, *b*) solutions corresponding to Models (4.7) and (4.10) have identical damping at the beginning, and after that solution corresponding to Model (4.10) begins to decay faster. For great initial angles (Fig. 4.17, *c*) Model (4.7) provides greater damping at the beginning of the process, than Model (4.10), but gradually with oscillations amplitude decrease the picture changes to the opposite one. Hence, consideration of only one additional degree of freedom corresponding to LV liquid vortex motion allows to describe the experimentally observed damping saturation effect for great oscillation amplitudes.

Note in conclusion, that the process of energy dissipation due to vortex formation at ribs edges is determined not by the oscillations amplitude, but by velocity of rib's motion relative to liquid. Therefore when extending the results discussed above to tanks of other radius we should take Strouhal Number (2.34) for the congruence criterion. It is necessary to take into consideration, that the boundary value of Strouhal number separating applicability areas of Models (4.7) and (4.10) is substantially less for transient processes than for steady-state harmonic oscillations. It is apparently connected with accelerated vortices stalling from the ribs' edges in the moment of motion start from rest.

Chapter 5

Some Dynamics Problems for Systems with Electromagnetic Actuators

However, true it may be that every generation and destruction proceeds from some principle, or even more than one, why does this happen and what is the cause?

Aristotle. *Metaphysics* *

In Chapter 1 we presented the mathematical model of eddy currents in HECF elements (Article 1.3), obtained within the scope of assumptions and constraints that, on one hand, make it possible to obtain a rigorous mathematical solution of the stated problem, and on the other hand, they result from specific features of application of AES and SC magnetic orientation and stabilization systems as well as electromagnetic levitation systems. In Chapter 4 devoted to experimental verification of the considered vortex models we showed by the example of the electromagnet being the principal element of the levitation system, that, by its structure, the suggested mathematical model allows a generalization for the cases of leakage flux existence, finite magnetic permeability of the core material and saturation effects. In that chapter the adequacy of skin effect mathematical model is corroborated, and it is shown that, for typical values of the electromagnetic levitation system parameters, the identification of the electromagnet with HECF core as an object of control can be performed with an accuracy sufficient for engineering calculations only taking into account eddy currents. Thus, eddy currents are in this case the 'principle' that determines appearance and disappearance of subtle dynamic effects in the closed-loop electromagnet – controller system for both steady-state oscillations and transient processes.

The principal purpose of this chapter is to reveal the mechanism, how that 'principle' acts, and to show the necessity of taking into account eddy currents when designing electromagnetic levitation system, for closed-loop system dynamics analysis problems as well as for air gap control law synthesis. By the example of studying a specific system using analysis methods presented in Chapter 3 we show the degree of the considered effects display and study the question of the measurements composition effect on the mathematical model completeness. It is shown, that based on the harmonic

---

* Aristotle's Metaphysics. Transl. with comment. and glossary by Hippocrates G.Apostle. Bloomington – London. Indiana University Press, 1966, (p. 18)

balance method also considered in Chapter 3, it is possible to reduce the set of integrodifferential equations describing skin effect to an equivalent set of ordinary differential equations allowing adequate synthesis of control laws, to which Chapter 7 is devoted.

5.1. Characteristic Properties of Electromagnet as an Object of Control. Requirements to the Air Gap Regulation System

Electromagnet is a basic element of electromagnetic levitation systems. At the same time the electromagnet is an object of control. Control voltage from the regulator implementing the control law, that affords suspension stability, is applied to it.

In the majority of works devoted to control modes of electromagnets in the suspension system and to the dynamics of the closed-loop system 'electromagnet – regulator' the electromagnet is represented by a series connection of an inductance coil and a resistor. We shall call this simplified model *traditional*, for it is adopted as a fundamental one and is published, e.g. in the works by K.H.Brock *et al.*, E.Gottzein, R.D.Thornton [14, 25 – 27, 96], the employees of the West-German firm MBB, which achieved the most success in magnetic levitation transport systems creation. A similar model is used in the works of other researchers, e.g. P.K.Sinha [91, 92] , V.I.Bocharov *et al.*, V.D.Nagorsky *et al.*, Yu.A.Nikitenko *et al.*, D.M.Veinberg *et al.* [9, 17, 71, 100]. In literature devoted to electromagnetic suspension a model is also presented, that describes eddying in terms of ordinary differential equations (see, e.g. T.I.Katsan, V.G.Lebedev, and A.I.Mytarev [35]).

The traditional model is admissible if the core is made of special laminated electrical sheet steels (LECF core). In a number of engineering applications, however, the core is made of ordinary structural steels having high electrical conductivity (HECF core), that appears advantageous for economic and structural reasons. In that case it is necessary to take into consideration unsteady effects associated with development of eddy currents in the conducting material of the core. It is the main feature, that is to be taken into account when synthesizing the control law and analyzing the closed-loop system dynamics.

The second feature consists in the fact, that the electromagnet as an object of control is an essentially non-linear system. The main non-linearities are the non-linear variation laws of the attraction force, inductance, and parameters characterizing skin effect with respect to the air gap, coil current, and eddy current, as well as saturation-type non-linearity of the static characteristic for the regulator's power part. These non-linear effects are not readily amenable to rigorous mathematical description, therefore analysis and synthesis of those systems is to be carried out using the experimental information.

When synthesizing control laws it is necessary to take into account characteristic properties of air gap sensors, accelerometers, current transducers, transducers of inductance in the air gap, whose signals are used for generating control actions in accordance with the adopted control law and set of measurements. One should take into consideration the following drawbacks of accelerometers:

1) the presence of an accelerometer causes cross connection of regulation channels, if electromagnetic actuators are used for attitude control of the vehicle;

2) with high-capacity power plants on board, an accelerometer becomes a source of rather a high level vibrational interference in the stabilization system operation, caused mainly by vibrations perpendicular to its sensitivity axis. This causes an unfavourable effect on the closed-loop system noise immunity.

The control law constructed based on indications of the air gap sensor and current transducer is free of those disadvantages, although it is characterized by smaller stability margins.

We shall list the main requirements to a regulation system for the air gap between the electromagnetic actuator and the ferromagnetic rail, that follow from functioning features specific for systems of the considered class.

1. Possibility of the control law simple and reliable implementation, that maintains stable levitation allowing for the probable straggling of the object's and regulator's parameters. To meet that requirement, it is desirable to design the control law based on the simplest set of measurements, namely air gap sensor and current transducer indications.

2. Provision of conditions for stable lift of the electromagnet resting on supports, i.e. the conditions of 'flow up' from the gap value of s_0 to the nominal gap value of $s^\circ (s_0 > s^\circ)$. Normally s° is a value of about 5 to 15 mm, and $s_0 = 1.5 \ldots 2\, s^\circ$.

3. Upkeep of the air gap with 1 to 2 mm accuracy in the presence of perturbations from the ferromagnetic rail. Typical perturbations are instantaneous changes in the ferromagnetic rail position of about 1 mm in the inertial space and the rail harmonic oscillations with the amplitude of about 1 mm and frequency of about 3 to 5 Hz.

4. Provisions for high level noise immunity of the regulation system.

5. Partitioning of the whole levitated load into two parts connected with elasto-viscous ties (two-mass system), with sprung mass partial frequency of about 1 to 1.5 Hz and the logarithmic decrement of the oscillations about 1 to 1.2. It is done in order for the high-frequency perturbations (those having frequency exceeding 1.5 Hz) tracked by the levitation system, which operating frequencies are 15 to 20 Hz, not to be transferred to the sprung mass. Due to that, the motions of the electromagnet and of the load are «dynamically isolated», the mass of the load being usually several times greater than the electromagnet mass. As a result, it becomes possible to extend significantly the system bandwidth not raising the control loop gain factor, to increase, thus, stability margins, noise immunity, and to reduce perturbation counteraction power consumption.

5.2. Mathematical Model of the Two-Mass System 'Controlled Electromagnet – Mass with Elasto-Viscous Suspension'

Dynamics of an electromagnet with an elasto-viscously suspended mass can be described in the first-order approximation with the following set of nonlinear integrodifferential equations being a generalization of mathematical Model (1.71) for the case considering core saturation effect and small flux closing outside the core, i.e. assuming $L = L(s, I + J) = M(s, I + J)$:

$$\ddot{\eta}_1 = -\omega_1^2(\eta_1 - \eta_2) - \varepsilon_1(\dot{\eta}_1 - \dot{\eta}_2) - g + \frac{1}{m_1}F,$$

$$\ddot{\eta}_2 = -\omega_2^2(\eta_2 - \eta_1) - \varepsilon_2(\dot{\eta}_2 - \dot{\eta}_1) - g, \tag{5.1}$$

$$\frac{d}{dt}[L(I + J)] + RI = V, \quad \frac{d}{dt}[L(I + J)] + \alpha\gamma^\circ \int_{-\infty}^{t} \frac{d}{d\tau}[\alpha J(\tau)]\frac{d\tau}{\sqrt{t - \tau}} = 0,$$

$$s = -\eta_1 + \eta_c(t),$$

where η_2 is the sprung mass coordinate in the inertial reference system; η_1 is the coordinate of the electromagnet pole upper edge; ω_2^2 is the partial frequency square of the sprung mass oscillations with motionless electromagnet; $\varepsilon_1, \varepsilon_2$ are damping coefficients; ω_1^2 is the partial frequency square of the electromagnet oscillations with motionless sprung mass; g is the acceleration of gravity; m_1 is the electromagnet mass; F is the electromagnet lifting force; V is the voltage applied to the electromagnet coil. Other symbols completely correspond to those introduced in Chapter 1.

Experimental investigations results presented in Chapter 4 completely corroborate adequacy of the considered mathematical model for the entire range of frequency, air gap, and current variations, interesting for practice. Between α coefficient and experimentally determined parameter $c_\gamma(s, I + J)$ there exists a relationship, somewhat more general than (1.88):

$$c_\gamma(s, I + J) = \gamma^\circ \alpha^2(s, I + J),$$

i.e.

$$\alpha = \left(\frac{c_\gamma(s, I + J)}{\gamma^\circ}\right)^{1/2}.$$

Thus, electromagnet lifting force F, self-induction coefficient L, and α coefficient are nonlinear functions of air gap and summary variable $I + J$:

$$F = F(s, I + J), \quad L = L(s, I + J), \quad \alpha = \alpha(s, I + J). \tag{5.2}$$

These nonlinear characteristics are presented in Chapter 4. By virtue of this, stabilization system is essentially nonlinear for great variations of variables in such modes as 'floating up' and control in response to standard disturbances. To investigate levitation system dynamics in those modes computer simulation is used. This question is considered in detail in Chapter 7.

Steady suspension mode corresponds to nominal air gap s°, nominal current I°, nominal voltage V°, $\mathring{\eta}_1 = -s^\circ$, $\mathring{\eta}_2$ and eddy current $J = 0$. These parameters are to satisfy the following relationships:

$$F^\circ = F(s^\circ, I^\circ) = (m_1 + m_2)\, g = m_1 g \left(1 + \frac{\omega_1^2}{\omega_2^2}\right),$$

$$V^\circ = I^\circ R, \quad \mathring{\eta}_2 = \mathring{\eta}_1 - \frac{g}{\omega_2^2}, \tag{5.3}$$

where m_2 is the sprung weight mass. Let us introduce variations of air gap Δs, current ΔI, voltage ΔV, as well as variations $\Delta\eta_1$ and $\Delta\eta_2$:

$$\Delta s = -\Delta\eta_1 + \eta_c(t), \quad \Delta I = I - I^\circ, \quad \Delta V = V - V^\circ,$$

$$\Delta\eta_1 = \eta_1 - \mathring{\eta}_1, \quad \Delta\eta_2 = \eta_2 - \mathring{\eta}_2 .$$

Variation of the variable describing eddy current J, coincides with the variable itself. Control variation of voltage ΔV is generated in accordance with the selected control law.

Let us write down the linearized model of the object in terms of of the nominal (programmed) mode variations:

$$\Delta\ddot{\eta}_1 = -\,\omega_1^2(\Delta\eta_1 - \Delta\eta_2) - \varepsilon_1(\Delta\dot{\eta}_1 - \Delta\dot{\eta}_2) - a_{\eta\eta}\,\Delta s + a_{\eta\delta}\,(\Delta I + J\,),$$

$$\Delta\ddot{\eta}_2 = -\,\omega_2^2(\Delta\eta_2 - \Delta\eta_1) - \varepsilon_2(\Delta\dot{\eta}_2 - \Delta\dot{\eta}_1),$$

$$T\,(\Delta\dot{I} + \dot{J}\,) + \Delta I = \frac{1}{R}\Delta V + \frac{c_s'}{R}\Delta\dot{s}, \tag{5.4}$$

$$\tau_J(\Delta\dot{I} + \dot{J}\,) + \frac{1}{\sqrt{\pi}}\int_{-\infty}^{t}\frac{dJ(\tau)}{d\tau}\frac{d\tau}{\sqrt{t-\tau}} = \frac{\tau_J}{T}\frac{c_s'}{R}\Delta\dot{s}, \quad \Delta s = -\Delta\eta_1 + \eta_c(t),$$

where the linearization coefficients are as follows:

$$a_{\eta\eta} = -\frac{1}{m_1}\frac{\partial F}{\partial s}\bigg|_{\substack{s=s^\circ,\\ I+J=I^\circ}}, \quad a_{\eta\delta} = \frac{1}{m_1}\frac{\partial F}{\partial\,(I+J\,)}\bigg|_{\substack{s=s^\circ,\\ I+J=I^\circ}},$$

$$T = \frac{1}{R}\left[\,L(s^\circ, I^\circ) + \frac{\partial L}{\partial\,(I+J\,)}\bigg|_{\substack{s=s^\circ,\\ I+J=I^\circ}} I^\circ\right] = \frac{\mathring{L}}{R}, \tag{5.5}$$

$$c_s' = -I^\circ \frac{\partial L}{\partial s}\bigg|_{\substack{s=s^\circ,\\ I+J=I^\circ}}, \qquad \tau_J = \frac{TR}{c_\gamma \sqrt{\pi}}.$$

Coefficient $a_{\eta\eta}$ characterizes the degree of the electromagnet static instability, $a_{\eta\delta}$ characterizes controllability, and τ_J the degree of eddy currents influence.

Let us concentrate on details of linearization coefficients computation based on natural experiment data. As it is clear from Fig. 4.10, the electromagnet attraction force is approximated with high accuracy by a function of $\frac{I+J}{s}$, up to the values of about $2\ \mathrm{A \cdot mm^{-1}}$, irrespective of the air gap, and has fairly good linearity with respect to this argument (see Article 1.6). It is convenient to use this property for linearization coefficients according to formulae

$$a_{\eta\eta} = -\frac{1}{m_1}\frac{\partial F(\varphi)}{\partial \varphi}\frac{\partial \varphi}{\partial s}\bigg|_{\substack{s=s^\circ,\\ I+J=I^\circ}} = \frac{1}{m_1}\frac{\partial F(\varphi)}{\partial \varphi}\bigg|_{\varphi^\circ}\frac{I^\circ}{s^{\circ 2}},$$

$$a_{\eta\delta} = -\frac{1}{m_1}\frac{\partial F(\varphi)}{\partial \varphi}\frac{\partial \varphi}{\partial (I+J)}\bigg|_{\substack{s=s^\circ,\\ I+J=I^\circ}} = -\frac{1}{m_1}\frac{\partial F(\varphi)}{\partial \varphi}\bigg|_{\varphi^\circ}\frac{1}{s^\circ}, \tag{5.6}$$

where $\varphi = (I+J)/s$, $\varphi^\circ = I^\circ/s^\circ$.

It is simple to choose the operating point from the curves presented in Fig. 4.10. Suppose, it is necessary, for instance, to suspend the load with the weight $(m_1 + m_2)\, g$ at the air gap s°. To determine the nominal value of current I°, it is necessary to place the operating point (OP) in the curve corresponding to air gap s°, then to determine φ°, and then I° according to the formula

$$I^\circ = \varphi^\circ s^\circ. \tag{5.7}$$

This is the scheme of linearization coefficients $a_{\eta\eta}$ and $a_{\eta\delta}$ calculation. Constants T and τ_J are determined during frequency tests described in Chapter 4. Coefficient c_s' can be calculated by differentiation of function $L(s)$ with respect to air gap s. This method provides low accuracy, however. For a more accurate determination we can use the following relationships. Let us write down electromagnetic field energy:

$$W = L\frac{(I+J)^2}{2}. \tag{5.8}$$

The electromagnet force

$$F = -\frac{\partial W}{\partial s} = -\frac{\partial L}{\partial s}\frac{(I+J)^2}{2}, \tag{5.9}$$

and linearization coefficient

$$a_{\eta\delta} = \frac{1}{m_1} \frac{\partial F}{\partial (I+J)}\bigg|_{I^\circ, s^\circ} = -\frac{1}{m_1}\frac{\partial L}{\partial s}\bigg|_{I^\circ, s^\circ} I^\circ. \tag{5.10}$$

Comparing (5.10) and the expression for c_s' in Formulae (5.5) we obtain

$$c_s' = m_1 a_{\eta\delta}, \tag{5.11}$$

that coincides with (1.88) up to denominations.

Taking into account that in the operating frequency interval $\omega >> \omega_2$ and $\eta_2 << \eta_1$ we can reduce the set of Equations (5.4) for synthesizing control law by eliminating generalized coordinate η_2 describing the sprung mass dynamics. As it is shown below, in Chapter 7, the system corresponding to the sprung mass motion has the characteristics proximate to partial ones, and with all that it has practically no influence on the stabilization system functioning in the specific operating frequency interval of about 15-20 Hz. Denoting $\Delta\eta_1 = \Delta\eta$ we obtain

$$\Delta\ddot{\eta} = -a_{\eta\eta}\Delta s + a_{\eta\delta}(\Delta I + J), \quad T(\Delta\dot{I} + \dot{J}) + \Delta I = \frac{1}{R}\Delta V + \frac{c_s'}{R}\Delta\dot{s},$$

$$\tau_J(\Delta\dot{I} + \dot{J}) + \frac{1}{\sqrt{\pi}}\int_{-\infty}^{t}\frac{dJ(\tau)}{d\tau}\frac{d\tau}{\sqrt{t-\tau}} = \frac{\tau_J}{T}\frac{c_s'}{R}\Delta\dot{s}, \tag{5.12}$$

$$\Delta s = -\Delta\eta + \eta_c(t).$$

The obtained set of Equations (5.12) is a foundation for electromagnetic levitation system control law synthesis.

Thus, we have performed the next stage of simplification reducing the two-mass system to a one-mass one. We have to remember, however, that mathematical Model (5.12) is adequate to (5.4) in the operating frequency interval only. When investigating system dynamics in the interval of secondary suspension characteristic frequencies, Equations (5.4) should be used.

Let us present the parameters of the two-mass system, which is chosen as an object of investigation. By this example we are going to present quantitative estimates of the investigated effects influence and to demonstrate applicability of analytic and numerical methods to analyzing and synthesizing systems of that class. Those parameters are:

- unsprung mass $m_1 = 120$ kg;
- sprung mass $m_2 = 180$ kg;

$\omega_2^2 = 39.5\ \mathrm{s}^{-2}$, $\varepsilon_2 = 2.4\ \mathrm{s}^{-1}$ – those characteristics correspond to suspension partial frequency 1 Hz with logarithmic decrement 1.2;

$\omega_1^2 = 59.2\ \mathrm{s}^{-2}$, $\varepsilon_1 = 3.6\ \mathrm{s}^{-1}$.

The following relationship exists between partial characteristics of the sprung and unsprung masses:

$$\frac{\omega_1^2}{\omega_2^2} = \frac{\varepsilon_1}{\varepsilon_2} = \frac{m_2}{m_1} = 1.5,$$

R resistance (for $t = 20°$ C) is 1.1 Ω , $g = 9.81$ m·s^{-2}. The electromagnet generates the lifting force of 300 kgf for air gap $s° = 7$ mm and current $I° = 15.6$ A; $\overset{\circ}{\eta}_1 = -0.007$ m, $\overset{\circ}{\eta}_2 = -0.256$ m, the average voltage output of the power amplifier $\overset{\circ}{V} = 17.2$ V, the maximum voltage available from the amplifier, $V_{max} = 80$ V.

The power amplifier is implemented in double-sided pulse-duration modulation scheme (PDM) based on powerful first class transistors TK 100. The characteristic carrier frequency of PDM is 1-2 kHz, that far exceeds the stabilization system operating frequencies (15-20 Hz). Therefore, in its relation to signals within the operating frequency range, the converter can be represented by an instantaneous element with a nonlinear static characteristic of level limitation ('saturation') type. It has parameters shown in Fig. 5.1, where V_l denotes the voltage corresponding to a converter with boundless linear zone.

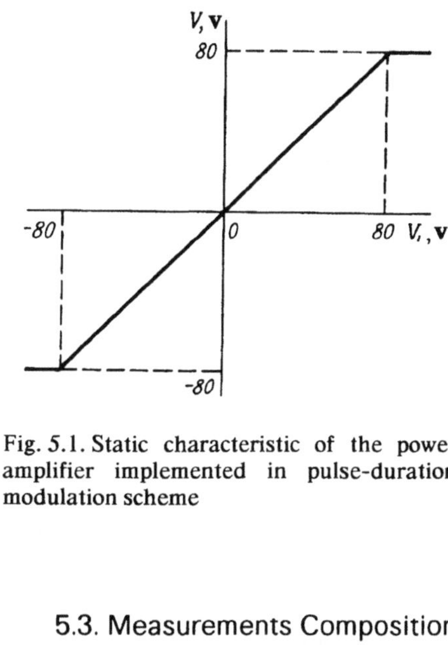

Fig. 5.1. Static characteristic of the power amplifier implemented in pulse-duration modulation scheme

For the chosen operation mode the linearization coefficients calculated according to the technique described above have the following values:

$$a_{\eta\eta} = 5200 \text{ s}^{-2},\ a_{\eta\delta} = 2.34 \text{ m}\cdot\text{A}^{-1}\cdot\text{s}^{-2},$$
$$T = 0.25 \text{ s},\quad c_s' = 280 \text{ V}\cdot\text{s}\cdot\text{m}^{-1},$$
$$\tau_J = 0.07\text{s}^{1/2}. \tag{5.13}$$

Let us proceed now to consideration of eddy currents influence on the closed-loop system properties within the scope of mathematical Model (5.12).

5.3. Measurements Composition Influence on Required Completeness of Controlled Electromagnet Mathematical Model

As it was shown in the previous Article, the basis for synthesizing control law is the linearized mathematical Model (5.12) of the one-mass system. Applying Laplace transform to it with zero initial conditions we obtain the following set of equations in the image plane:

$$p^2\Delta\eta\ (p) = -\ a_{\eta\eta}\Delta s\ (p) + a_{\eta\delta}\,[\Delta I\ (p) + J\ (p)],$$

$$T\,p\ [\Delta I\ (p) + J\ (p)] + \Delta I\ (p) = \frac{1}{R}\,\Delta V\ (p) + d_{\eta\eta}\ p\Delta s\ (p), \tag{5.14}$$

$$\tau_J\ p\ [\Delta I\ (p) + J\ (p)] + \sqrt{p}\ J\ (p) = b'_{\eta\eta}\ p\ \Delta s\ (p),\quad \Delta s\ (p) = -\Delta\eta\ (p) + \eta_c(p),$$

where

$$a'_{\eta\eta} = \frac{c'_s}{R}, \quad b'_{\eta\eta} = \frac{\tau_J}{T} a'_{\eta\eta},$$

and, unlike Chapter 3, Laplace images are denoted with the same letters as the original functions.

Consider the question about degree of eddy currents influence on closed-loop system stability depending on the chosen measurements composition (see [52, 67]). By *closed-loop* we further mean system 'object of control (electromagnet) – controller (control law)'. Suppose, and it is the most general case, control is generated based on air gap sensor, accelerometer, and current transducer signals, i.e.

$$\frac{\Delta V(p)}{R} = \Im_s(p)\Delta s(p) - \Im_{\ddot{\eta}}(p)\Delta\ddot{\eta}(p) - \Im_I(p)\Delta I(p), \qquad (5.15)$$

where $\Im_s(p)$, $\Im_{\ddot{\eta}}(p)$ and $\Im_I(p)$ are some linear differential operators implementing the control law. Signs at these operators in Expression (5.15) are selected in order to carry out the negative feedback principle with positive coefficients of transfer functions corresponding to the operators.

Let us write down the characteristic equation of system (5.14) closed by control Law (5.15):

$$T p^3 + \tau_J p^2 \sqrt{p} + p^2 + (a_{\eta\delta} a'_{\eta\eta} - T a_{\eta\eta})\, p +$$

$$+ (a_{\eta\delta} b'_{\eta\eta} - a_{\eta\eta}\tau_J)\sqrt{p} - a_{\eta\eta} + a_{\eta\delta}\,[\Im_s(p) + p^2 \Im_{\ddot{\eta}}(p)] +$$

$$+ \Im_I(p)\,[(p^2 - a_{\eta\eta})(\tau_J\sqrt{p} + 1) + a_{\eta\delta} b'_{\eta\eta}\sqrt{p}\,] = 0. \qquad (5.16)$$

We omitted multiplier $q = \sqrt{p}$ from the left side of characteristic Equation (5.16), because this multiplier produces a zero root in terms of variable $\sqrt{p}$ (see Formula (3.21a) and the corresponding note on allowability of zero roots).

Suppose control is generated based on air gap sensor and accelerometer signals. In this case, as quantitative estimate shows, terms $\tau_J p^2 \sqrt{p}$ and $(a_{\eta\delta} b'_{\eta\eta} - a_{\eta\eta}\tau_J)\sqrt{p}$ of characteristic Equation (5.16) are negligible compared to the others at operating frequencies of the stabilization system. This is characteristic not only of the considered example, but of a variety of systems of that class. Therefore we can neglect eddy currents influence on closed-loop system stability for such composition of measurements, although these effects are not negligible in the sense of electromagnet current I variation.

The situation is absolutely different from the viewpoint of stability, if control is generated using current transducer signal. It is clear from (5.16), that operator $\Im_I(p)$ as if amplifies eddy currents influence on closed-loop system stability. Besides that, $\Im_I(p)$ also increases the influence of static instability characterized by $a_{\eta\eta}$ coefficient. A control law based on current value will therefore have essentially lesser stability margins than the one based on air gap and acceleration measurements.

Let us consider the problem of control law design without current correction. It follows from the closed-loop system characteristic Equation (5.16) analysis, that in this case we can synthesize control law based on the mathematical model derived from (5.12) by ignoring eddy currents influence, i.e. for $J \equiv 0$:

$$\Delta\ddot{\eta} = -\,a_{\eta\eta}\Delta s + a_{\eta\delta}\Delta I\ , T\Delta\dot{I} + \Delta I = \frac{1}{R}\Delta V + a'_{\eta\eta}\,\Delta\,\dot{s},$$

$$\Delta s = -\Delta\eta + \eta_c(t)\ . \qquad (5.17)$$

This model is usually cited, and we decided to call it traditional.

Let us consider one of the possible control laws based on air gap and acceleration measurements, which is carried over from works by K.H.Brock, E.Gottzein *et al.* [14, 25 – 27]. Such investigation is expedient from the viewpoint of obtaining a primary standard of already existing system, which can be compared to various kinds of control laws synthesized herein. The structure of this control law is presented in Fig. 5.2.

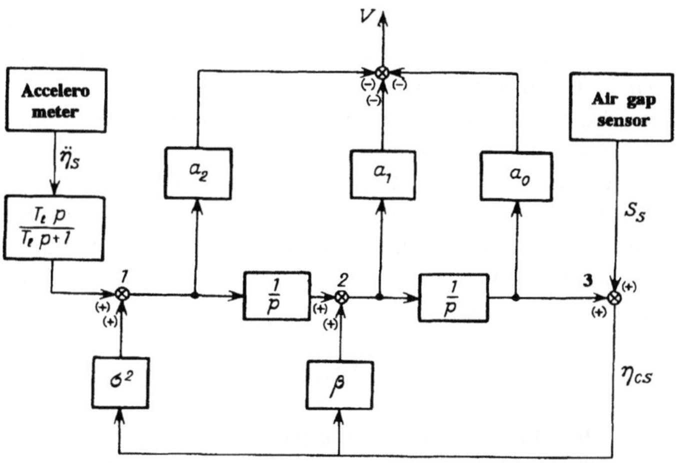

Fig. 5.2. Structural scheme of control law with an accelerometer.

The principle of constructing this control law is fairly clear from the presented scheme; it is as follows. Accelerometer signal $\ddot{\eta}_s$ passes through a low-frequency filter with transfer function $T_l\,p/(T_l\,p + 1)$ included to filter off the constant component connected with accelerometer zero setting error as well as technological misalignments when aligning accelerometer sensitivity axis.

Within the frequency interval we are interested in, this filter does not affect the system functioning. After double integration $\ddot{\eta}_s$ signal converts to η_s. By adding in the third summer η_s signal with air gap measurements s_s we obtain signal $\eta_{cs}(t)$ corresponding to the ferromagnetic rail disturbance $\eta_c(t)$. When conditions $\ddot{\eta}_s = \ddot{\eta}$, $s_s = s$ are fulfilled and with no disturbance $\eta_c(t) = 0$, there are no signals in feedback loops closed by summers 1 and 2 as well. Thus we carry out stabilization in variations

$\Delta\eta$, $\Delta\dot{\eta}$, $\Delta\ddot{\eta}$, the control law looking as follows (we neglect low frequency filter influence):

$$\frac{\Delta V}{R} = -a_0\Delta\eta - a_1\Delta\dot{\eta} - a_2\Delta\ddot{\eta},\ s_s = s,\ \ddot{\eta}_s = \ddot{\eta},\ \eta_c(t) = 0. \tag{5.18}$$

Feedback loops with parameters β and σ^2 are introduced to obtain the necessary response characteristics of the system with respect to ferromagnetic rail disturbance $\eta_c(t)$.

Thus we managed to separate the problems of providing stability and filtration.

After simple transformations we obtain the control law transfer function:

$$\frac{\Delta V}{R} = \frac{1}{1 + T_1 p + T_2 p^2}\Big\{-\frac{T_l p}{T_l p + 1}[a_0T_2 + (a_1T_2 + a_2T_1)p + a_2T_2p^2]\,\Delta\ddot{\eta} + \\ + [a_0 + (a_1 + a_0T_1)p + (a_2 + a_1T_1)p^2]\Delta s\Big\}, \tag{5.19}$$

where $T_1 = \beta/\sigma^2$, $T_2 = 1/\sigma^2$. One can easily make sure that control Law (5.19) converts to (5.18) for ideal functioning of accelerometer and air gap sensor, absence of disturbances ($\eta_c \equiv 0$), and if we assume $T_l p/(T_l p + 1) \cong 1$.

To illustrate the properties of control Law (5.19) we present results of the following closed-loop system characteristics calculation:

1) amplitude response $A_{\eta\eta c}(\omega)$ of transfer function $W_{\eta\eta c}(p) = \eta(p)/\eta_c(p)$ characterizing the quality of control in response to low-frequency disturbances of the ferromagnetic rail and filtering off high-frequency ones (see Fig. 5.3);

2) stability domain in the plane of parameters $a_{\eta\eta}$ and $a_{\eta\delta}$ deviations from their nominal (calculated) values (Fig.5.3), i.e.

$$\Delta a_{\eta\eta} = \frac{a_{\eta\eta} - \mathring{a}_{\eta\eta}}{\mathring{a}_{\eta\eta}},\quad \Delta a_{\eta\delta} = \frac{a_{\eta\delta} - \mathring{a}_{\eta\delta}}{\mathring{a}_{\eta\delta}},$$

where $\mathring{a}_{\eta\eta}$, $\mathring{a}_{\eta\delta}$ are the nominal values of the parameters;

3) the family of control law stability domains in the plane of a_0, a_1, with parameter a_2 (Fig. 5.4).

Note, that all the calculations herein have been carried out for two modifications of the mathematical model, the simplified one, (5.17) and the 'rigorous' one, (5.12). Obtained results match so well, that they would be practically indistinguishable in the figures.

It is clear from the results presented, that the considered control Law (5.19) provides good filtering off for high-frequency components of ferromagnetic rail disturbances $\eta_c(t)$, allowing control in response to low-frequency components. At the same time the closed-loop system has sufficient stability margins overlapping possible errors in electromagnet mathematical model parameters determination and in specification of controller parameters. Presence of accelerometer causes several disadvantages, however. These are low noise immunity with respect to vibrational disturbances and possibility of high-degree coupling between stabilization channels.

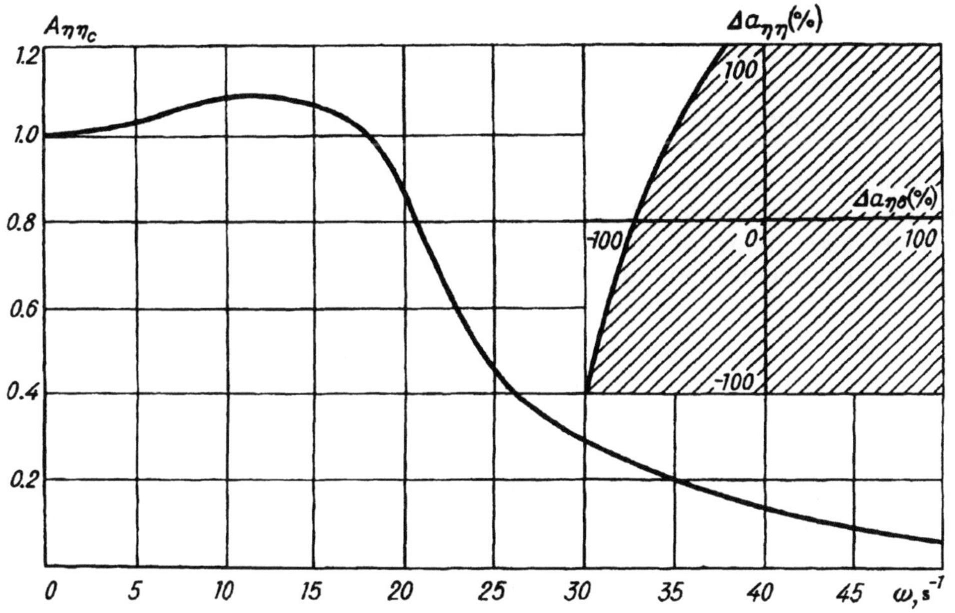

Fig. 5.3. Amplitude response characteristic and stability domain in coordinates $\Delta a_{\eta\eta}$, $\Delta a_{\eta\delta}$ for the control law with an accelerometer. Stability is at the hatched side.

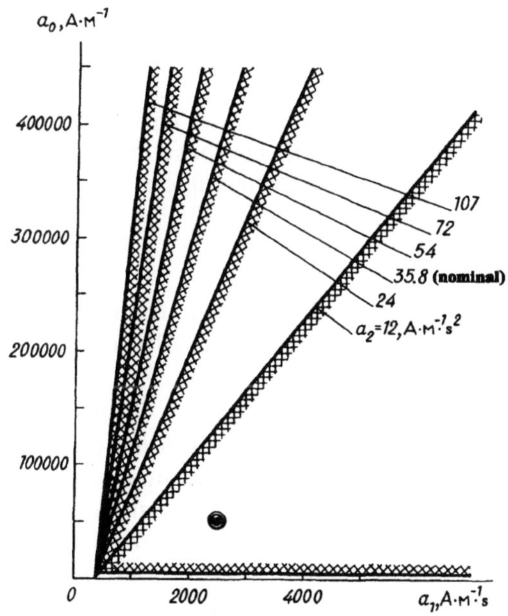

Fig. 5.4. Stability domains in parameters a_0, a_1, a_2 for the control law with an accelerometer. Stability is at the hatched side. ⊛ denotes the nominal value

Because of these disadvantages a more preferable from the practical point of view is still a control law not using accelerometer indications, particularly, the algorithm based on air gap sensor and current transducer indications.

Thus, if there is no current correction, we ignore eddy currents influence when synthesizing control law and use the traditional (simplified) mathematical Model (5.17).

With current transducer indications present among the set of measurements, the situation is entirely different. It is necessary to take into account eddy currents when analyzing closed-loop system properties as well as when synthesizing control law. This question is considered below in detail.

5.4. Eddy Currents Influence on Closed-Loop System Dynamics. Reduction of Integrodifferential Equations to an Equivalent Set of Differential Equations

Mathematical Model (5.12) of controlled electromagnet with HECF core, taking into account skin effect, should be used for the closed-loop system analysis in the interval of its characteristic (operating) frequencies, based on analytical methods presented in Chapter 3. When investigating dynamics taking into consideration secondary suspension, the model of (5.4) form is necessary. Final analysis (test calculation) taking into account nonlinear properties of the controller and object of control is carried out by Equations (5.1) integration with the help of the numerical method presented in Chapter 3. This regards to the analysis, however.

As to the synthesis, it is necessary to note, that there are no universal methods of synthesizing control laws for such objects, similar to methods for systems described by ordinary differential equations, developed in automatic control theory. In this connection a question emerges of the possibility to reduce the set of integrodifferential equations to a set of ordinary differential equations equivalent to it in some sense, that allows to synthesize the control law. For this purpose (see Chapter 3) let us apply the harmonic balance method to transform the integral term

$$c_\gamma \int_{-\infty}^{t} \frac{dJ(\tau)}{d\tau} \frac{d\tau}{\sqrt{t-\tau}}$$

in the vicinity of closed-loop system characteristic frequency Ω (see V.G.Lebedev, A.I.Mytarev, B.I.Rabinovich and R.V.Linkov, Y.V.Urman [52, 57]).

There are two ways to perform this operation.

1) An approximate one, when the integral in the third equation of (5.12) is replaced by an active resistance R^*equivalent to it in the sense of energy dissipation during the characteristic period (taking into account only one term in Formula (3.46)). In dynamics of solids with liquid filling this transformation corresponds physically to introduction of some equivalent damping. We shall discuss it in a more detail in the next chapter. For an electromagnet such operation corresponds to modelling skin effect in a HECF core by a short-circuit loop with internal resistance R_*^* which is inductively

coupled with the electromagnet coil. This operation is carried out, for instance, in transformers engineering for estimating power losses due to eddy currents. This model was already investigated in Chapter 4 (Article 4.5) when identifying the electromagnet parameters.

2) We perform a rigorous operation of averaging the integral over the same characteristic period. For objects having cavities filled with liquid this corresponds to taking into account not only damping, but also associated moment of inertia, caused by vortex motion (both terms in Formula (3.46)).

The first way of Equations (5.12) transformation differs from the second one in neglecting some additional inductance L^*.

The first way corresponds to the following set of linear differential equations:

$$\Delta\ddot{\eta} = -\, a_{\eta\eta}\Delta s + a_{\eta\delta}(\Delta I + J\,),\;\; L^{\circ}\,(\Delta\dot{I} + \dot{J}\,) + R\Delta I = \Delta V + c_s'\,\Delta\dot{s},$$
$$L^{\circ}\,(\Delta\dot{I} + \dot{J}\,) + R^{*}J = c_s'\,\Delta\dot{s},\;\; \Delta s = -\Delta\eta + \eta_c(t). \tag{5.20}$$

Expression for R^* is obtained by approximate application of harmonic balance method that is reduced to consideration of component $g_D\, x(t)$ of Formula (3.46) only, neglecting component $g_L\, \dot{x}(t)$. Including Expression (3.47) for g_D coefficient, we obtain

$$R^{*} = c_{\gamma}\left(\frac{\pi\Omega}{2}\right)^{1/2}. \tag{5.21}$$

After dividing the second equation of (5.20) by R and the third one by R^*, we obtain

$$\Delta\ddot{\eta} = -\, a_{\eta\eta}\Delta s + a_{\eta\delta}(\Delta I + J\,),\;\; T\,(\Delta\dot{I} + \dot{J}\,) + \Delta I = \frac{\Delta V}{R} + a_{\eta\eta}'\,\Delta\dot{s},$$
$$T_J\,(\Delta\dot{I} + \dot{J}\,) + J = c_{\eta\eta}'\,\Delta\dot{s},\quad \Delta s = -\Delta\eta + \eta_c(t), \tag{5.22}$$

where

$$T_J = \frac{L^{\circ}}{R^{*}} = \tau_J\left(\frac{2}{\Omega}\right)^{1/2},\; c_{\eta\eta}' = \frac{c_s'}{R^{*}} = a_{\eta\eta}'\,\frac{T_J}{T}.$$

The second way corresponds to the following set of equations:

$$\Delta\ddot{\eta} = -\, a_{\eta\eta}\Delta s + a_{\eta\delta}(\Delta I + J\,),\;\; L^{\circ}\,(\Delta\dot{I} + \dot{J}\,) + R\Delta I = \Delta V + c_s'\,\Delta\dot{s},$$
$$(L^{\circ} + L^{*})\,\dot{J} + L^{\circ}\Delta\dot{I} + R^{*}J = c_s'\,\Delta\dot{s},\;\; \Delta s = -\Delta\eta + \eta_c(t). \tag{5.23}$$

Additional inductance L^*, whose effect in harmonic oscillations mode is equivalent to increase of eddy current loop inertia, is determined, considering (3.47) from the relationship

$$L^* = c_\gamma \left(\frac{\pi}{2\Omega}\right)^{1/2}. \tag{5.24}$$

After dividing the second equation of (5.23) by R and the third one by R^* we obtain:

$$\Delta\ddot{\eta} = -a_{\eta\eta}\Delta s + a_{\eta\delta}(\Delta I + J), \quad T(\Delta\dot{I} + \dot{J}) + \Delta I = \frac{\Delta V}{R} + a'_{\eta\eta}\Delta\dot{s}, \tag{5.25}$$

$$T_{J_1}\dot{J} + T_J\Delta\dot{I} + J = c_s'\Delta\dot{s}, \quad \Delta s = -\Delta\eta + \eta_c(t),$$

where

$$T_{J_1} = T_J + \frac{1}{\Omega}.$$

When LECF core is used (laminated core, special electrical-sheet steels), skin effect is poorly pronounced, and as it was already mentioned, mathematical Model (5.17) is valid. In this case we can ensure the necessary stability margins, frequency properties, and quality of transient processes of the closed-loop system by the following simple control law:

$$\frac{\Delta V}{R} = a_0\Delta s_s + a_1\Delta\dot{s}_s - K_I\Delta I_s, \quad K_I > 0, \tag{5.26}$$

where Δs_s and ΔI_s are differences of air gap sensor and current transducer indications s_s and I_s, respectively, and nominal values of those quantities s° and I°.

Let us use this simple control law to estimate the influence of eddy currents on dynamic properties of the closed-loop system designed ignoring them. At the same time let us try to find the best of the considered approximations for the set of integrodifferential equations by a set of differential equations.

Consider, for instance, a modification of control Law (5.26), which provides three repeated real roots $p_i = -100\ \mathrm{s}^{-1}$ (i=1, 2, 3) for closed-loop System (5.17) with initial Data (5.13). It has the following coefficient values: $a_0 = 273\ 000\ \mathrm{A\cdot m^{-1}}$; $a_1 = 3500\ \mathrm{A\cdot m^{-1}\cdot s}$; $K_I = 74$. Transformation of Equations (5.12) to the form of (5.22) and (5.25) is performed by harmonic balance method at frequency $\Omega = |p_i| = 100\ \mathrm{s}^{-1}$. For four modifications of the mathematical model (namely, (5.12), (5.17), (5.22), and (5.25)) with control Law (5.26) the following closed-loop system characteristics were calculated:

1. Amplitude response of transfer function $W_{\eta\eta c}(p) = \eta(p)/\eta_c(p)$. The results are presented in Fig. 5.5, where different line types correspond to different mathematical model modifications.

2. Stability domains in the plane of deviations from nominal values of T and τ_J, i.e.

$$\Delta T = \frac{T - T^{\circ}}{T^{\circ}}, \quad \Delta \tau_J = \frac{\tau_J - \tau_J^{\circ}}{\tau_J^{\circ}},$$

where T° and τ_J° are the nominal (calculated) values. Those parameters were chosen for control because they are the least stable. Stability domains shown in Fig. 5.6 were calculated based on stability Criterion (3.20) presented in Chapter 3.

3. Real frequency response $P_{\eta\eta c}(\omega)$ of transfer function $W_{\eta\eta c}(p)$ presented in Fig. 5.7.

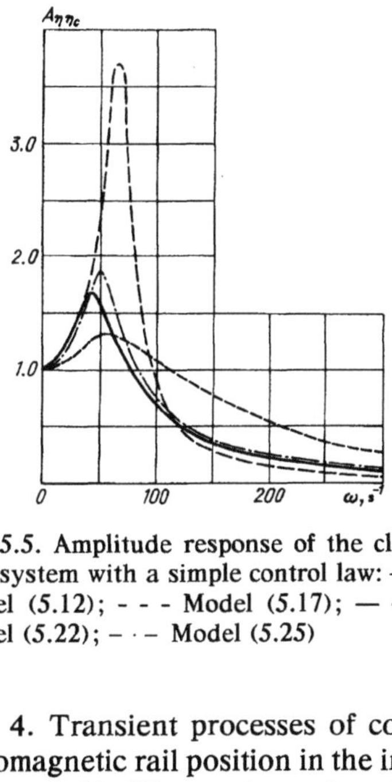

Fig. 5.5. Amplitude response of the closed-loop system with a simple control law: —— Model (5.12); - - - Model (5.17); — — — Model (5.22); – · – Model (5.25)

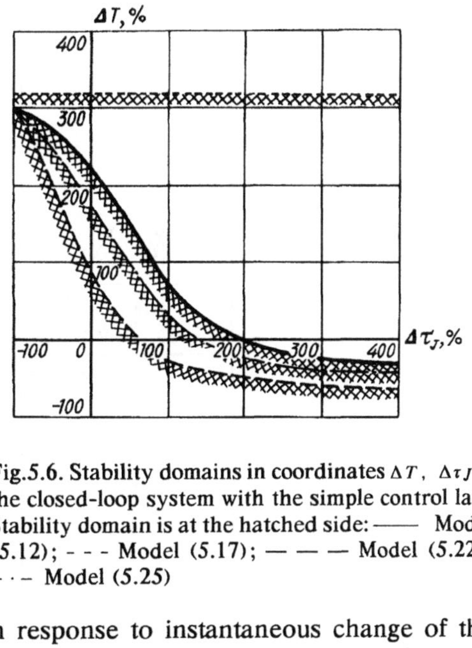

Fig.5.6. Stability domains in coordinates ΔT, $\Delta\tau_J$ of the closed-loop system with the simple control law. Stability domain is at the hatched side: —— Model (5.12); - - - Model (5.17); — — — Model (5.22); – · – Model (5.25)

4. Transient processes of control in response to instantaneous change of the ferromagnetic rail position in the inertial space, described by Heaviside unity function $\eta_c(t) = \delta_1(t)$. We construct them based on real frequency response characteristics, taking advantage of the known formula (see V.A.Besekersky and G.P.Popov [8])

$$\eta(t) = \frac{2}{\pi} \int_0^{\infty} \frac{P_{\eta\eta c}(\omega) \sin \omega t}{\omega} d\omega, \tag{5.27}$$

that lets us avoid direct numerical solution of integrodifferential equations. Results are presented in Fig. 5.8.

With the help of unit-step function response obtained according to Formula (5.27) and using Duhamel integral we can solve the initial set of Equations (5.4) for arbitrary disturbance composition. But a more advantageous way is to apply the efficient universal algorithm for numerical solution of integrodifferential equations presented in Chapter 3. Its efficiency is illustrated in Chapters 6 and 7 by the examples of particular systems modelling.

Joint analysis of the obtained results brings us to the following conclusions:

1. Comparison of mathematical model modifications (5.12) and (5.17) in frequency response, stability domains, and transient processes convinces that, when current transducer indications are included in the set of measurements, eddy currents effect in HECF core is great enough to be taken into account while analyzing and synthesizing the closed-loop system. Neglecting that effect can cause stability margins exaggeration, that is very dangerous, since for significant deviation of real parameter

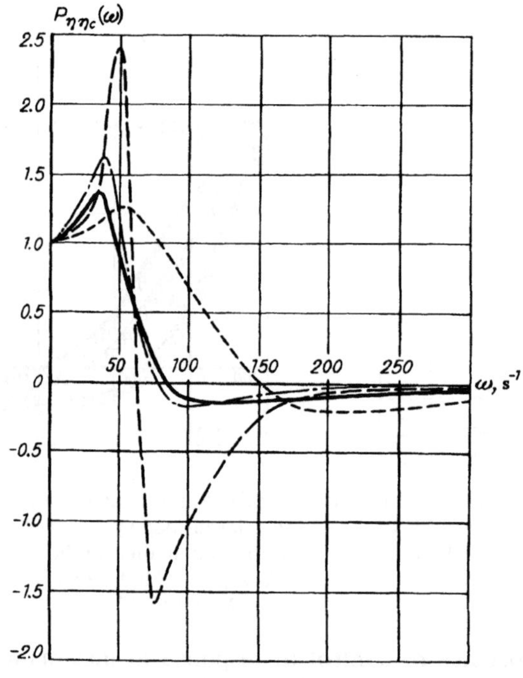

Fig. 5.7. Real frequency response of the closed-loop system with the simple control law: —— Model (5.12); - - - Model (5.17); — — — Model (5.22); – · – Model (5.25)

values from the calculated ones an unforeseen loss of system stability or major distortion of dynamic properties can take place. Hence, it is necessary to take into consideration eddy currents effect not only for accurate identification of the mathematical model parameters, but also for reliable synthesis of the control law.

2. Comparison of Systems (5.12) and (5.25) shows that rigorous consideration of dissipative R^* and inertial L^* components of the integral by the harmonic balance method (rigorous averaging of the integral) lets us obtain the mathematical model of controlled electromagnet taking into account unsteady skin effect. The model has the

form of a set of ordinary differential equations. It fairly well describes closed-loop system dynamics in the operating frequency range and can be put in the foundation of control law synthesis. Note, that good matching of real frequency response characteristics being the kernel of Integral (5.27) ensures practical coincidence of transient processes calculated based on mathematical Models (5.12) and (5.25), in spite of the different forms of solution. For System (5.12) solution has the form of (3.22) (see Chapter 3), and for System (5.25) it has the following form

$$\eta(t) = \sum_{i=1}^{n} C_i \exp(p_i t),$$

where p_i are the roots of the closed-loop system characteristic equation; n is the order of the system.

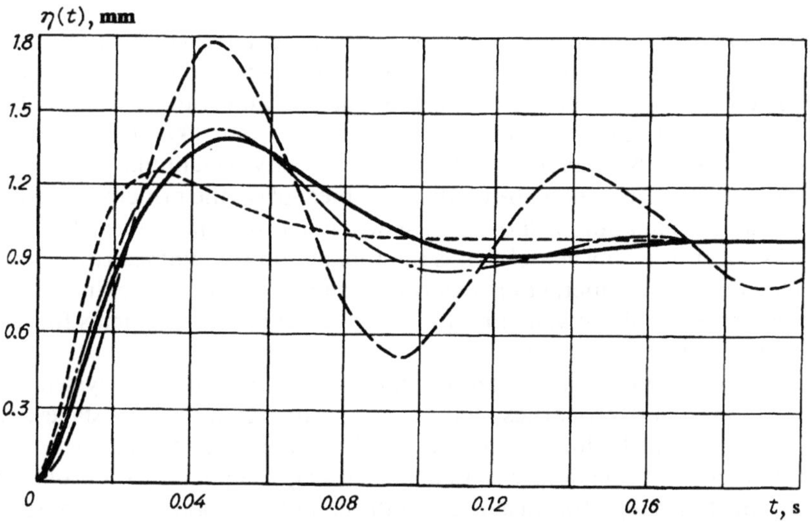

Fig. 5.8. Transient processes in the closed-loop system with the simple control law: —— Model (5.12); - - - Model (5.17); — — — Model (5.22); – · – Model (5.25)

3. Comparison of calculation results based on mathematical Models (5.12) and (5.22) shows that the mathematical model obtained by the harmonic balance method ignoring inertial term L^* produces a picture much worse than the real one. That can cause unreasonably high stability margins in real systems. It follows that traditional modelling of skin effect in HECF core by a short-circuit loop is not accurate enough for closed-loop systems analysis and synthesis problems solution. Drawing on the analogy with dynamics of stabilized objects containing cavities partially filled with LV liquid, we can expect similar results in their dynamic properties determined based on mathematical models for the controlled object of various completeness degree. That is the subject of the next chapter.

Chapter 6

Some Dynamics Problems for a Spacecraft with Compartments Partially Filled with Liquid

Thus, master-artists are considered wither not in virtue of their ability to do something, but in virtue of having the theory and knowing the causes.

Aristotle. *Metaphysics**

'Ability to do something' is however a necessary feature; without it it is impossible to solve fairly complicated problems of SC dynamics discussed in the present chapter. Those problems are at the junction of hydrodynamics and solid dynamics. This very difficulty, however, brings forward 'having the theory' and 'knowing the causes' of specific dynamic effects occurring in dynamics of a body containing vortex LV liquid.

The mathematical model of the controlled object plays the principal role in correct description and explanation of those effects when investigating closed-loop 'object – controller' system. It is this model that is the foundation for synthesizing stabilization system allowing for counteracting external disturbances and keeping generalized coordinates and generalized velocities of the object within certain limits for the entire interval of the object's movement.

Because of SC complexity as an object of control, stabilization system synthesis and closed-loop system properties analysis are performed in several stages, as a rule (see V.D.Arens, S.M.Fedorov, and M.S.Khitrik [4]). A mathematical model of different structure and complexity is used at each stage, depending on problems being solved. At the first step, for instance, controllability margins, static amplification coefficient and time constant of differentiation are determined. They are needed for control in response to specified external disturbances. At this stage the most simplified mathematical model is used, representing the object of control by a solid, ignoring mobility of liquid in compartments and elastic properties of the body.

After the initial approximation of stabilization system, we proceed to study the question of its interaction with additional degrees of freedom caused by liquid mobility and elastic deformations of the body. At this stage designers usually seek to afford stability 'in the small' (in the sense of Liapunov). To achieve this goal they introduce correcting filters and even additional sensors into the stabilization system, or make

---

* Aristotle's Metaphysics. Transl. with comment. and glossary by Hippocrates G. Apostle. Bloomington-London. Indiana University Press, 1966, p. 13.

construction changes in the object of stabilization (introduce special damping devices, change the object's centre of gravity position, *etc.*).

In case the problem cannot be solved within the scope of allowable design solutions, they seek for stability 'in the great', i.e. steady limit cycles with allowable oscillation amplitudes. And the concluding stage is perturbed motion analysis within the scope of the most complete mathematical model of the object, closed by the synthesized stabilization system, taking into account possible straggling of parameters values and calculated perturbation forces.

The reader can get acquainted with the methods of synthesizing and analyzing stabilization systems in specialized literature devoted to that problem, e. g. V. D. Arens, S. M. Fedorov, M. S. Khitrick [4] or I. M. Sidorov, L. Ye. Goncharova, V. G. Lebedev [90]. The present chapter considers only aspects of the complete problem solution, associated with vortex motions of LV liquid. Mathematical models used are of various degrees of accuracy in describing complex physical phenomena of liquid interaction with damping devices.

6.1. Stabilizability and Dynamic Stability of Spacecrafts Having Compartments Partially Filled with Liquid

Let us consider possible mechanisms of SC instability development in the object-controller system described by the set of Equations (2.120). At the first stage we exclude vortex effects from consideration and ignore the influence of SC centre of mass stabilization system, i.e. assume $\Im_2(\zeta_s) \equiv 0$. The last assumption is based on the fact, that characteristic frequencies of side motion are, as a rule, an order lower than those for angular motion. Besides that, suppose LV liquid energy dissipation can be described by the model of viscous damping proportional to generalized velocity. With these assumptions we obtain from (2.120)

$$
\begin{gathered}
\ddot{\zeta} + a_{\zeta\psi}\,\psi + \sum_{n=1}^{N} a_{\zeta s_n}\,\ddot{s}_n + a_{\zeta\delta}\,\delta_\psi = 0, \\
\ddot{\psi} + \sum_{n=1}^{N} \left(a_{\psi s_n}\,\ddot{s}_n + a'_{\psi s_n}\, s_n\right) + a_{\psi\delta}\,\delta_\psi = 0, \qquad (6.1) \\
\ddot{s}_n + \varepsilon_n \dot{s}_n + \omega_n^2\, s_n + a_{s_n\zeta}\,\ddot{\zeta} + a_{s_n\psi}\,\ddot{\psi} + a'_{s_n\psi}\,\psi = 0, \\
\delta_\psi = \Im_1(\psi), \quad n = 1, 2, \ldots, N,
\end{gathered}
$$

where ε_n are the coefficients of dissipative forces.

Note that in (6.1) coordinates ζ are related to the other coordinates via their second derivative $\ddot{\zeta}$. Calculating $\ddot{\zeta}$ from the first equation and eliminating it from N last Equations (6.1) we obtain a set of $N + 1$ equations describing angular motion of the object. Let us write it down in matrix form

$$
\mathbf{A}\ddot{\mathbf{x}} + \mathbf{B}\dot{\mathbf{x}} + \mathbf{C}\mathbf{x} + \mathbf{D}\delta_\psi = 0, \quad \delta_\psi = \Im_1(\psi), \qquad (6.2)
$$

where A, B, C are matrices of $(N+1)\times(N+1)$ dimensions representing inertial, dissipative, and positional forces, respectively; D is a column vector of coefficients preceding the control coordinate; x is an $(N+1)$ – dimensional vector of generalized coordinates $\{\psi, s_n\}$.

Suppose characteristic equation

$$\det(\mathrm{A}\lambda + \mathrm{C}) = 0 \tag{6.3}$$

has simple real negative roots; then, as it is well known (see, e.g. F.R.Gantmacher [23]), there exists a transformation

$$\mathrm{x} = \mathrm{G}\mathrm{y}, \quad \mathrm{G} = \|g_{ij}\|, \quad i, j = 1, 2, \ldots, N+1, \quad g_{ii} = 1, \tag{6.4}$$

that reduces matrices A and C to diagonal form simultaneously. Perform in (6.2) change of variables according to (6.4) and neglect nondiagonal components in the obtained dissipative force matrix. It results in

$$\ddot{y}_i + \varepsilon_{id}\,\dot{y}_i + \omega_{id}^2\, y_i + b_{i\delta}\,\delta_\psi = 0, \quad i = 1, 2, \ldots, N+1,$$

$$\delta_\psi = \Im_1(\psi), \; \psi = \sum_{j=1}^{N+1} g_{1j}\, y_j. \tag{6.5}$$

Here ω_i denotes natural frequencies of the open-loop system. We assume further, that $\Im_1(p)$ operator is a slowly varying function of its argument in the sense, that in the vicinity of open-loop system natural frequencies it can be substituted by amplitude and phase response

$$\Im_1(p) \approx A(\omega) e^{i\psi(\omega)}. \tag{6.6}$$

Characteristic equation of System (6.5) can gain the form

$$\begin{vmatrix} \omega_1^2 + \varepsilon_{1d}\, p + p^2 + b_{1\delta}\,\Im_1(p) & -g_{12}(p^2 + \omega_{1d}^2) & \ldots & -g_{1N+1}(p^2 + \omega_{1d}^2) \\ b_{2\delta}\,\Im_1(p) & \omega_{2d}^2 + \varepsilon_{2d}\, p + p^2 & 0 & 0 \\ \cdots & \cdots & \cdots & \cdots \\ b_{N+1\delta}\,\Im_1(p) & 0 & \ldots & \omega_{N+1d}^2 + \varepsilon_{N+1d}\, p + p^2 \end{vmatrix} = 0. \tag{6.7}$$

To calculate root p_n corresponding to n th equation of System (6.5) it is enough in the first-order approximation to consider the characteristic determinant composed of the first and n th equations of System (6.5):

$$\begin{vmatrix} \omega_1^2 + \varepsilon_{1d} p + p^2 + b_{1\delta}\,\Im_1(p) & -g_{1n}(p^2 + \omega_{1d}^2) \\ b_{n\delta}\,\Im_1(p) & \omega_{nd}^2 + \varepsilon_{nd}\, p + p^2 \end{vmatrix} = 0. \tag{6.8}$$

Let us write down root p_n corresponding to oscillations of n th oscillator in the form

$$p_n = i\omega_{nd} + \Delta p_n, \quad \Delta p_n = \alpha_n + i\Delta\omega_n. \tag{6.9}$$

Let us assume, that closing of the object control loop by the stabilization system causes a small shift of the natural frequency and that real part of the root is small in comparison with the natural frequency of oscillations. Assume besides, that the frequency of the object's oscillations as a solid body ω_1 is small compared to ω_n, i.e. the following conditions are satisfied:

$$\left|\frac{\alpha_n}{\omega_{nd}}\right| << 1, \quad \left|\frac{\Delta\omega_n}{\omega_{nd}}\right| << 1, \quad \left|\frac{\omega_{1d}}{\omega_{nd}}\right| << 1. \tag{6.10}$$

Then substituting (6.9) into (6.8) and using (6.6) we obtain, up to the introduced small quantities, expressions for the real and imaginary parts of the root corresponding to n th associated oscillator:

$$\alpha_n = -\frac{\varepsilon_{nd}}{2} - \frac{g_{1n}\, b_{n\delta}\, A\,(\omega_{nd})\, \sin\varphi(\omega_{nd})}{2\omega_{nd}},$$

$$\Delta\omega_n = \frac{g_{1n} b_{n\delta}\, A\,(\omega_{nd})\, \cos\varphi(\omega_{nd})}{2\omega_{nd}}. \tag{6.11}$$

It follows from Formula (6.11) for the real part of the root for the characteristic equation, that stabilization system influence on the object's oscillations at n th oscillator frequency is specified by expression

$$\alpha_{n\delta} = -\frac{g_{1n} b_{n\delta}\, A\,(\omega_{nd})\, \sin\varphi(\omega_{nd})}{2\omega_{nd}}. \tag{6.12}$$

If it is possible to choose phase response $\varphi(\omega)$ of stabilization system to fulfil at frequency ω_n condition

$$g_{1n}\, b_{n\delta}\, \sin\varphi(\omega_{nd}) > 0, \tag{6.13}$$

control effect in this case appears a stabilizing one. Condition (6.13) is usually called *phase stabilization condition* (see [63]).

If amplitude response $A\,(\omega)$value at ω_n frequency is such that condition

$$\left|\frac{g_{1n}\, b_{n\delta}\, A\,(\omega_{nd})}{\omega_{nd}}\right| < \varepsilon_{nd}, \tag{6.14}$$

is fulfilled, the real part of α_n root is negative irrelevant of $\sin\varphi\,(\omega_{nd})$ sign. Condition (6.14) is called *amplitude stabilization condition.*

An object, for which Condition (6.13) is fulfilled at frequencies of all the associated oscillators for one sign of $\sin\varphi\,(\omega)$ (positive or negative), is called *stabilizable*. The stabilizability feature is determined by parameters related to the object of control only, hence it can be expressed via construction and configuration characteristics of the object.

Consider a simple example of a SC with a rigid body and two compartments containing liquid. The set of equations describing perturbed motion of such an object according to 'floating cover' scheme can be obtained from (2.91), when components describing control action are extracted from P_z and M_{Gy}. Besides that, let us change from point O to metacentre, neglect vortex motions of liquid at this stage of research, and divide each equation by the corresponding diagonal element of the kinetic energy matrix. Resulting is the following mathematical model being a particular case of (2.120):

$$\ddot{\zeta} + a_{\zeta\psi}\,\psi + \sum_{n=1}^{2} a_{\zeta s_n}\,\ddot{s}_n = -\,a_{\zeta\delta}\,\delta_\psi\,, \qquad \ddot{\psi} + \sum_{n=1}^{2} a_{\psi s_n}\,\ddot{s}_n = -\,a_{\psi\delta}\,\delta_\psi\,,$$

$$\ddot{s}_n + \omega_n^2\, s_n + a_{s_n\zeta}\,\ddot{\zeta} + a_{s_n\psi}\,\ddot{\psi} = 0, \quad n = 1, 2, \quad \delta_\psi = \Im_1(\psi). \tag{6.15}$$

The characteristic equation has the form

$$\Phi(p^2) + \Im_1(p)\,\Phi_1(p^2) = 0,$$

$$\Phi(p^2) = a_0\,p^4 + a_1\,p^2 + a_2, \quad \Phi_1(p^2) = b_0\,p^4 + b_1\,p^2 + b_2. \tag{6.16}$$

For coefficients a_i, b_i we obtain the following expressions:

$$a_0 = 1 - \nu_1 - \nu_2, \quad a_1 = \omega_1^2\,(2 - \nu_1 - \nu_3), \quad a_2 = \omega_1^2\,(1 + \nu_3), \tag{6.17}$$

$$b_0 = -\,a_{\psi\delta}\,(-1 + \nu_4), \quad b_1 = \omega_2^2\,a_{\psi\delta}\,(\nu_4 - 2), \quad b_2 = \omega_2^2\,a_{\psi\delta},$$

where, in turn,

$$\nu_1 = \sum_{n=1}^{2} (a_{s_n\psi}\,a_{\psi s_n} + a_{s_n\zeta}\,a_{\zeta s_n}),$$

$$\nu_2 = \sum_{n=1}^{2} (a_{s_n\psi}\,a_{s_2\zeta} - a_{s_2\psi}\,a_{s_n\zeta})\,(a_{\psi s_n}\,a_{\zeta s_n} - a_{\psi s_2}\,a_{\zeta s_n}),$$

$$\nu_3 = \frac{a_{\zeta\psi}}{\omega_1^2}\sum_{n=1}^{2} a_{\psi s_n}\,a_{s_n\zeta}, \quad \nu_4 = \sum_{n=1}^{2} a_{s_n\zeta}\,a_{\zeta s_n} - \frac{a_{\zeta\psi}}{a_{\psi\delta}}\,a_{s_n\zeta}\,a_{\psi s_n}.$$

Represent the root of characteristic Equation (6.16) in the form

$$p_n = \alpha_n + i\,(\omega_n + \Delta\,\omega_n),$$

then, if Conditions (6.10) are obeyed, we obtain

$$\alpha_n = -\,\frac{\Phi'(-\omega_n^2)}{\Phi_1(-\omega_n^2)}\,A(\omega_n)\,\sin\varphi\,(\omega_n). \tag{6.18}$$

It follows from (6.18), that the object is stabilizable if one of inequalities

$$\Phi'(-\omega_n^2)\,\Phi_1(-\omega_n^2) > 0, \tag{6.19a}$$

$$\Phi'(-\omega_n^2)\,\Phi_1(-\omega_n^2) < 0 \tag{6.19b}$$

is satisfied for $n = 1, 2$ simultaneously. Conditions (6.19) are equivalent to the condition of alternating zeroes for functions $\Phi(-\omega^2)$ and $\Phi_1(-\omega^2)$. For alternating of zeroes for polynomials $\Phi(-\omega^2)$ and $\Phi_1(-\omega^2)$ it is necessary and sufficient (see F.R.Gantmacher [23]) for polynomial

$$F(\lambda) = \Phi(\lambda^2) + \varepsilon\,\lambda\Phi_1(\lambda^2) \tag{6.20}$$

to be a Hurwitz polynomial for $\varepsilon = 1$ or $\varepsilon = -1$.

Write down Hurwitz determinants for Polynomial (6.20)

$$\Delta_1 = a_0 > 0, \quad \varepsilon\Delta_2 = \varepsilon\begin{vmatrix} a_0 & a_1 \\ b_0 & b_1 \end{vmatrix} > 0, \tag{6.21}$$

$$\varepsilon\Delta_3 = \varepsilon\begin{vmatrix} a_0 & a_1 & a_2 \\ b_0 & b_1 & b_2 \\ 0 & a_0 & a_1 \end{vmatrix} > 0, \quad \varepsilon^2\Delta_4 = \varepsilon^2\begin{vmatrix} b_0 & b_1 & b_2 & 0 \\ a_0 & a_1 & a_2 & 0 \\ 0 & b_0 & b_1 & b_2 \\ 0 & a_0 & a_1 & a_2 \end{vmatrix} > 0.$$

Thus, the criterion of stabilizability for an object with characteristic equation of (6.16) is positive Hurwitz determinants for $\varepsilon = 1$ or $\varepsilon = -1$. To obey at least one group of Inequalities (6.21) it is sufficient to obey condition

$$\Delta_4 > 0, \tag{6.21a}$$

which appears to be the criterion of stabilizability for the considered class of objects.

Parallel to Condition (6.21a) we should consider the condition of open-loop system characteristic equation roots being real (with respect to frequency square). This characteristic equation is as follows:

$$\Phi(-\omega^2) = a_0\,\omega^4 + a_1\,\omega^2 + a_2 = 0.$$

The condition has the form

$$a_1^2 - 4a_0\,a_2 > 0. \tag{6.22}$$

Inequality (6.22) violation means severe instability of flutter type, i.e. intrinsic dynamic instability.

Ample literature is devoted to stabilizability and intrinsic dynamic instability criteria formulation and expressing via construction and configuration characteristics for the object of control (see, e.g. [63, 76, 86]). Here we cite only the result obtained by I.M.Sidorov and I.P.Korotaeva [89], who formulated criterion (6.21) for a SC with cylindric tanks of equal diameters. The formulation is carried out in the plane of two dimensionless parameters z_n $(n = 1, 2)$ (Fig. 6.1) characterizing distances from the metacentre to free surfaces of liquid, related to radius of inertia:

$$l = [(J^\circ + J)/(m^\circ + m)]^{1/2}.$$

Tank number is counted from the tail and parameters $z_n$in the opposite direction ($z_n > 0$ for the lower tank).

Non-stabilizability domain is shown by single hatching in Fig. 6.1. The domains of intrinsic dynamic instability are shown by cross-hatching. When liquid level in cavities varies, the corresponding point in the plane (z_1, z_2) can move from the domain of stabilizability to the domain of non-stabilizability and back. Analysis of stabilizability criterion makes it possible to purposely change (within the scope of technical possibilities) the object's construction parameters in order to improve dynamic properties of the closed-loop system.

Let us consider some ways of ensuring closed-loop system stability, resulting from the analysis of stabilizability and intrinsic dynamic instability criteria and the approximate Formula (6.11) for the real part of the root corresponding to associated oscillator's oscillations.

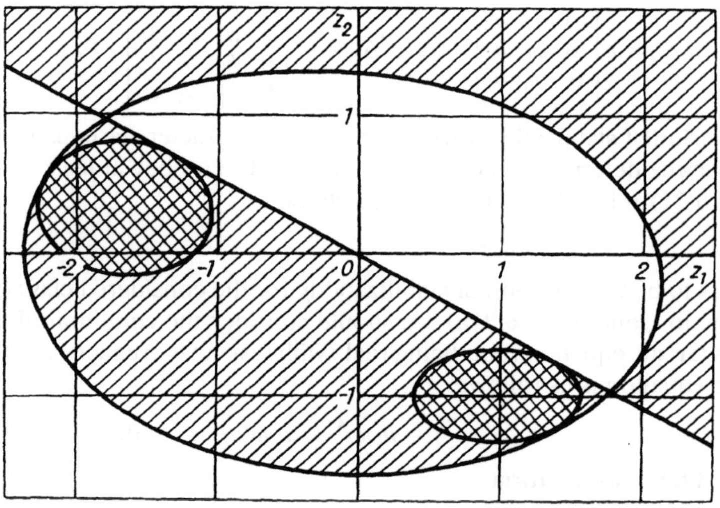

Fig. 6.1. Domains of stability (single hatching) and of intrinsic dynamic instability (cross hatching) drawn in coordinates z_1, z_2

If Inequality (6.22) is violated, i.e. the SC is characterized by intrinsic dynamic instability as an object of control, it is impossible to choose such an operator $\Im(\psi)$ within the scope of stabilization system of the considered structure for the closed-loop system to be stable. In this case it is necessary to change inertial and centring characteristics for the object of control so, as Condition (6.22) to be obeyed throughout the time interval of system functioning. Another way is to install special damping devices in compartments containing liquid in order to dissipate the energy of liquid oscillations (see G.N.Mikishev [62]).

In the case when the object of control is characterized by intrinsic dynamic stability, the most efficient way to ensure closed-loop system stability is phase stabilization, as it follows from Formula (6.11). Choosing phase response $\varphi(\omega)$ such, that for each frequency ω_n of the associated oscillators the second component of Formula (6.11) is positive, we can make real parts of the characteristic equation roots be located in the left half-plane, irrespective of amplitude response $A(\omega_n)$ values and damping coefficient ε_n describing energy dissipation in liquid.

For all the attractiveness of phase stabilization, it is usually practically infeasible. It is connected with the fact that eigenvalue spectrum is very dense, and therefore at close frequencies conflicting requirements are sometimes applied to phase response. For one frequency value phase lead is required, and for another one, close to that one, phase lag is required, and such conflicting requirements alternate. Besides that, it is necessary to take into account possible straggling of determined frequencies and their time variations. All those factors together are inhibitory to the implementation of phase stabilization of all the associated oscillators. Usually phase stabilization conditions are obeyed for some of natural frequencies only and not for the entire time interval of system functioning.

For associated oscillators frequencies, for which phase stabilization conditions are not satisfied, it is necessary to ensure amplitude stabilization Conditions (6.14) satisfaction. The situation here is more subtle, since the play consists in selection of relationships between the damping coefficients and amplitude response values at the frequencies of the associated oscillators. It is clear from (6.14), that in order to ensure amplitude stabilization we should tend to decrease amplification coefficient $A(\omega)$ and to increase the damping coefficient. Methods of decreasing the amplification coefficient are described in literature devoted to stabilization systems design (see e.g. V.D.Arens, S.M.Fedorov, M.S.Khitrik and I.M.Sidorov, L.Ye.Goncharova, V.G.Lebedev [4, 90]). Here it is important for us to note, that $A(\omega)$ value cannot be zero, and therefore it is bounded from below.

In order to increase LV liquid oscillations energy dissipation (the damping coefficient), special damping devices are placed into the cavities containing liquid. They can have the shape of radial ribs, when it is necessary to damp liquid oscillations over a long time interval, or the shape of coaxial rings, when excessive damping is necessary for particular levels only of the liquid free surface. The effect of those devices having sharp edges is that when liquid moves, vortices are generated at the edges. Those vortices accomplish dissipation of the energy replenished by the stabilization system. Note that energy dissipation rate depends not only on the ribs geometry, but on liquid oscillations amplitude too, or, to be exact, on the velocity of liquid motion relative to the rib. Hence, in the case, when the stabilization system is destabilizing at the associated oscillator's frequency, continuous oscillations establish in the closed-loop system. Their amplitude is determined by the balance of energy inflow and dissipation. From the considerations of stabilization system functioning accuracy, those steady-state oscillations amplitudes are to be bounded by specific values.

It is clear from above, that one of principal items in providing object – controller closed-loop system stability is an adequate description for the mechanism of LV liquid interaction with the damping device, i.e. development of an appropriate mathematical model for the phenomenon and analysis of that model structure and parameters influence on closed-loop system stability. Chapters 2 and 4 are devoted to the first question, a part of Chapter 3 to the second one, in a more detail it is discussed below.

Presence of highly developed in-tank devices causes two fundamental dynamic features of the considered objects:

1) nonlinearity of coefficients characterizing degree of in-tank devices influence;

2) additional degrees of freedom describing liquid dynamics in its vortex motion.

Let us consider in sequence the influence of those two factors on stabilized objects stability. Begin with investigation of nonlinear liquid oscillations.

6.2. Simplified Mathematical Models of Perturbed Motion for a Spacecraft Having Compartments Partially Filled with Liquid

Mathematical models of a stabilized SC with cavities partially filled with LV liquid in the form of (2.119), (2.120) (yawing plane) and (2.122), (2.123) (roll plane) are the most complete (at the level of applications to practical problems) in consideration of liquid vortex motion and nonlinear effects connected with presence of special damping devices inside fuel tanks.

The following simplified equations of perturbed motion can be obtained from Equations (2.119), (2.122) by the technique used in Chapter 2 (see Equations (2.91)) within the scope of the mathematical model most often used (see [63, 76, 86]).

$$(m^\circ + m)\ddot{\xi} + \sum_{n=1}^{N} \lambda_n \ddot{s}_n = P_z,$$

$$(J^\circ + J)\ddot{\psi} + \sum_{n=1}^{N} \left(\lambda_{Gn} \ddot{s}_n + \gamma_n \int_{-\infty}^{t} \frac{\ddot{s}_n(\tau)\, d\tau}{\sqrt{t-\tau}} + j\lambda_n s_n \right) + \beta \int_{-\infty}^{t} \frac{\ddot{\psi}(\tau)\, d\tau}{\sqrt{t-\tau}} = M_{Gy},$$

$$(6.23)$$

$$\mu_n \left(\ddot{s}_n + \omega_n^2 s_n + \beta_n \int_{-\infty}^{t} \frac{\ddot{s}_n(\tau)\, d\tau}{\sqrt{t-\tau}} \right) + \lambda_n \ddot{\xi} + \lambda_{Gn} \ddot{\psi} + j\lambda_n \psi +$$

$$+ \gamma_n \int_{-\infty}^{t} \frac{\ddot{\psi}(\tau)\, d\tau}{\sqrt{t-\tau}} = 0, \quad n = 0, 2, \ldots, N.$$

$$(I^\circ + I)\ddot{\varphi} + \beta^* \int_{-\infty}^{t} \frac{\ddot{\varphi}(\tau)\, d\tau}{\sqrt{t-\tau}} = M_{Gx}. \tag{6.24}$$

In these equations J, I are associated moments of inertia; β, β^* are coefficients characterizing ribs influence:

$$J = \sum_{n=1}^{N} J_n, \quad I = \sum_{n=1}^{N} I_n, \quad \beta = \sum_{n=1}^{N} \beta_n^\circ, \quad \beta^* = \sum_{n=1}^{N} \beta_n^*. \tag{6.25}$$

Let us rewrite Equations (6.23) and (6.24) so as to make them similar to Equations (2.120) and (2.123). That is convenient for further analysis.

$$\ddot{\zeta} + a_{\zeta\psi}\,\psi + \sum_{n=1}^{N} a_{\zeta sn}\,\ddot{s}_n + a_{\zeta\delta}\,\delta_\psi = \widetilde{P}_z(t),$$

$$\ddot{\psi} + \sum_{n=1}^{N}\left(\hat{a}_{\psi sn}\,\ddot{s}_n + a''_{\psi sn}\int_{-\infty}^{t}\frac{\ddot{s}_n(\tau)\,d\tau}{\sqrt{t-\tau}} + \hat{a}'_{\psi sn}\,s_n\right) +$$

$$+ a'_{\psi\psi}\int_{-\infty}^{t}\frac{\ddot{\psi}(\tau)\,d\tau}{\sqrt{t-\tau}} + \hat{a}_{\psi\delta}\,\delta_\psi = \hat{M}_{Gy}(t), \qquad (6.26)$$

$$\ddot{s}_n + \omega_n^2\,s_n + \beta_n\int_{-\infty}^{t}\frac{\ddot{s}_n(\tau)\,d\tau}{\sqrt{t-\tau}} + a_{sn\zeta}\,\ddot{\zeta} + a_{sn\psi}\,\ddot{\psi} +$$

$$+ a'_{sn\psi}\,\psi + a_{sn\vartheta n}\int_{-\infty}^{t}\frac{\ddot{\psi}(\tau)\,d\tau}{\sqrt{t-\tau}} = 0,$$

$$\delta_\psi = \Im_1(\psi) - \Im_2(\zeta_S), \quad \zeta_S = \zeta - (x_S - x_G)\,\psi, \qquad n = 1, 2, \ldots, N.$$

$$\ddot{\varphi} + \beta''_{\chi n}\int_{-\infty}^{t}\frac{\ddot{\varphi}(\tau)\,d\tau}{\sqrt{t-\tau}} + \hat{a}_{\varphi\delta}\,\delta_\varphi = \hat{M}_{Gx}(t), \quad \delta_\varphi = \Im_3(\varphi). \qquad (6.27)$$

Coefficients of Equations (6.26) and (6.27) are related to coefficients of Equations (2.120) and (2.123) as follows:

$$\hat{a}_{\psi sn} = \frac{a_{\psi sn}}{\kappa_\vartheta}, \qquad a''_{\psi sn} = \frac{a_{\psi\vartheta n}\,a_{\vartheta nsn}}{\kappa_\vartheta},$$

$$\hat{a}'_{\psi sn} = \frac{a'_{\psi sn}}{\kappa_\vartheta}, \qquad a'_{\psi\psi} = \frac{\sum_{n=1}^{N} a_{\psi\vartheta n}\,\beta_{\vartheta n}}{\kappa_\vartheta}, \qquad \hat{a}_{\psi\delta} = \frac{a_{\psi\delta}}{\kappa_\vartheta}, \qquad (6.28)$$

$$\beta''_{\chi n} = \frac{\sum_{n=1}^{N} a_{\varphi\chi n}\,\beta_{\chi n}}{\kappa_\chi}, \qquad \hat{a}_{\varphi\delta} = \frac{a_{\varphi\delta}}{\kappa_\chi}.$$

And the right sides are related as follows:

$$\hat{M}_{Gy}(t) = \frac{\widetilde{M}_{Gy}(t)}{\kappa_\vartheta}, \qquad \kappa_\vartheta = 1 - \sum_{n=1}^{N} a_{\psi\vartheta n},$$

$$(6.29)$$

$$\hat{M}_{Gx}(t) = \frac{\widetilde{M}_{Gx}(t)}{\kappa_\chi}, \qquad \kappa_\chi = 1 - \sum_{n=1}^{N} a_{\varphi\chi n}.$$

Mathematical Models (6.23), (6.24) or (6.26), (6.27) are applicable in the case, when the influence of effects connected with LV liquid vortex motion is faint, for objects with load-bearing units playing the damping role, or for objects having dampers, when liquid oscillation amplitudes do not exceed some threshold value. The simpler Equations (6.26), (6.27) are not used in practical computations too. It is apparently due to absence of well developed methods for analyzing and synthesizing systems described by nonlinear integrodifferential equations. One of the problems is to determine legitimacy bounds of such an approach and to show when simplified models are applicable, and when analysis and synthesis of systems should be performed based on more complete models. In practical applications based on harmonic balance method the set of integrodifferential equations is reduced to a set of ordinary differential equations (see Chapters 3, 4).

Suppose the object of consideration is the class of solutions to Equations (6.23) and (6.24), having the type of quasi-harmonic oscillations with slowly varying amplitude, i.e.

$$\psi = \psi_a \sin \omega_a t, \quad s_n = s_{na} \sin \omega_a t, \quad \varphi = \varphi_a \sin \omega^a t, \tag{6.30}$$

where ω_a and ω^a are characteristic frequencies of yawing plane and roll plane oscillations, respectively. By virtue of Formula (3.46), (3.47), ignoring inertial components of integral terms (i.e. kinetic energy of liquid vortex motion), we obtain

$$\int_{-\infty}^{t} \frac{\ddot{\psi}(\tau)\, d\tau}{\sqrt{t-\tau}} \cong \left(\frac{\pi\omega_a}{2}\right)^{1/2} \dot{\psi}(t),$$

$$\int_{-\infty}^{t} \frac{\ddot{s}_n(\tau)\, d\tau}{\sqrt{t-\tau}} \cong \left(\frac{\pi\omega_a}{2}\right)^{1/2} \dot{s}_n(t), \quad \int_{-\infty}^{t} \frac{\ddot{\varphi}(\tau)\, d\tau}{\sqrt{t-\tau}} \cong \left(\frac{\pi\omega^a}{2}\right)^{1/2} \dot{\varphi}(t). \tag{6.31}$$

Using (6.31), we transform the set of integrodifferential equations to the traditional set of ordinary differential equations. We shall compare to it the results of investigations based on more complete models:

$$(m^\circ + m)\ddot{\xi} + \sum_{n=1}^{N} \lambda_n \ddot{s}_n = P_z.$$

$$(J^\circ + J)\ddot{\psi} + \sum_{n=1}^{N} (\lambda_{Gn} \ddot{s}_n + \gamma_n' \dot{s}_n + j\lambda_n s_n) + \beta' \dot{\psi} = M_{Gy}, \tag{6.32}$$

$$\mu_n (\ddot{s}_n + \beta_n' \dot{s}_n + \omega_n^2 s_n) + \lambda_n \ddot{\xi} + \lambda_{Gn} \ddot{\psi} + j\lambda_n \psi + \gamma_n' \dot{\psi} = 0,$$
$$n = 1, 2, \ldots, N.$$

$$(I^\circ + I)\ddot{\varphi} + \beta^{*\prime} \dot{\varphi} = M_{Gx}, \tag{6.33}$$

where

$$\beta' = \beta\left(\frac{\pi\omega_a}{2}\right)^{1/2}, \quad \gamma_n' = \gamma_n\left(\frac{\pi\omega_a}{2}\right)^{1/2}, \quad \beta_n' = \beta_n\left(\frac{\pi\omega_a}{2}\right)^{1/2}, \quad \beta^{*\prime} = \beta^*\left(\frac{\pi\omega^a}{2}\right)^{1/2}. \tag{6.34}$$

Taking into account the controller and dividing the left and right sides of each equation by the coefficients at the second derivative of the corresponding generalized coordinate, we obtain:

$$\ddot{\zeta} + a_{\zeta\psi}\,\psi + \sum_{n=1}^{N} a_{\zeta s_n}\,\ddot{s}_n + a_{\zeta\delta}\,\delta_\psi = \widetilde{P}_z(t),$$

$$\ddot{\psi} + \sum_{n=1}^{N} (\hat{a}_{\psi s_n}\,\ddot{s}_n + \gamma_n'\,\dot{s}_n + \hat{a}_{\psi s_n}'\,s_n) + \beta'\dot{\psi} + \hat{a}_{\psi\delta}\,\delta_\psi = \widehat{M}_{Gy}(t), \qquad (6.35)$$

$$\ddot{s}_n + \beta_n'\dot{s}_n + \omega_n^2\,s_n + a_{s_n\zeta}\,\ddot{\zeta} + a_{s_n\psi}\,\ddot{\psi} + a_{s_n\psi}'\,\psi + \hat{\gamma}_n''\dot{\psi} = 0,$$

$$\delta_\psi = \Im_1(\psi) - \Im_2(\zeta_S),\quad \zeta_S = \zeta - (x_S - x_G)\,\psi, \qquad n = 1, 2, \ldots, N,$$

$$\ddot{\varphi} + \hat{\beta}^{*\prime}\dot{\varphi} + \hat{a}_{\varphi\delta}\,\delta_\varphi = \widehat{M}_{Gx}(t),\quad \delta_\varphi = \Im_3(\varphi), \qquad (6.36)$$

where

$$\hat{\gamma}_n' = \frac{\gamma_n'}{\overset{\circ}{J} + J},\quad \hat{\beta}' = \frac{\beta'}{\overset{\circ}{J} + J},\quad \hat{\beta}^{*\prime} = \frac{\beta^{*\prime}}{\overset{\circ}{I} + I},\quad \hat{\gamma}_n'' = \frac{\gamma_n'}{\mu_n}. \qquad (6.37)$$

In the equations presented above, the coefficients, which characterize the degree of ribs influence considering the hypothesis introduced in Chapter 2 of wave motions prevailing over vortex ones ($l\dot{\vartheta}_n << \dot{s}_n$), are nonlinear functions of generalized velocities describing wave motion of liquid:

$$\beta' = \sum_{n=1}^{N} \hat{\overset{\circ}{\beta}}_n \sqrt{|\dot{\tilde{s}}_n|} \left(\frac{\pi\omega_a}{2}\right)^{1/2},\quad \gamma_n' = \hat{\gamma}_n \sqrt{|\dot{\tilde{s}}_n|} \left(\frac{\pi\omega_a}{2}\right)^{1/2},$$

$$\beta_n' = \hat{\beta}_n \sqrt{|\dot{\tilde{s}}_n|} \left(\frac{\pi\omega_a}{2}\right)^{1/2},\quad \beta^{*\prime} = \hat{\beta}^* \sqrt{|\dot{\varphi}|} \left(\frac{\pi\omega^a}{2}\right)^{1/2},\quad \dot{\tilde{s}}_n = \frac{\dot{s}_n}{l}. \qquad (6.38)$$

Here l is the characteristic size.

Coefficients at the first derivatives in Equations (6.32), (6.33), (6.35), (6.36) can be considered as damping coefficients characterizing energy dissipation. As it is clear from (6.38), these coefficients continuously vary in nonlinear manner with the variation of the appropriate coordinate. For the considered class of solutions having quasi-harmonic oscillation type we can, however, introduce an equivalent linear damping, being the function of liquid oscillations amplitude. To do this let us, following V.M.Rogovoi and S.V.Cheremnykh [86], introduce nonlinear damping force for harmonic oscillations

$$F_D = K_D\sqrt{|\dot{s}|}\,\dot{s}\,|_{s=s_a\sin\omega_a t} = K_D\,\omega_a\sqrt{\omega_a s_a}\,\sqrt{s_a}\,\sqrt{|\cos\omega_a t|}\cos\omega_a t \qquad (6.39)$$

and an equivalent linear damping force:

$$F_D' = \kappa K_D\sqrt{\dot{s}_a}\,\dot{s}\,|_{s=s_a\sin\omega_a t} = \kappa K_D\,\omega_a\sqrt{\omega_a s_a}\cos\omega_a t. \qquad (6.40)$$

Equivalence is in the sense of equal works for virtual displacement over the characteristic period of oscillations. Let us equate the works of those forces for the characteristic period $T = 2\pi/\omega_a$

$$\int_0^T F_D\, ds = \int_0^T F_D\, \dot{s}\, dt = \int_0^T F_D'\, ds = \int_0^T F_D'\, \dot{s}'\, dt. \tag{6.41}$$

We obtain (see [86])

$$\kappa = \frac{k_s^*}{k_s} = \frac{8.08}{5\sqrt{\pi}} \approx 0.91, \tag{6.42}$$

where k_s and k_s^* are coefficients specified by Formula (2.49). It follows, for instance that linear damping coefficient ε_n equivalent to β_n' is specified as follows:

$$\varepsilon_n = \kappa\,\widehat{\beta}_n \sqrt{|\dot{\tilde{s}}_{na}|} \left(\frac{\pi\omega_a}{2}\right)^{1/2}. \tag{6.43}$$

All the other dissipation coefficients are calculated similarly.

In Formula (6.43) the coefficient connecting quantities ε_n and $\dot{\tilde{s}}_{na}$ in harmonic oscillations mode coincides with that obtained in Chapter 4 $\kappa_\varepsilon = \kappa\sqrt{\pi/2} = 1.147$.

6.3. Self-Sustained Oscillations in the Closed-Loop System 'Spacecraft – Liquid – Controller'

As it was already mentioned above, in certain cases we do not manage to provide closed-loop system stability 'in the small' at the frequencies of liquid wave motions with damping of the order of LV liquid linear damping. Those cases are: violation of phase stabilization conditions and great dynamic amplification coefficient of the controller, intrinsic dynamic instability of the SC. Under such circumstances liquid oscillations dampers are usually applied. Especially complicated is the problem of amplitude stabilization of an object characterized by intrinsic dynamic instability. For normal amplitudes of liquid oscillations the object of control becomes essentially nonlinear, and the only possible stable mode is the mode of self-sustained oscillations. The stabilization problem is reduced to providing steady self-sustained oscillations with amplitudes not exceeding the maximum acceptable ones. It is solved by rational selection and distribution of damping devices in fuel compartments.

Let us consider, following [34], specific dynamic features of stabilized SC with compartments (tanks) partially filled with liquid, connected with nonlinear character of its wave motions damping. In order to clarify the system features caused by this factor particularly, let us neglect additional degrees of freedom corresponding to vortex motions of liquid. We shall confine first to the class of motions in the yaw plane, described by mathematical Model (6.26). We shall make another simplification, namely ignore all the elements in the matrix of coefficients at the integrals, which elements play a secondary role, except for the main coefficients β_n. Then the equations of perturbed motion for the considered object with angular and lateral stabilization systems are written as follows:

$$\begin{gathered} \ddot{\zeta} + a_{\zeta\psi}\,\psi + \sum_{n=1}^{N} a_{\zeta s n}\,\ddot{s}_n + a_{\zeta\delta}\,\delta = \widetilde{P}_z(t), \\ \ddot{\psi} + \sum_{n=1}^{N} (\hat{a}_{\psi s n}\ddot{s}_n + \hat{a}'_{\psi s n}\,s_n) + \hat{a}_{\psi\delta}\,\delta = \widehat{M}_{Gy}\,(t), \\ \ddot{s}_n + \beta_n \int_{-\infty}^{t} \frac{\ddot{s}_n(\tau)\,d\tau}{\sqrt{t-\tau}} + \omega_n^2 s_n + a_{sn\zeta}\,\ddot{\zeta} + a_{sn\psi}\,\ddot{\psi} + a'_{sn\psi}\,\psi = 0, \\ \delta = \Im_\psi\,(\psi) - \Im_\zeta\,(\zeta), \qquad n = 1, 2, \ldots, N, \end{gathered} \tag{6.44}$$

where, according to denominations accepted in (6.26),

$$\Im_\psi = \Im_1 + \Im_2\,(x_S - x_G), \qquad \Im_\zeta = \Im_2. \tag{6.45}$$

The coefficients at the integrals are, as in (6.38), nonlinear functions in relative velocity of liquid oscillations $\dot{\tilde{s}}_n = \dot{s}_n/r_0$, where r_0 is the characteristic size (usually it is the maximum radius of liquid free surface). Thus the object of control is described by a set of nonlinear integrodifferential equations.

In practice usually the assumption of (6.30) is used, that enables reduction of (6.44) to a set of ordinary differential equations with some equivalent damping coefficients ε_n specified by Formula (6.43), in which it is additionally assumed that $\omega_a = \omega_n$:

$$\begin{gathered} \ddot{\zeta} + a_{\zeta\psi}\,\psi + \sum_{n=1}^{N} a_{\zeta s n}\,\ddot{s}_n + a_{\zeta\delta}\,\delta = \widetilde{P}_z(t), \\ \ddot{\psi} + \sum_{n=1}^{N} (\hat{a}_{\psi s n}\ddot{s}_n + \hat{a}'_{\psi s n}\,s_n) + \hat{a}_{\psi\delta}\,\delta = \widehat{M}_{Gy}\,(t), \\ \ddot{s}_n + \varepsilon_n \dot{s}_n + \omega_n^2 s_n + a_{sn\zeta}\,\ddot{\zeta} + a_{sn\psi}\,\ddot{\psi} + a'_{sn\psi}\,\psi = 0, \\ \delta = \Im_\psi\,(\psi) - \Im_\zeta\,(\zeta), \qquad n = 1, 2, \ldots, N. \end{gathered} \tag{6.46}$$

Mathematical Model (6.46) corresponds to consideration of dissipative part only for the integral terms. When taking into account inertial component too, we should introduce into Equations (6.44) an equivalent relative associated mass of liquid $\widetilde{\mu}_n$ (referred to associated mass of liquid μ_n corresponding to consideration of liquid potential motion only), which together with the damping coefficient is calculated considering harmonic linearization coefficients obtained in Chapter 4. It is calculated for $\omega_a = \omega_n$ according to the following formulae:

$$\varepsilon_n = \kappa\,\widehat{\beta}_n \sqrt{\tilde{s}_{na}} \left(\frac{\pi\omega_a}{2}\right)^{1/2} = \kappa_\varepsilon \widehat{\beta}_n\, \omega_n \sqrt{\tilde{s}_{na}},$$

$$\tilde{\mu}_n = \kappa_\mu\, \widehat{\beta}_n \sqrt{\tilde{s}_{na}} = \kappa_{\mu\varepsilon} \frac{\varepsilon_n}{\omega_n}, \quad \kappa_\mu = 0.765, \quad \kappa_{\mu\varepsilon} = \frac{\kappa_\mu}{\kappa_\varepsilon} = 0.667. \tag{6.47}$$

The mathematical model of the object-controller closed-loop system (SC with a controller) is represented by the following set of differential equations:

$$\ddot{\zeta} + a_{\zeta\psi}\,\psi + \sum_{n=1}^{N} a_{\zeta s_n}\,\ddot{s}_n + a_{\zeta\delta}\,\delta = \tilde{P}_z(t),$$

$$\ddot{\psi} + \sum_{n=1}^{N} (\widehat{a}_{\psi s_n}\ddot{s}_n + \widehat{a}'_{\psi s_n} s_n) + \widehat{a}_{\psi\delta}\delta = \widehat{M}_{Gy}(t), \tag{6.48}$$

$$(1 + \tilde{\mu}_n)\,\ddot{s}_n + \varepsilon_n\,\dot{s}_n + \omega_n^2\, s_n + a_{s_n\zeta}\,\ddot{\zeta} + a_{s_n\psi}\,\ddot{\psi} + a'_{s_n\psi}\,\psi = 0,$$

$$\delta = \Im_\psi(\psi) - \Im_\zeta(\zeta), \qquad n = 1, 2, \ldots, N.$$

Mathematical Model (6.48) differs from the traditional Model (6.46) in the presence of an additional relative associated mass of liquid. Those models coincide for relatively weak influence of the damping devices (small damping), i.e. while the following relationship characterizing equivalent LV liquid is true:

$$\tilde{\mu}_n = \kappa_{\mu\varepsilon} \frac{\varepsilon_n}{\omega_n} \ll 1. \tag{6.49}$$

We shall assume the latter at the initial stage as the criterial estimate for the influence of the additional associated liquid mass, connected with its vortex motion.

Consideration of the dissipation part only in the integral term, i.e. ignoring $\tilde{\mu}_n$ coefficient, corresponds in the equivalent electrodynamics problem to description of skin effect in a HECF core by a short-circuit loop with an equivalent active resistance R^* inductively coupled with the electromagnet coil (see (5.20)). Retaining $\tilde{\mu}_n$ corresponds to consideration of additional inductance L^* in mathematical Model (5.23) of the electromagnet.

Let us consider one of methods for investigating self-sustained oscillations of object – controller closed-loop system, caused by nonlinearity of liquid wave motions damping in the tanks (see [34]). It is characterized by high informativeness, simplicity of computer implementation, and requires comparatively small computation time. This method is based on determination of periodical modes by constructing stability domains in the space of damping coefficients. Let damping devices be installed in m from N compartments, and the stability domain is bounded in m -dimensional space of $\varepsilon_1, \ldots, \varepsilon_m$ by the surface

$$\Phi_m(\varepsilon_1, \ldots, \varepsilon_m) = 0. \tag{6.50}$$

Each point of Surface (6.50) corresponds to frequency ω_a, the imaginary part for one of pure imaginary roots in the closed-loop system characteristic equation.

Suppose, that self-sustained oscillations at frequency ω_a develop in the system. Since the damping coefficients nonlinearity is 'weak' and the system has high selectivity at the frequencies close to partial frequencies ω_n of liquid oscillations, we can assume that the self-sustained oscillations are proximate to harmonic ones. Particularly,

$$\psi(t) = \psi_a \sin \omega_a t. \tag{6.51}$$

Liquid oscillations amplitudes have the values, for which point $\varepsilon_{1a}, \ldots, \varepsilon_{ma}$ (where $\varepsilon_{na} = \tilde{\varepsilon}_n \sqrt{\tilde{s}_{na}}$) lies in Surface (6.50), $n = 1, 2, \ldots, m$.

Let us apply Laplace transform to Equations (6.46) for zero initial conditions and present the relationship of images $\Psi(p)$ and $S_n(p)$ (considering $\psi(t)$ as an external disturbance) in the form

$$\frac{S_n(p)}{\Psi(p)} = W_{s_n\psi}(p) = \frac{\Delta_{sn}(p)}{\Delta^\circ(p)}, \quad n = 1, 2, \ldots, m, \tag{6.52}$$

where

$$\Delta^\circ(p) = \begin{vmatrix} p^2 & a_{\zeta s1}\, p^2 & \ldots & a_{\zeta sn}\, p^2 & \ldots & a_{\zeta sN}\, p^2 & a_{\zeta\delta} \\ a_{s1\zeta}\, p^2 & p^2+\varepsilon_1\, p+\omega_1^2 & \ldots & 0 & \ldots & 0 & 0 \\ \ldots & \ldots & \ldots & \ldots & \ldots & \ldots & \ldots \\ a_{sn\zeta}\, p^2 & 0 & \ldots & p^2+\varepsilon_n\, p+\omega_n^2 & \ldots & 0 & 0 \\ \ldots & \ldots & \ldots & \ldots & \ldots & \ldots & \ldots \\ a_{sN\zeta}\, p^2 & 0 & \ldots & 0 & \ldots & p^2+\varepsilon_N\, p+\omega_N^2 & 0 \\ \Im_\zeta(p) & 0 & \ldots & 0 & \ldots & 0 & 1 \end{vmatrix}, \tag{6.53}$$

$$\Delta_{sn}(p) = \begin{vmatrix} p^2 & a_{\zeta s1}\, p^2 & \ldots & -a_{\zeta\psi} & \ldots & a_{\zeta sn}\, p^2 & a_{\zeta\delta} \\ a_{s1\zeta}\, p^2 & p^2+\varepsilon_1\, p+\omega_1^2 & \ldots & -\hat{a}_{s1\psi}\, p^2 - \hat{a}_{s1\psi} & \ldots & 0 & 0 \\ \ldots & \ldots & \ldots & \ldots & \ldots & \ldots & \ldots \\ a_{sn\zeta}\, p^2 & 0 & \ldots & -\hat{a}_{sn\psi}\, p^2 - \hat{a}_{sn\psi} & \ldots & 0 & 0 \\ \ldots & \ldots & \ldots & \ldots & \ldots & \ldots & \ldots \\ a_{sN\zeta}\, p^2 & 0 & \ldots & -\hat{a}_{sN\psi}\, p^2 - \hat{a}_{sN\psi} & \ldots & p^2+\varepsilon_N\, p+\omega_N^2 & 0 \\ \Im_\zeta(p) & 0 & \ldots & \Im_\psi(p) & \ldots & 0 & 1 \end{vmatrix}. \tag{6.54}$$

In harmonic oscillations mode with frequency ω_a

$$\frac{s_{na}}{\psi_a} = \left.\frac{A_{sn}(\omega_a)}{A^\circ(\omega_a)}\right|_{\varepsilon_n = \varepsilon_{na}}. \tag{6.55}$$

Here $A_{s_n}(\omega)\,|_{\varepsilon_n=\varepsilon_{na}}$ and $A^\circ(\omega)\,|_{\varepsilon_n=\varepsilon_{na}}$ are amplitude response values of operators $\Delta_{s_n}(p)$ and $\Delta^\circ(p)$ at frequency ω_a, calculated for ε_n value satisfying (6.50).

When investigating steady-state self-sustained oscillations, we can use not the transfer function, but the controller's frequency response, that is more acceptable in practice and allows to extend the described method to systems with essentially nonlinear control laws, particularly relay-type ones, and to systems with an on-board computer included in the control loop. Dividing ε_{na} by ε_{1a} for $n = 2, ..., m$ we obtain

$$\kappa_n = \frac{\varepsilon_{na}}{\varepsilon_{1a}} = \frac{\widetilde{\varepsilon}_n}{\widetilde{\varepsilon}_1}\left(\frac{A_{s_n}(\omega_a)}{A_{s_1}(\omega_a)}\right)^{1/2}\Bigg|_{\varepsilon_n=\varepsilon_{na}}. \tag{6.56}$$

Thus the operating point $\varepsilon_{1a}, \ldots, \varepsilon_{ma}$ lies in the intersection of Surface (6.50) and Ray (6.56).

There are several solutions for the set of Equations (6.50) and (6.56) corresponding to periodical modes, that will take place depending on the initial conditions. It is natural to suppose that Surface (6.50) points, through which Ray (6.56) enters the stability domain, corresponds to stable modes, and the points, through which it exits, correspond to unstable modes.

Having the operating points we obtain self-sustained oscillations amplitudes for all the generalized coordinates, particularly

$$s_{na} = \frac{r_0\,\varepsilon_{na}^2}{\omega_a\,\widetilde{\varepsilon}_n^2}, \quad \psi_a = \frac{A^\circ(\omega_a)}{A_{s_n}(\omega_a)}\Bigg|_{\varepsilon_n=\varepsilon_{na}} s_{na}, \quad \dot{\psi}_a = \omega_a\,\psi_a. \tag{6.57}$$

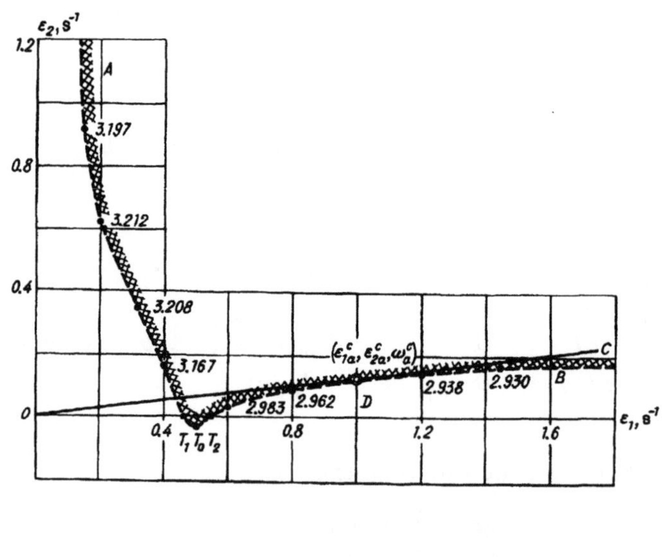

Fig. 6.2. Stability domain boundary in coordinates $\varepsilon_1, \varepsilon_2$. Stability is at the hatched side; ⊖ denotes the critical point

Note, that the operating point location at Surface (6.50) depends on $\eta_{n1} = \tilde{\varepsilon}_n/\tilde{\varepsilon}_1$ value only, which is a construction parameter. By its optimal choice we can optimally distribute damping devices among fuel compartments.

Let us for a methodological example consider a SC with an analogue controller and two tanks, one of toroidal shape and the other of lens-like shape (see [34]). These shapes of tanks are the least favourable in dynamic sense for the case of servo force presence, because they encourage appearance of intrinsic dynamic instability, that is just what happens in the considered case. To compensate it rather a high level of damping is required.

Fig. 6.2 shows stability domain in $\varepsilon_1, \varepsilon_2$ plane constructed by D – partitioning method in complex plane ε_2 with respect to the grid of ε_1 values. Generally speaking, stability domain can be constructed by the method of scanning in $\varepsilon_1, \varepsilon_2$ plane and calculation of the characteristic equation roots. The stability domain boundary

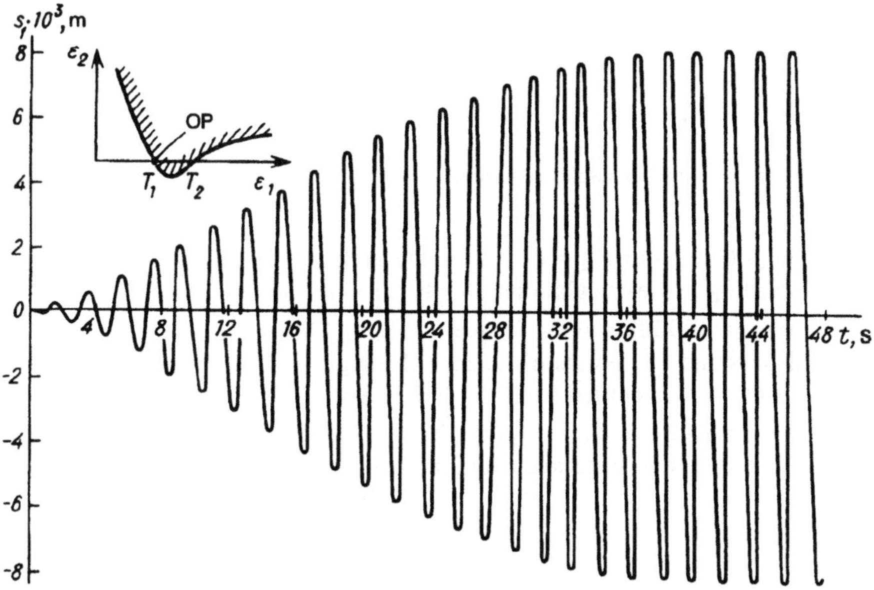

Fig. 6.3. The process of steady self-sustained oscillations development, s_1 is the coordinate of liquid wave motions in the 1st cavity. OP - operating point

corresponds to two pure imaginary complex conjugated roots $\pm\, i\omega_a$ (numbering in ω_a is shown in Fig. 6.2), while the other roots are located in the left half plane of complex variable p. It is clear from stability domains, that it is impossible to ensure closed-loop system stability without installing special damping devices in the tanks. This results from intrinsic dynamic instability of the object. The system appears unstable 'in the small', and the only possible stable mode is self-sustained oscillations mode (stability 'in the great'). It is also clear that a stable periodic mode can be afforded by installing dampers in the first tank only. The presence of a damper in the second tank only does not help to achieve system stability even 'in the great'.

If there is a damper in the first tank only, steady self-sustained oscillations develop

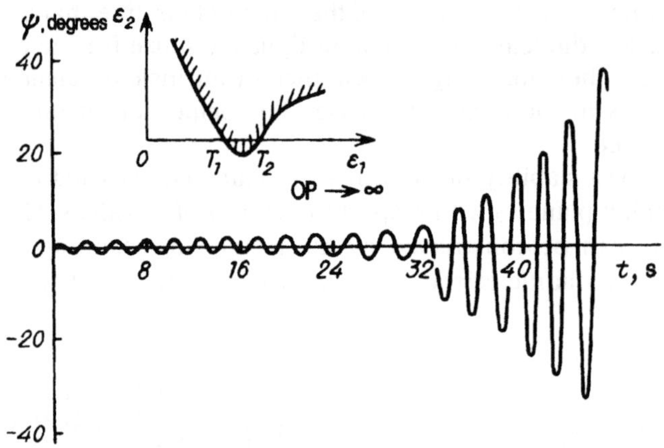

Fig. 6.4. The process of unstable oscillations development; ψ is the yaw angle

at such an amplitude, that $\varepsilon_{1a} = \widetilde{\varepsilon}_1 \sqrt{\dot{s}_{1a}}$ value matches T_1 point. For instance, with $\widetilde{\varepsilon}_1 = 4.07$ according to the calculations, steady self-sustained oscillations establish with the amplitude $s_{1a} = 0.0088$m at the frequency $\omega_a = 3.00\ \mathrm{s}^{-1}$.

Fig. 6.3 presents the results of numerical integration for Equations (6.46). The initial conditions were specified for $\psi_0 = 0.57°$. According to simulation results $s_{1a} = 0.0086$ m; $\omega_a = 2.92\ \mathrm{s}^{-1}$; $\psi_a = 2°$, that is an evidence of the approximate method's high accuracy. Those self-sustained oscillations will establish in the system, however, if at the initial moment ε_1 value (for $\varepsilon_2 = 0$) lies to the left of T_1 point ('soft excitation'). If we provide the system with the initial conditions, when that value is to the right of T_2 point ('hard excitation'), oscillations with increasing amplitude will begin to develop in the system. The system appears unstable 'in the great'. The results of appropriate simulation are presented in Fig. 6.4. Hence point T_1 corresponds to a stable periodic cycle, and T_2 to an unstable one.

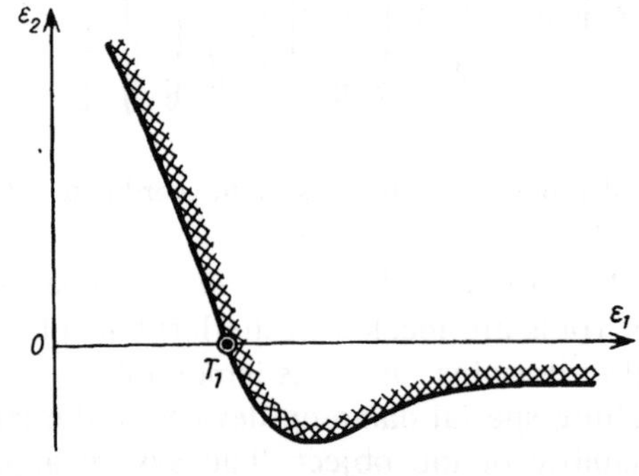

Fig. 6.5. Stability domain boundary in ε_1, ε_2 coordinates, where the mode of instability 'in the great' is absent. Stability domain is at the hatched side

Thus, we can install damping devices in the first tank only, when either the system has a stability domain of the type shown in Fig. 6.5, or there are surely no initial disturbances exceeding some threshold value.

Let us present an example of periodic modes calculation, when damping devices are installed in both tanks. The calculations was performed by successive approximation method according to the following algorithm.

(1) The first approximation point $\varepsilon_{1a}^{(1)}$, $\varepsilon_{2a}^{(1)}$, $\omega_a^{(1)}$ was specified in curve AB (see Fig. 6.2).

(2) Curve AB (6.50) with ray OC (6.56) intersection point was found: $\varepsilon_{1a}^{(2)}$, $\varepsilon_{2a}^{(2)}$, $\omega_a^{(2)}$.

(3) Steps (1) and (2) were repeated until the successive approximation matched the previous one with sufficient accuracy.

(4) Steady-state oscillations amplitudes were determined according to Formulae (6.57).

Stability boundary analysis in Fig. 6.2 shows that, depending on η_{21} value, the system can have either one stable solution at the intersection of ray OC and the curve interval AT_1, or two stable solutions: one at the OC and AT_1 intersection, and the other at the intersection of ray OC with curve interval T_2B. The η_{21}^c critical value is determined from (6.56) by substituting ε_{1a}^c, ε_{2a}^c, ω_a^c (critical point D), where the ray is tangent to the Curve (6.50):

$$\eta_{21}^c = \frac{\varepsilon_{2a}^c \sqrt{A_{s1}(\omega_a^c)}}{\varepsilon_{1a}^c \sqrt{A_{s2}(\omega_a^c)}}, \quad \varepsilon_1 = \varepsilon_{1a}^c, \quad \varepsilon_2 = \varepsilon_{2a}^c. \tag{6.58}$$

For $\eta_{21} > \eta_{21}^c$ the system has one stable periodic mode, for $\eta_{21} < \eta_{21}^c$ it has two such modes. For the given system $\varepsilon_{1a}^c = 1.0$, $\varepsilon_{2a}^c = 0.13$, $\omega_a = 3.00\ \mathrm{s}^{-1}$, and $\eta_{21}^c = 0.078$. For $\widetilde{\varepsilon_1} = 4.07$, $\widetilde{\varepsilon_2} = 4.07$, $\eta_{21} = 1 > \eta_{21}^c$ the limit cycle parameters are as follows: $s_{1a} = 0.004096$ m, $s_{2a} = 0.003855$ m, $\omega_a = 3.21\ \mathrm{s}^{-1}$, $\psi_a = 0.86°$. According to numerical simulation results, $s_{1a} = 0.004$ m, $s_{2a} = 0.00372$ m, $\omega_a = 3.2\ \mathrm{s}^{-1}$, $\psi_a = 0.84°$.

For $\widetilde{\varepsilon_1} = 16.3$, $\widetilde{\varepsilon_2} = 0.815$, $\eta_{21} = 0.05 < \eta_{21}^c$ there are two possible stable modes, which amplitude difference is by a factor of 30: Mode 1: $\psi_a = 0.13°$, Mode 2: $\psi_a = 3.9°$. During numerical simulation the system was brought to Mode 1 specifying the initial condition of $\psi_0 = 0.057°$ (then $\psi_a = 0.14°$), and to Mode 2 at $\psi_0 = 5.7°$, $\psi_a = 3.8°$.

Between the two found stable periodic modes there must exist the third, unstable one, corresponding to a separatrix which separates the domains of periodic modes attraction. The solution corresponding to the separatrix is obviously situated at the ray OC intersection with the curve interval T_0D (see Fig. 6.2).

Note that the above algorithm for periodic modes determination in the vicinity of the separatrix is numerically unstable. To determine the unstable mode a modified algorithm of [34] can be used. It is described below applied to a SC with two tanks partially filled with liquid.

Stability Surface (6.50) can be plotted in coordinates ε_1 and Γ_2, where

$$\Gamma_2 = \frac{\varepsilon_2}{\varepsilon_1}. \tag{6.59}$$

Stability domain boundary is then obtained in the form

$$\Phi_2(\Gamma_2, \varepsilon_1) = 0. \tag{6.60}$$

Solutions corresponding to the periodic modes, both the stable and unstable one, are situated at the intersection of Curve γ_2 (6.56) and Surface (6.60).

Fig. 6.6 presents an example of calculation for the case considered above

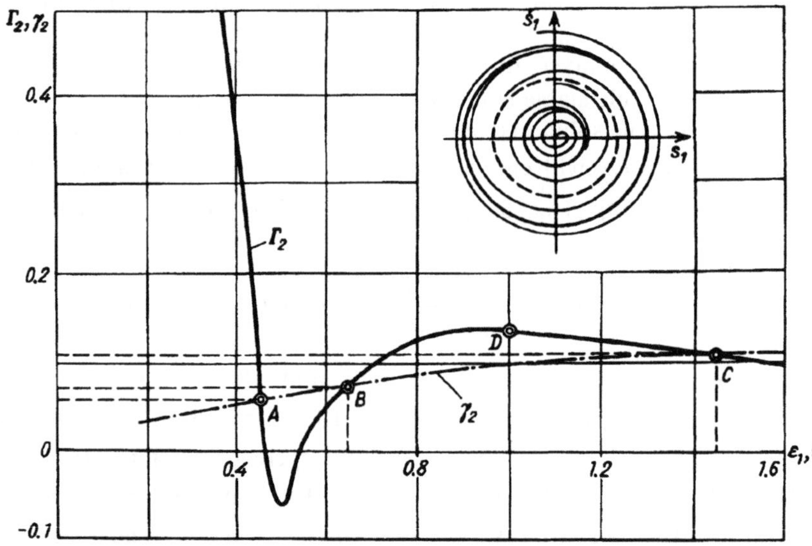

Fig. 6.6. Limit cycle determination; *A*, *C* denote the stable limit cycles, *B* denotes the unstable one (the separatrix).

($\eta_{21} = 0.05$). Point *A* corresponds to the first stable periodic mode, point *C* to the second one, and point *B* to the separatrix. The figure also shows the phase picture of the oscillation process in the first tank (s_1, $\dot{s}_1$).

This method also facilitates determination of the critical value η_{21}^c, for which curves Γ_2 and γ_2 are tangent to each other at point *D*.

For stabilization of the considered object characterized by intrinsic dynamic instability it is obviously necessary to provide a sufficiently high level of damping, when the coefficient characterizing the additional associated mass calculated by Formula (6.47) is not a small quantity having the order of 0.1 − 0.2.

To estimate the influence of the additional associated mass on stability and on required damping, let us consider stability domains in the plane of damping coefficients $\varepsilon_1, \varepsilon_2$. Those domains were obtained by calculating roots of the characteristic equation with respect to variable $q = \sqrt{p}$ for the closed-loop system written in the form of (6.44) with scanning in parameters $\beta_{1,2}$ = const corresponding to a particular amplitude of the harmonic mode, followed by conversion to equivalent damping coefficients according to Formula (6.47). For comparison stability domains were calculated with the same parameters based on mathematical Models (6.48) and (6.46). The latter was used above to illustrate dynamic peculiarities connected with nonlinearity of damping coefficients for liquid wave motions. The corresponding stability domains are presented in Fig. 6.7.

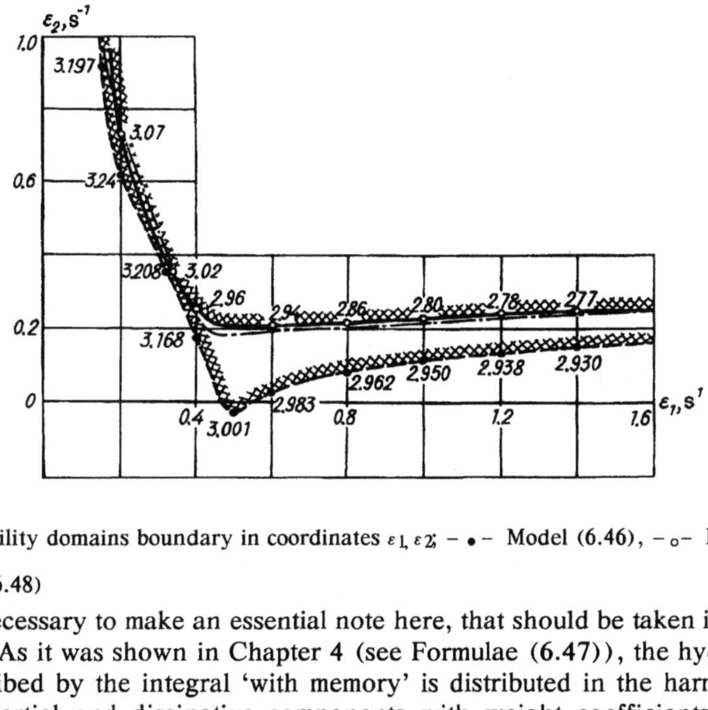

Fig. 6.7. Stability domains boundary in coordinates $\varepsilon_1, \varepsilon_2$; –•– Model (6.46), –o– Model (6.44), –·– Model (6.48)

It is necessary to make an essential note here, that should be taken into account further on. As it was shown in Chapter 4 (see Formulae (6.47)), the hydrodynamic force described by the integral 'with memory' is distributed in the harmonic mode between inertial and dissipative components with weight coefficients κ_ε and κ_μ, differing in a half.

Therefore application of the technique assuming $\beta_{1,2}$ = const with further transformation of the integral terms results in inaccurate account for the inertial component of the hydrodynamic force produced by the damping elements. The inertial component is thus taken into account with the same weight coefficient κ_ε as the dissipative one, i.e. it is set half as great as the real value.

However, as comparison of stability domains for Models (6.44) and (6.48) shows, even for relatively strong impact of the additional associated mass its effect on the accuracy of the final result is insignificant. That enables usage of Model (6.44) instead of (6.48) for solution of appropriate problems. To calculate a stability domain based on Model (6.48) it is necessary to organize a multi-step computational process with limit cycle frequency ω_a refinement, while Model (6.44) allows to obtain the frequency at the first step.

Let us explain the sense of stability domain boundary numbering for Model (6.44). The characteristic equation has two roots at the point lying in this boundary:

$$q_{1,2} = \alpha \pm i\alpha. \tag{6.61}$$

As it was shown in Chapter 3, for the steady state (when the transient components corresponding to the located roots to the left from stability Bound (3.20) according to Law (3.22) decay) those roots correspond to undamped harmonic oscillations at frequency

$$\omega_a = 2\alpha^2. \tag{6.62}$$

According to it is the stability domain boundary numbered.

Fig. 6.7 shows not only the proximity of the stability domains calculated based on mathematical Models (6.44) and (6.48), but also the possibility to reduce rigorous account of the integral term to account of additional associated mass based on Model (6.48) for that level of required damping. Therefore new dynamic features of the system, that stability analysis using Model (6.44) allows to reveal, should be referred basically to the additional associated mass effect.

From the quantitative point of view, the difference is as follows. Firstly, the required damping grows noticeably. It is necessary to afford stable self-sustained oscillations. It becomes about half as high, that being converted to oscillation amplitude corresponds to two-fold increase. Secondly, the possible limit cycles frequencies decrease by 5-6 %.

From the qualitative point of view, the system described by Model (6.44) differs in that it is impossible to afford stability by installing oscillation dampers in the first tank only. With no dampers in the second tank the system appears unstable both 'in the small' and 'in the great'. Also less distinct is the system's feature to have two stable limit cycles with the same parameters, the amplitudes of those limit cycles differing in magnitude by an order.

Essential qualitative and quantitative difference in results of analysis based on the traditional model and on the model using integrodifferential equations shows that it is prohibitive to neglect kinetic energy of liquid vortex motions in the considered case. This causes the necessity of a more accurate account of effects connected with vortex formation at the edges of in-tank construction elements including account of additional degrees of freedom characterizing LV liquid vortex motion. There are reasons to assume that such an approach is about to reveal essentially new forms of instability.

A mathematical description of the considered objects dynamics taking into account the mentioned effects is presented in Chapter 2 in the form of nonlinear integrodifferential equations including equations for additional generalized coordinates ϑ_n (the yawing plane) and χ_n (the roll plane) that are an integral description of the vortexed liquid mass dynamics (n is tank number).

Let us discuss this aspect of the problem in a more detail, and analyze the effect of additional degrees of freedom on object – controller closed-loop system stability by way of a specific example.

6.4. SC Stability in the Yaw Plane with Account of Potential and Vortex Motions of Liquid in Tanks

The object considered in the previous article has intrinsic dynamic instability that is a result of servo force presence and tanks of complicated shape (toroidal and lens-like). All dynamic features connected with vortex motion of liquid and nonlinear character of the corresponding hydrodynamic forces are to display in this object in full scale. It is necessary to solve complicated boundary-value problems to calculate the coefficients of such object's refined mathematical model.

We further concentrate on analyzing a simpler object, namely a spacecraft with a gimbaled LPRE and two cylindric tanks of the same radius r_0 each having 8 radial ribs installed with $0.2r_0$ width and length through the entire tank depth. For the lower tank (subscript 1) phase stabilization conditions formulated above are obeyed, and for the upper one (subscript 2) those conditions are violated. Therefore the object is unstable at the frequencies of liquid wave motions, the instability being connected with the controller destabilizing effect. Both tanks have equal partial frequencies of liquid oscillations, that, as it is shown below, makes it expedient to install dampers in both tanks simultaneously.

If the hypothesis of approximately uniform vorticity distribution over the liquid volume is right, and with dominating role of wave motions of liquid $r_0\ |\dot{\vartheta}_n| << |s_n|$ for the considered object it is possible to calculate all the hydrodynamic coefficients including those connected with vortex motions of liquid according to formulae presented in Chapter 2.

Let us write down the mathematical model of perturbed motion in the yaw plane of a stabilized SC with two tanks. The model takes into account additional degrees of freedom corresponding to vortex motion of liquid. We obtain this model from Equations (2.120). For simplicity we consider a SC with a yaw stabilization controller without lateral stabilization. In this case lateral motion of the SC is independent. Having expressed the corresponding acceleration $\ddot{\xi}$ from the first equation and substituting it into the other equations we obtain, disturbances $\widetilde{P}_z(t)$ and $\widetilde{M}_{Gy}(t)$ being assumed absent:

$$\ddot{\psi} + a_{\psi\vartheta_1}\ddot{\vartheta}_1 + a_{\psi s_1}\ddot{s}_1 + a'_{\psi s_1}s_1 + a_{\psi\vartheta_2}\ddot{\vartheta}_2 + a_{\psi s_2}\ddot{s}_2 + a'_{\psi s_2}s_2 = -a_{\psi\delta}\,\delta,$$

$$\ddot{\vartheta}_1 + \ddot{\psi} + \beta_{\vartheta_1}\int_{-\infty}^{t}\frac{\ddot{\vartheta}_1(\tau)\,d\tau}{\sqrt{t-\tau}} + a_{\vartheta_1 s_1}\int_{-\infty}^{t}\frac{\ddot{s}_1(\tau)\,d\tau}{\sqrt{t-\tau}} = 0,$$

$$(1 - a_{s1\zeta}\, a_{\zeta s1})\, \ddot{s}_1 + \omega_1^2 s_1 + \beta_1 \int_{-\infty}^{t} \frac{\ddot{s}_1(\tau)\, d\tau}{\sqrt{t-\tau}} - a_{s1\zeta}\, a_{\zeta s2}\, \ddot{s}_2 + a_{s1\psi}\, \ddot{\psi} +$$

$$+ (a'_{s1\psi} - a_{s1\zeta}\, a_{\zeta\psi})\, \psi + a_{s1\vartheta_1} \int_{-\infty}^{t} \frac{\ddot{\vartheta}_1(\tau)\, d\tau}{\sqrt{t-\tau}} - a_{s1\zeta}\, a_{\zeta\delta}\, \delta = 0, \qquad (6.63)$$

$$\ddot{\vartheta}_2 + \ddot{\psi} + \beta_{\vartheta 2} \int_{-\infty}^{t} \frac{\ddot{\vartheta}_2(\tau)\, d\tau}{\sqrt{t-\tau}} + a_{\vartheta 2 s2} \int_{-\infty}^{t} \frac{\ddot{s}_2(\tau)\, d\tau}{\sqrt{t-\tau}} = 0,$$

$$(1 - a_{s2\zeta}\, a_{\zeta s2})\, \ddot{s}_2 + \omega_2^2 s_2 + \beta_2 \int_{-\infty}^{t} \frac{\ddot{s}_2(\tau)\, d\tau}{\sqrt{t-\tau}} - a_{s2\zeta}\, a_{\zeta s1}\, \ddot{s}_1 + a_{s2\psi}\, \ddot{\psi} +$$

$$+ (a'_{s2\psi} - a_{s2\zeta}\, a_{\zeta\psi})\, \psi + a_{s2\vartheta 2} \int_{-\infty}^{t} \frac{\ddot{\vartheta}_2(\tau)\, d\tau}{\sqrt{t-\tau}} - a_{s2\zeta}\, a_{\zeta\delta}\, \delta = 0, \qquad \delta = \Im_\psi(\psi).$$

For yaw stabilization controller consider a servo drive with a correction unit implementing a feedback in the yaw angle and its first derivative. The appropriate control law is written in operator form as follows:

$$(Tp + 1)\, \Delta(p) = \frac{K_\psi + K_{\dot{\psi}}\, p}{1 + T_\psi\, p}\, \Psi(p), \qquad (6.64)$$

where $\Delta(p)$ and $\Psi(p)$ are Laplace transforms of functions $\delta(t)$ and $\psi(t)$; K_ψ, $K_{\dot{\psi}}$ are control law coefficients; T is the servo drive time constant; T_ψ is the real differentiation loop time constant necessary to provide the system noise immunity.

Allowing for hypothesis $|r_0 \dot{\vartheta}_{1,2}| << |s_{1,2}|$, the coefficients at the integral terms of Equations (6.63) are nonlinear functions of liquid wave motions velocity and are defined by the formulae

$$\beta_{\vartheta j} = \hat{\beta}_{\vartheta j} \sqrt{|\tilde{s}_j|}, \quad a_{\vartheta j\, sj} = \hat{a}_{\vartheta j\, sj} \sqrt{|\tilde{s}_j|}, \quad \beta_j = \hat{\beta}_j \sqrt{|\tilde{s}_j|},$$

$$a_{sj\vartheta j} = \hat{a}_{sj\vartheta j} \sqrt{|\tilde{s}_j|}, \quad \tilde{s}_j = \frac{\dot{s}_j}{r_0}, \quad j = 1,2. \qquad (6.65)$$

Coefficients with a hat sign ($\wedge$) in Formulae (6.65) are independent of amplitude. It is obvious from (6.65) that all the nonlinear coefficients can be expressed in terms of two ones, β_1 and β_2, containing all the information on the studied object nonlinearity:

$$\beta_{\vartheta j} = K_{\vartheta j\vartheta j}\, \beta_j, \quad a_{\vartheta j\, sj} = K_{\vartheta j\, sj}\, \beta_j, \quad a_{sj\vartheta j} = K_{sj\vartheta j}\, \beta_j, \quad j = 1,2, \qquad (6.66)$$

where

$$K_{\vartheta j\, \vartheta j} = \frac{\hat{\beta}_{\vartheta j}}{\hat{\beta}_j}, \quad K_{\vartheta j sj} = \frac{\hat{a}_{\vartheta j sj}}{\hat{\beta}_j}, \quad K_{sj\vartheta j} = \frac{\hat{a}_{sj\vartheta j}}{\hat{\beta}_j}, \quad j = 1,2. \qquad (6.67)$$

Table 6.1

Coefficients of the Perturbed Motion Equations

$a_{\zeta\psi}$	$a_{\zeta\delta}$	$a_{\zeta s1}$	$a_{\zeta s2}$	$a_{\psi\vartheta1}$	$a_{\psi\vartheta2}$	$a_{\psi s1}$	$a_{\psi s2}$
7.63	7.63	0.06206	0.04558	0.172164	0.060068	0.003802	- 0.0094
$a'_{\psi s1}$	$a'_{\psi s2}$	$a_{\psi\delta}$	ω_1^2	$a_{s1\zeta}$	$a_{s1\psi}$	$a'_{s1\psi}$	ω_2^2
0.03071	0.02255	2.622	7.024	1.6801	1.6138	12.27	7.024
$a_{s2\zeta}$	$a_{s2\psi}$	$a'_{s2\psi}$	$\hat{a}_{\vartheta1s1}$	$\hat{a}_{s1\vartheta1}$	$\hat{\beta}_1$	$\hat{\beta}_{\vartheta1}$	$\hat{a}_{\vartheta2s2}$
1.5408	- 4.9211	11.7563	- 0.00512	- 0.35221	0.1005	0.0341	- 0.0093
$\hat{a}_{s2\vartheta2}$	$\hat{\beta}_2$	$\hat{\beta}_{\vartheta2}$	$K_{\vartheta1s1}$	$K_{s1\vartheta1}$	$K_{\vartheta2s2}$	$K_{s2\vartheta2}$	$K_{\vartheta1\vartheta1}$
- 0.29126	0.1005	0.0593	- 0.00509	- 3.505	- 0.0925	- 0.29126	0.339
$K_{\vartheta2\vartheta2}$	K_{ψ}	$K_{\dot{\psi}}$	T	T_{ψ}			
0.59	0.76	0.76	0.1	0.05			

Table 6.1 presents the coefficients for the considered hypothetical object of control described by (6.63), controller (6.64) and coefficients characterizing nonlinearity, specified by (6.65), (6.66), (6.67).

Let us perform harmonic linearization of the nonlinear coefficients at frequency ω_a and use Formulae (6.47). Integral terms of the type of

$$J_{\vartheta1} = \beta_{\vartheta1} \int_{-\infty}^{t} \frac{\ddot{\vartheta}_1(\tau)\, d\tau}{\sqrt{t-\tau}}, \quad \beta_{\vartheta1} = \hat{\beta}_{\vartheta1} \sqrt{|\tilde{s}_1|}$$

being represented by a sum of inertial and dissipative components similar to that in Chapter 4, is an ambiguous operation even in harmonic mode and it depends on relative phasing of coordinates ϑ_1 and s_1 and relation of their amplitudes. Therefore further on we assume that inertial and dissipative components are related, as before, with the weight $\kappa_{\mu\varepsilon}$ (see (6.47)). It is shown below, that such an assumption affects the final results accuracy insignificantly. Taking into account relations of nonlinear Coefficients (6.65), we shall reduce the set of integrodifferential equations to an equivalent set of ordinary differential equations similar to (6.48). Those ordinary differential equations are nonlinear due to nonlinear dependence of damping coefficients ε_1, ε_2 on liquid oscillations amplitude:

$$\ddot{\psi} + a_{\psi\vartheta1}\ddot{\vartheta}_1 + a_{\psi s1}\ddot{s}_1 + a'_{\psi s1}s_1 + a_{\psi\vartheta2}\ddot{\vartheta}_2 + a_{\psi s2}\ddot{s}_2 + a'_{\psi s2}s_2 = -a_{\psi\delta}\delta,$$

$$\left(1 + \frac{\kappa_{\mu\varepsilon}K_{\vartheta1\vartheta1}\varepsilon_1}{\omega_a}\right)\ddot{\vartheta}_1 + \ddot{\psi} + K_{\vartheta1\vartheta1}\varepsilon_1\dot{\vartheta}_1 + \frac{\kappa_{\mu\varepsilon}K_{\vartheta1s1}\varepsilon_1}{\omega_a}\ddot{s}_1 + K_{\vartheta1s1}\varepsilon_1\dot{s}_1 = 0,$$

$$\left(1 - a_{s1\zeta}a_{\zeta s1} + \frac{\kappa_{\mu\varepsilon}\varepsilon_1}{\omega_a}\right)\ddot{s}_1 + \varepsilon_1\dot{s}_1 + \omega_1^2 s_1 - a_{s1\zeta}a_{\zeta s2}\ddot{s}_2 + a_{s1\psi}\ddot{\psi} +$$

$$+ (a'_{s1\psi} - a_{s1\zeta}a_{\zeta\psi})\psi + \frac{\kappa_{\mu\varepsilon}K_{s1\vartheta1}\varepsilon_1}{\omega_a}\ddot{\vartheta}_1 + K_{s1\vartheta1}\varepsilon_1\dot{\vartheta}_1 - a_{s1\zeta}a_{\zeta\delta}\delta = 0, \quad (6.68)$$

$$\left(1+\frac{\kappa_{\mu\varepsilon} K_{\vartheta_2\vartheta_2}\varepsilon_2}{\omega_a}\right)\ddot{\vartheta}_2+\ddot{\psi}+K_{\vartheta_2\vartheta_2}\varepsilon_2\dot{\vartheta}_2+\frac{\kappa_{\mu\varepsilon} K_{\vartheta_2 s_2}\varepsilon_2}{\omega_a}\ddot{s}_2+K_{\vartheta_2 s_2}\varepsilon_2\dot{s}_2=0,$$

$$\left(1-a_{s_2\zeta}a_{\zeta s_2}+\frac{\kappa_{\mu\varepsilon}\varepsilon_2}{\omega_a}\right)\ddot{s}_2+\varepsilon_2\dot{s}_2+\omega_2^2 s_2-a_{s_2\zeta}a_{\zeta s_1}\ddot{s}_1+a_{s_2\psi}\ddot{\psi}+$$

$$+\left(a'_{s_2\psi}-a_{s_2\zeta}a_{\zeta\psi}\right)\psi+\frac{\kappa_{\mu\varepsilon}K_{s_2\vartheta_2}\varepsilon_2}{\omega_a}\ddot{\vartheta}_2+K_{s_2\vartheta_2}\varepsilon_2\dot{\vartheta}_2-a_{s_2\zeta}a_{\zeta\delta}\delta=0,$$

$$(Tp+1)\,\Delta(p)=\frac{K_\psi+K_{\dot{\psi}}\,p}{1+T_\psi\,p}\,\Psi(p).$$

Let us parallel to that write down the mathematical model of the considered system, derived from Equations (6.68) for $\dot{\vartheta}_1=\dot{\vartheta}_2=-\dot{\psi}$ after additionally neglecting inertial components while transforming the appropriate integrals:

$$(1-a_{\psi\vartheta_1}-a_{\psi\vartheta_2})\ddot{\psi}-(a_{\psi\vartheta_1}K_{\vartheta_1\vartheta_1}\varepsilon_1+a_{\psi\vartheta_2}K_{\vartheta_2\vartheta_2}\varepsilon_2)\dot{\psi}+a_{\psi s_1}\ddot{s}_1-$$

$$-a_{\psi\vartheta_1}K_{\vartheta_1 s_1}\varepsilon_1\dot{s}_1+a'_{\psi s_1}s_1+a_{\psi s_2}\ddot{s}_2-a_{\psi\vartheta_2}K_{\vartheta_2 s_2}\varepsilon_2\dot{s}_2+a'_{\psi s_2}s_2=-a_{\psi\delta}\delta,$$

$$(1-a_{s_1\zeta}a_{\zeta s_1})\ddot{s}_1+\varepsilon_1\dot{s}_1+\omega_1^2 s_1-a_{s_1\zeta}a_{\zeta s_2}\ddot{s}_2+a_{s_1\psi}\ddot{\psi}-$$

$$-K_{s_1\vartheta_1}\varepsilon_1\dot{\psi}+(a'_{s_1\psi}-a_{s_1\zeta}a_{\zeta\psi})\psi-a_{s_1\zeta}a_{\zeta\delta}\delta=0, \qquad (6.69)$$

$$(1-a_{s_2\zeta}a_{\zeta s_2})\ddot{s}_2+\varepsilon_2\dot{s}_2+\omega_2^2 s_2-a_{s_2\zeta}a_{\zeta s_1}\ddot{s}_1+a_{s_2\psi}\ddot{\psi}-$$

$$-K_{s_2\vartheta_2}\varepsilon_2\dot{\psi}+(a'_{s_2\psi}-a_{s_2\zeta}a_{\zeta\psi})\psi-a_{s_2\zeta}a_{\zeta\delta}\delta=0,$$

$$(Tp+1)\,\Delta(p)=\frac{K_\psi+K_{\dot{\psi}}\,p}{1+T_\psi\,p}\,\Psi(p).$$

Let us perform a comparative analysis of the three mathematical models: (6.63), (6.68), and (6.69). Let us compare those models based on stability domains in the plane of coefficients ε_1, ε_2, constructed by scanning in parameters ε_1 and ε_2 and calculating the roots of the closed-loop system characteristic equation.

For Model (6.63) the roots were calculated with respect to variable $q=\sqrt{p}$, then boundary values of β_1, β_2 were obtained according to stability Criterion (3.20), and from those the required damping coefficients were derived with the help of formulae

$$\varepsilon_{1,2}=\left(\frac{\pi\omega_a}{2}\right)^{1/2}\beta_{1,2}\,. \qquad (6.70)$$

System (6.68) characteristic equation in variable has two zero roots, the presence of which is connected with choice of variables. If we introduce $\Omega_1=\dot{\vartheta}_1$ and $\Omega_2=\dot{\vartheta}_2$ as variables, those roots disappear. Hence we can ignore them when analyzing stability. System (6.63) in variable $q=\sqrt{p}$ has 6 zero roots. Two pairs of zero roots are of the same nature as System (6.68) zero roots, and the other two roots are due to the specific features of solution representation for systems of such class, just like for the system

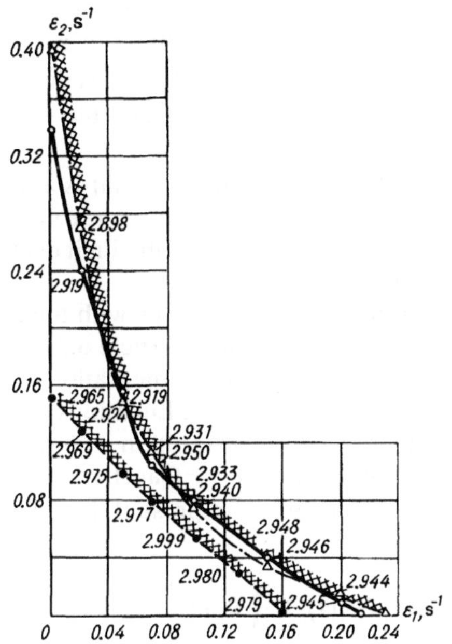

Fig. 6.8. Stability domain boundaries in coordinates ε_1, ε_2, when the drive inertia is taken into account. Stability is at the hatched side; -o- Model (6.63); -•- Model (6.69); -Δ- Model (6.68)

with electromagnetic elements, discussed in Chapter 5 (see Article 5.3). Those 6 zero roots were also neglected during stability analysis. The frequency corresponding to undamped harmonic oscillations of System (6.63) was calculated according to Formula (6.62).

Results of the calculation are presented in Fig. 6.8. The hatching is at the stability domain side. The curve numbering is the frequency of undamped harmonic oscillations described by Systems (6.68) and (6.69) (the imaginary part of two pure imaginary complex conjugated roots $\pm i\omega_a$). The upper number refers to the upper curve.

When calculating stability domains for Systems (6.63) and (6.69), frequency for each point was refined by the successive approximations method until it coincided with the frequency shown in the graph. The initial approximation was $\omega_a = \omega_1 = \omega_2 = 2.65\ \mathrm{s}^{-1}$.

Analysis of the results obtained brings to the following major conclusions:

1. Let damping be uniformly distributed among the tanks at liquid wave motion frequencies within the scope of mathematical Model (6.63) considering the effect of additional degrees of freedom corresponding to liquid vorticity uniform over the volume of each tank. Then to afford closed-loop system stability the required damping is 15-20% greater than that corresponding to Model (6.69). The maximum difference for $\varepsilon_1 = 0$ reaches a factor of 2.2. This indicates the necessity to analyze additional degrees of freedom more completely, those degrees of freedom being connected with liquid vortex motions, particularly to estimate the effect of vorticity nonuniformity, i.e. effect of higher harmonics in the vortex field generated by the dampers.

2. Mathematical Model (6.68) yields results coinciding with the calculations based on Model (6.63) with accuracy sufficient for practice. That corroborates the possibility to take into account inertial and dissipative components of all the integral terms with weight coefficients κ_ε, κ_μ. Just as for the electromagnetic levitation system, it allows to reduce the set of integrodifferential Equations (6.63) to the set of ordinary differential Equations (6.68), for which well developed methods of analysis and control law synthesis are applicable.

Note that stability domain calculated for Model (6.44), which is not shown in Fig. 6.8, practically coincides with that obtained based on Model (6.69). This is an evidence of small effect of the integral terms inertial components and coefficient $a'_{\psi\psi}$. At the same time such an approximation requires a certain caution, i.e. a check is desirable in each particular case.

3. Required values of the damping coefficients are about 0.1 s^{-1}, that is at least two orders greater than values due to liquid viscosity only. Thus the considered system, although being unstable 'in the small', is stable 'in the great' because the limit cycle is implemented, caused by nonlinear damping of liquid oscillations.

To corroborate the possibility of calculating limit cycle amplitudes with the help of stability domains construction in the damping coefficients plane (Article 6.3) let us consider perturbed motion of the studied object by way of numerical integration of the initial nonlinear integrodifferential Equations (6.63). We shall solve Equations (6.63) according to the scheme presented in Chapter 3, by its stepwise resolution with respect to variable $\ddot{\psi}$, $\ddot{\vartheta}_1$, $\ddot{s}_1$, $\ddot{\vartheta}_2$, $\ddot{s}_2$ followed by transformation to Liapunov – Cauchy form. For that purpose let us assume that the system has no pre-history, and present the integral terms in the form

$$
\begin{aligned}
&\int_{-\infty}^{t} \frac{\ddot{\vartheta}_1(\tau)\, d\tau}{\sqrt{t-\tau}} = \int_{0}^{t} \frac{\ddot{\vartheta}_1(\tau)\, d\tau}{\sqrt{t-\tau}} \cong \sqrt{h}\left\{G_{\vartheta 1}(t) + \frac{4}{3}\ddot{\vartheta}_1(t)\right\},\\
&\int_{-\infty}^{t} \frac{\ddot{s}_1(\tau)\, d\tau}{\sqrt{t-\tau}} = \int_{0}^{t} \frac{\ddot{s}_1(\tau)\, d\tau}{\sqrt{t-\tau}} \cong \sqrt{h}\left\{G_{s1}(t) + \frac{4}{3}\ddot{s}_1(t)\right\},\\
&\int_{-\infty}^{t} \frac{\ddot{\vartheta}_2(\tau)\, d\tau}{\sqrt{t-\tau}} = \int_{0}^{t} \frac{\ddot{\vartheta}_2(\tau)\, d\tau}{\sqrt{t-\tau}} \cong \sqrt{h}\left\{G_{\vartheta 2}(t) + \frac{4}{3}\ddot{\vartheta}_2(t)\right\},\\
&\int_{-\infty}^{t} \frac{\ddot{s}_2(\tau)\, d\tau}{\sqrt{t-\tau}} = \int_{0}^{t} \frac{\ddot{s}_2(\tau)\, d\tau}{\sqrt{t-\tau}} \cong \sqrt{h}\left\{G_{s2}(t) + \frac{4}{3}\ddot{s}_2(t)\right\},
\end{aligned}
\tag{6.71}
$$

where $G_{\vartheta 1}(t)$, $G_{s1}(t)$, $G_{\vartheta 2}(t)$, $G_{s2}(t)$ are calculated for discrete moments of time 0, h, . . ., ih, . . ., nh with functions $\ddot{\vartheta}_1(t)$, $\ddot{s}_1(t)$, $\ddot{\vartheta}_2(t)$, $\ddot{s}_2(t)$ by Formulae (3.55).

Substituting Expressions (6.71) into initial Equations (6.63), in view of (6.65), (6.66), (6.67), we obtain

$$\ddot{\psi} + a_{\psi\vartheta 1}\ddot{\vartheta}_1 + a_{\psi s1}\ddot{s}_1 + a_{\psi\vartheta 2}\ddot{\vartheta}_2 + a_{\psi s2}\ddot{s}_2 + a'_{\psi s1}s_1 + a'_{\psi s2}s_2 = -a_{\psi\delta}\,\delta,$$

$$\ddot{\psi} + \left(1 + \frac{4}{3}K_{\vartheta 1\vartheta 1}\beta_1\sqrt{h}\right)\ddot{\vartheta}_1 + \frac{4}{3}K_{\vartheta 1 s1}\beta_1\sqrt{h}\,\ddot{s}_1 =$$

$$= -K_{\vartheta 1\vartheta 1}\beta_1\sqrt{h}\,G_{\vartheta 1} - K_{\vartheta 1 s1}\beta_1\sqrt{h}\,G_{s1},$$

$$a_{s1\psi}\ddot{\psi} + \frac{4}{3}K_{s1\vartheta 1}\beta_1\sqrt{h}\,\ddot{\vartheta}_1 + \left(1 - a_{s1\zeta}a_{\zeta s1} + \frac{4}{3}\beta_1\sqrt{h}\right)\ddot{s}_1 - a_{s1\zeta}a_{\zeta s2}\ddot{s}_2 =$$

$$= -\left(a'_{s1\psi} - a_{s1\zeta}a_{\zeta\psi}\right)\psi - \omega_1^2 s_1 - \beta_1\sqrt{h}G_{s1} - K_{s1\vartheta_1}\beta_1\sqrt{h}G_{\vartheta_1} + a_{s1\zeta}\,a_{\zeta\delta}\,\delta, \tag{6.72}$$

$$\ddot{\psi} + \left(1 + \frac{4}{3}K_{\vartheta_2\vartheta_2}\beta_2\sqrt{h}\right)\ddot{\vartheta}_2 + \frac{4}{3}K_{\vartheta_2 s2}\beta_2\sqrt{h}\,\ddot{s}_2 =$$

$$= -K_{\vartheta_2\vartheta_2}\beta_2\sqrt{h}\,G_{\vartheta_2} - K_{\vartheta_2 s2}\beta_2\sqrt{h}\,G_{s2},$$

$$a_{s2\psi}\ddot{\psi} + a_{s2\zeta}\,a_{\zeta s1}\,\ddot{s}_1 + \frac{4}{3}K_{s2\vartheta_2}\beta_2\sqrt{h}\,\ddot{\vartheta}_2 + \left(1 - a_{s2\zeta}\,a_{\zeta s2} + \frac{4}{3}\beta_2\sqrt{h}\right)\ddot{s}_2 =$$

$$= -\left(a'_{s2\psi} - a_{s2\zeta}\,a_{\zeta\psi}\right)\psi - \omega_2^2 s_2 - \beta_2\sqrt{h}G_{s2} - K_{s2\vartheta_2}\beta_2\sqrt{h}G_{\vartheta_2} + a_{s2\zeta}\,a_{\zeta\delta}\,\delta,$$

$$(Tp + 1)\,\Delta(p) = \frac{K_\psi + K_{\dot{\psi}}\,p}{1 + T_\psi\,p}\,\Psi(p), \quad \beta_1 = \hat{\beta}_1\sqrt{|\tilde{s}_1|}, \quad \beta_2 = \hat{\beta}_2\sqrt{|\tilde{s}_2|}.$$

For numerical integration System (6.72) should be transformed to Liapunov – Cauchy form at each integration step, since the matrix of coefficients at the second derivatives is nonlinear with respect to quantities $\dot{\tilde{s}}_1$ and $\dot{\tilde{s}}_2$ and is continuosly changing with time. Besides that, it is necessary to remember the whole 'history of the process' starting with the moment $t = 0$. That causes the solution of the stated problem to require great amounts of computer memory and computational time, and the smaller is the integration step, the greater are those quantities.

After transformation to Liapunov – Cauchy form the set of equations can be integrated by one of standard methods. As a matter of fact the computations were performed by Runge-Kutta method with 4-point approximation.

Consider a simpler system with inertialess drive and ideal differentiating circuit, i.e. assume $T = 0$ and $T_\psi = 0$. The controller equation in that case is

$$\delta = K_\psi \psi + K_{\dot{\psi}}\,\dot{\psi}, \tag{6.73}$$

with the same coefficients K_ψ, $K_{\dot{\psi}}$ as for control Law (6.64). Setting the time constants equal to zero, i.e. performing a transformation to Equation (6.73), lets us increase the integration step of Equations (6.63) by approximately an order and hence save the computer memory and decrease the integration time appropriately.

The study performed for the vehicle with controller described by (6.64) was repeated for the considered vehicle with controller described by (6.73). In other words, we calculated stability domains in the plane of ε_1, ε_2 for models with various degrees of completeness, and numerical integration was performed additionally. Results of this calculation, presented in Fig. 6.9 with the same denominations as in Fig. 6.8, completely corroborate the qualitative conclusions made from investigation of the system with control Law (6.64). It should be noted, that drive and differentiation circuit inertia makes the problem of oscillations amplitude stabilization at liquid wave motions frequencies essentially more complicated. Required damping for a system with inertialess controller in case of the same dampers is smaller, than for a system considering the controller's inertia, by a factor of 2-3, and for amplitude by a factor of 4-9.

For the system described by Equations (6.63) with parameters presented in Table 6.1 and controller described by (6.73) we performed numerical simulation of self-sustained oscillations reaching the steady state. The simulation results are presented in Fig. 6.10. Integration was performed with the step h = 0.05 s. Initial conditions were

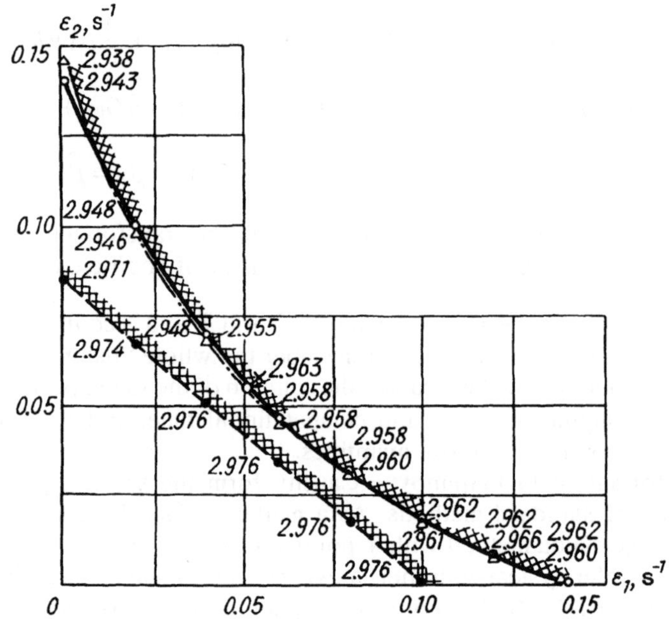

Fig. 6.9. Stability domain boundary in ε_1, ε_2 coordinates for inertialess drive. The hatching is at the stability side of the boundary; – o – Model (6.63); – • – Model (6.69); – Δ – Model (6.68), – × – numerical integration of Model (6.63)

specified for coordinates $s_1(0) = s_2(0) = 0.045$ m such, as for the system to reach oscillation amplitudes proximate to those, for which the required damping is provided. In the opposite case the process of reaching the limit cycle would take too much time because of extremely low oscillation increments. It is clear, that by the 25th second the system reaches steady self-sustained oscillations with constant parameters $\omega_a =$ $= 2.639\ s^{-1}$, $\psi_a = 0.039°$, $\dot{\vartheta}_{1a} = 0.00211\ s^{-1}$, $s_{1a} = 0.0465$ m, $\dot{\vartheta}_{2a} = 0.00195\ s^{-1}$, $s_{2a} = 0.051$ m, $\delta_a = 0.0034°$.

During the process of reaching the limit cycle beating is well seen in the system. It is caused by close frequencies corresponding to liquid natural oscillations frequencies. Equivalent linear damping calculated by Formula (6.47) matches the one calculated using stability domains with high accuracy. In Fig. 6.9 the point denoted by

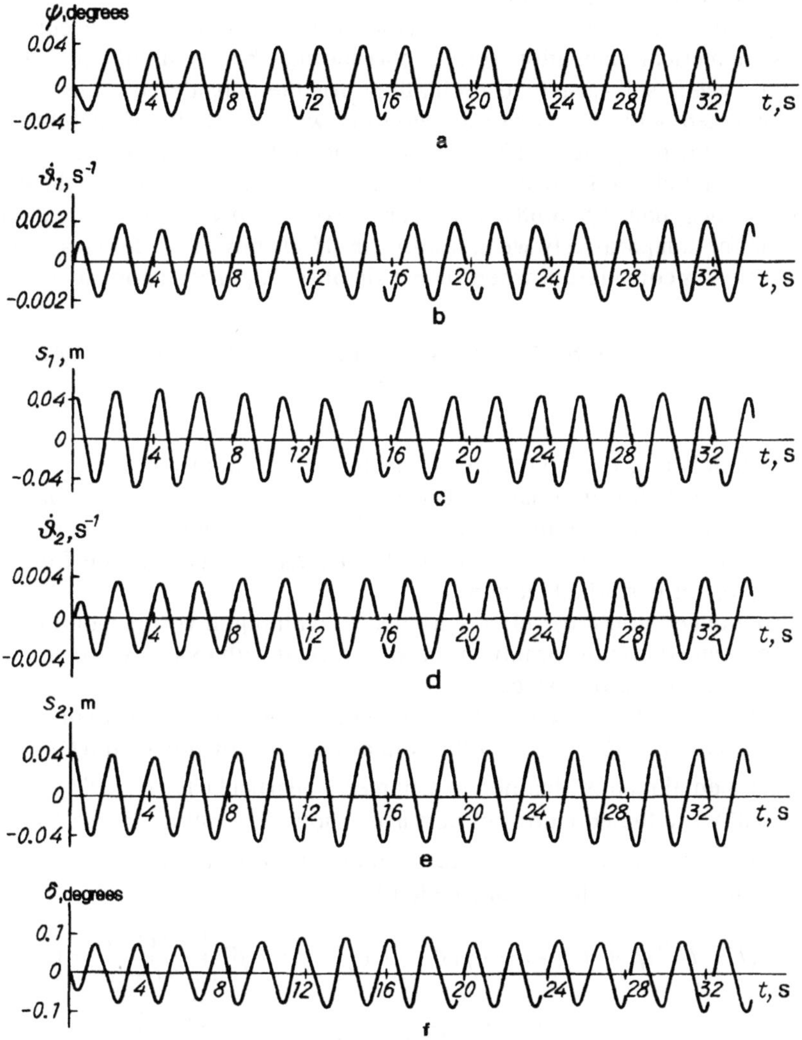

Fig 6.10. The process of reaching steady self-sustained oscillations: a) in the yaw angle ψ; b) in liquid vortex motions coordinate variation rate $\dot{\vartheta}_1$ in the first cavity; c) in liquid wave motions coordinate s_1 in the first cavity; d) in liquid vortex motions coordinate variation rate $\dot{\vartheta}_2$ in the second cavity; e) in liquid wave motions coordinate s_2 in the second cavity; f) in the control element deflection angle δ.

a cross corresponds to the damping calculated using the results of perturbed motion equations integration. Self-sustained oscillations frequency 2.963 is also shown there.

The following relationships are actual in the considered case: $r_0 \dot{\vartheta}_{1a} / s_{1a} = 0.034$, $r_0 \dot{\vartheta}_{2a} / s_{2a} = 0.028$.

It is clear from these relationships that the hypothesis of liquid wave motions dominating role in relative velocities formation is followed with high accuracy.

Thus, numerical simulation results corroborate the possibility of investigating closed-loop system self-sustained oscillations for the considered scale of nonlinear effects connected with liquid vortex motions with the help of stability domains construction in the damping coefficients plane, according to the scheme similar to that presented in Article 6.3. Moreover, Equations (6.63) can be replaced by the approximate ones, namely (6.68), which are incomparably simpler to deal with. This is all correct for comparatively weak nonlinear effects however. Below an example of a specific object is considered, when such a simplified approach appears inapplicable.

6.5. SC Stability in the Roll Plane for Non-Small Amplitudes of Angular Oscillations

For a concluding example let us consider an essentially nonlinear problem of roll-plane dynamics for an axisymmetric SC with two cylindric tanks, one of which contains 6 radial ribs having relative width $\bar{b} = 0.3$ with the simplest angular stabilization controller [50, 54]. We shall consider in this case the following two specific features of perturbed motion.

1. Pronounced nonlinear effects connected with liquid vortex motion.
2. Pronounced nonuniformity of the vortex field with extensive vortex formation at rib edges and stalling of vortices.

Let us abandon therefore now the hypothesis assuming the dominating role of the vortex field component uniform over the volume occupied by liquid, and take advantage of the experimental results obtained in Chapter 4 (particularly, $I^* \neq I^{(0)} - I$).

We shall write the equations of perturbed motion, resulting from (2.122), in the form of (4.10) (the mathematical model with empiric values I^* and β^*, considering the additional liquid 'vortex' degree of freedom):

$$(\overset{\circ}{I} + I^{(0)})\ddot{\varphi} + I^*\ddot{\chi} = -a_{\varphi\delta}\,\delta, \quad I^*(\ddot{\chi} + \ddot{\varphi}) + \beta^* \int_{-\infty}^{t} \frac{\ddot{\chi}(\tau)\,d\tau}{\sqrt{t-\tau}} = 0, \tag{6.74}$$

$$T\dot{\delta} + \delta = a_\varphi\,\varphi + a_{\dot{\varphi}}\,\dot{\varphi}, \quad \beta^* = \hat{\beta}^*\sqrt{|\chi|},$$

where T is the drive time constant, specifying its inertia; a_φ, $a_{\dot{\varphi}}$ are control law coefficients. Parallel to (6.74) we shall consider the mathematical model following from Model (4.7) with empiric value β:

$$(\overset{\circ}{I} + I)\,\ddot{\varphi} + \beta \int_{-\infty}^{t} \frac{\ddot{\varphi}(\tau)\,d\tau}{\sqrt{t-\tau}} = -a_{\varphi\delta}\delta, \tag{6.75}$$

$$T\dot{\delta} + \delta = a_\varphi\,\varphi + a_{\dot{\varphi}}\,\dot{\varphi}, \quad \beta = \hat{\beta}\sqrt{|\varphi|}.$$

Instead of (6.75), an even simpler mathematical model is usually considered, which results from (6.75) after the same operations as in the transformation of (6.24) to (6.33):

$$(I^{\circ} + I)\ddot{\varphi} + \beta'\dot{\varphi} = -a_{\varphi\delta}\,\delta, \tag{6.76}$$

$$T\dot{\delta} + \delta = a_{\varphi}\,\varphi + a_{\dot{\varphi}}\,\dot{\varphi}, \quad \beta' = \widehat{\beta}\sqrt{|\dot{\varphi}|}\left(\frac{\pi\omega^{a}}{2}\right)^{1/2}.$$

Assuming the characteristic velocity equal to tangential velocity of a rib edge with respect to liquid at rest and assigning subscript 'm' to the model and 'r' to the real, object, we obtain from Formula (2.34) the following values of Strouhal numbers:

$$\mathrm{Sh}_r = \left[\frac{\varphi^a\,\omega^a\,r_0}{V^*}(1-\widetilde{b})\right]_r, \quad \mathrm{Sh}_m = \left[\frac{\varphi^a\,\omega^a\,r_0}{V^*}(1-\widetilde{b})\right]_m, \tag{6.77}$$

where φ^a is the amplitude of the tank angular oscillations with frequency ω^a, r_0 is its radius. Let us assume the rib relative width $\widetilde{b}$ and quantity of ribs for the model and real object the same. Also assume, that characteristic velocities V^* determining occurrence of extensive vortex stalling in rib edges vicinity are approximately equal. We obtain the following:

$$\frac{\mathrm{Sh}_m}{\mathrm{Sh}_r} = \frac{[\varphi^a\,\omega^a\,r_0]_m}{[\varphi^a\,\omega^a\,r_0]_r}. \tag{6.78}$$

Particularly, equal Strouhal numbers correspond to angular velocities $\dot{\varphi}^a = \varphi^a\,\omega^a$ inversely proportional to tanks radii:

$$\dot{\varphi}^a_m = \dot{\varphi}^a_r\left(\frac{r_{0r}}{r_{0m}}\right) \tag{6.79}$$

This relationship specifies applicability fields of the mathematical models considered in Chapter 4, when appropriate results are carried over to the real vehicle.

On the other side, within self-similarity domain in Strouhal number dimensionless parameters $\widetilde{I}^*$, $\widetilde{\beta}^*$ and $\widetilde{\beta}$ do not change when we proceed from the model to the real vehicle (see Chapter 4.5). The corresponding values of I^*, β^*, $\widehat{\beta}$ for the real vehicle can be calculated by formulae

$$I_r^* = \widetilde{I}^*\rho_r\,r_{0r}^5, \quad \beta_r^* = \widetilde{\beta}^*\rho_r\,r_{0r}^5, \quad \widehat{\beta}_r = \widetilde{\beta}\,\rho_r\,r_{0r}^5. \tag{6.80}$$

Numerical values of coefficients in Equations (6.74)-(6.76) are presented in Table 6.2. Dimension for the moment of inertia is $\mathrm{kg\cdot m^2}$, damping coefficients β – $\mathrm{N\cdot m\cdot s^2\cdot rad^{-1.5}}$, $a_{\varphi\delta}$ – $\mathrm{N\cdot m}$, $a_{\dot{\varphi}}$ – s, a_{φ} are dimensionless.

Table 6.2

Coefficients of the Perturbed Motion Equations

I°	$I^{(0)}$	I	I^*	$\widehat{\beta}^*$	$\widehat{\beta}$	$a_{\varphi\delta}$	a_{φ}	$a_{\dot{\varphi}}$
7200	14400	3520	12960	26870	8957	192000	0.448	0.185

According to the introduced self-similarity Criterion (6.79), we can carry the results obtained in Chapter 4 over to the vehicle considered in this methodological example and determine angle φ amplitude ranges, for which one or the other model is preferable. In the interval $\varphi^a = 0-1.5°$ the considered processes should be described with Model (6.75), in the interval $\varphi^a = 1.5-3.0°$ Models (6.74) and (6.75) are equivalent, and for amplitudes $\varphi^a > 3°$ Model (6.74) should be used.

Let us consider stability domains for Models (6.74)-(6.76) in T, φ^a plane, where T is the drive time constant, φ^a are the amplitudes of the limit cycle, which takes place if conditions of system stability 'in the small' are violated. To construct those domains let us perform numerical integration of nonlinear differential Equations (6.76) and nonlinear integrodifferential Equations (6.74), (6.75) by the method described in

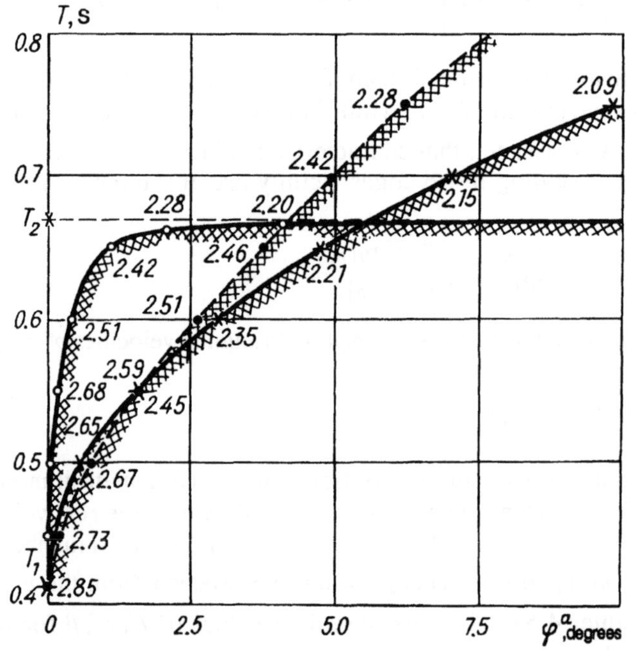

Fig. 6.11. Stability domain boundaries in coordinates T, φ^a. Stability is at the hatched side; - o - Model (6.74); - × - Model (6.75); - • - Model (6.76)

Chapter 3. To apply this method we shall introduce a new variable

$$\ddot{\chi}(t) = \sigma_\chi(t) \tag{6.81}$$

and write down the integral equation with respect to it, assuming the process pre-history in $\ddot{\chi}(t)$ absent:

$$\int_{-\infty}^{t} \frac{\ddot{\chi}(\tau)\, d\tau}{\sqrt{t-\tau}} = \int_{0}^{t} \frac{\sigma_\chi(\tau)\, d\tau}{\sqrt{t-\tau}} \cong \sqrt{h}\left\{G_\chi(t) + \frac{4}{3}\sigma_\chi(t)\right\}. \tag{6.82}$$

Having substituted (6.81) into (6.74), we obtain a set of equations that can be integrated by any of standard methods (Runge-Kutta, Adams-Stormer, *etc.*):

$$(I^\circ + I^{(0)})\,\ddot{\varphi} + \overset{*}{I}\sigma_\chi = -\,a_{\varphi\delta}\delta, \quad \left(I^* + \frac{4}{3}\beta^*\sqrt{h}\right)\sigma_\chi = -\,(I^*\ddot{\varphi} + \beta^*\sqrt{h}\,G_\chi), \tag{6.83}$$

$$\ddot{\chi} = \sigma_\chi, \quad T\dot{\delta} + \dot{\delta} = a_\varphi\,\varphi + a_{\dot{\varphi}}\,\dot{\varphi}, \quad \beta^* = \hat{\beta}^*\sqrt{|\chi|}.$$

Function $G_\chi(t)$ in Equations (6.83) is calculated by Formulae (3.55) at each integration step.

The obtained stability domains are presented in Fig 6.11. Double hatching is within the stability domain. The curves numbering corresponds to the limit cycle frequency.

The obtained stability domains analysis brings us to the following conclusions:

1. There are two critical values $T_1 = 0.415$ s and $T_2 = 0.67$ s of the drive time constant. For $0 < T < T_1$ the system is stable 'in the small'. For $T > T_1$ the system appears unstable 'in the small'. Stability condition violation is due to phase lag introduced to the system by the angular stabilization controller, while for the considered vehicle stability as a solid body phase lead is required.

2. In case of Models (6.75), (6.76) the system is stable 'in the great' for any $T > T_1$. The only possible stable mode is the mode of self-sustained oscillations with an amplitude monotonically increasing with parameter T. This mode is provided by nonlinear damping of liquid oscillations.

As it is clear from Fig. 6.11, up to amplitudes of about $\varphi^a = 4°$ the simplest mathematical Model (6.76) and Model (6.75) are practically equivalent. On the other side, mathematical Model (6.74) in the range of $\varphi^a = 0 - 2.5°$ corresponds to self-sustained oscillations with amplitudes almost an order smaller than those obtained using Models (6.75), (6.76).

3. In case of Model (6.74) the system is stable 'in the great' for $T_1 < T \leq T_2$. For $T > T_2$ the system is unstable 'in the great', i.e. there is no such value φ^a of oscillation amplitude, for which a stable limit cycle can take place.

It is impossible to reveal this extremely dangerous mode within the scope of mathematical Models (6.75), (6.76). From the technical point of view this mode means loss of SC stability. Thus, a more rigorous mathematical description allows to reveal not only quantitative, but also some essential qualitative effects.

Fig 6.12 shows the results of Equations (6.74), (6.75), and (6.76) numerical integration for $T = 0.8$ s ($T > T_2$). It is clear from the processes in Fig. 6.12 that, depending on the used mathematical model, either a stable limit cycle with quite admissible parameters takes place in the system (Models (6.75), (6.76)) or dangerous oscillations with increasing amplitude develop (Model (6.74)).

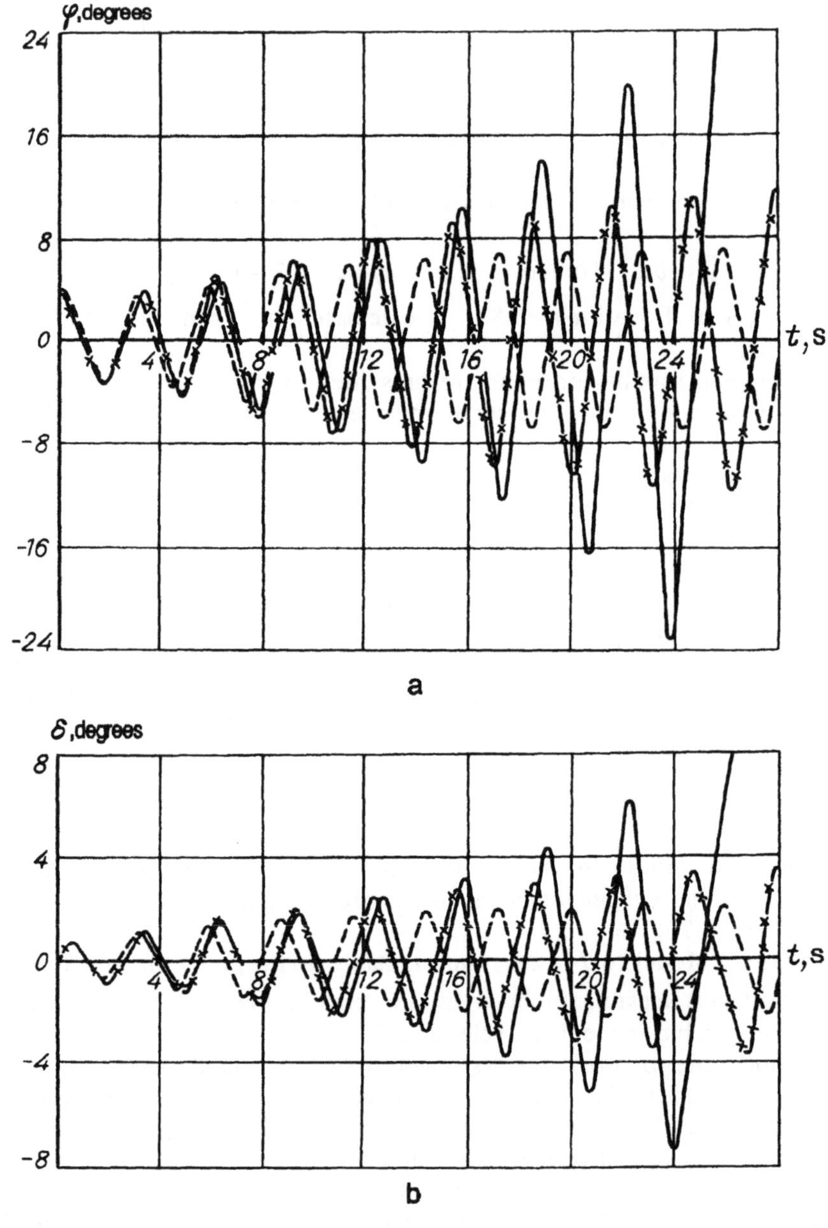

Fig. 6.12. SC motion around its longitudinal axis; φ is the roll angle; δ is the control deflection angle; ——— Model (6.74); – × – Model (6.75); – – – Model (6.76)

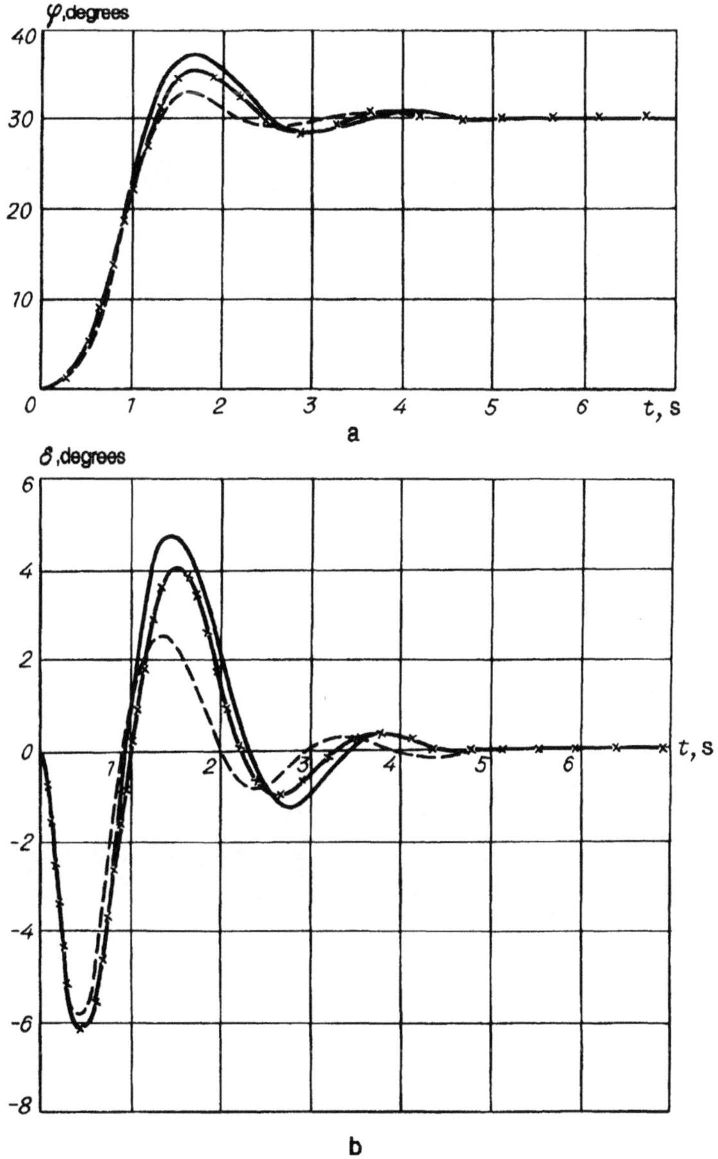

Fig. 6.13. SC turn in roll for a specified angle within the domain of stability 'in the small'; φ is the roll angle; δ is the control deflection angle; ——— Model (6.74); –×– Model (6.75); – – – Model (6.76)

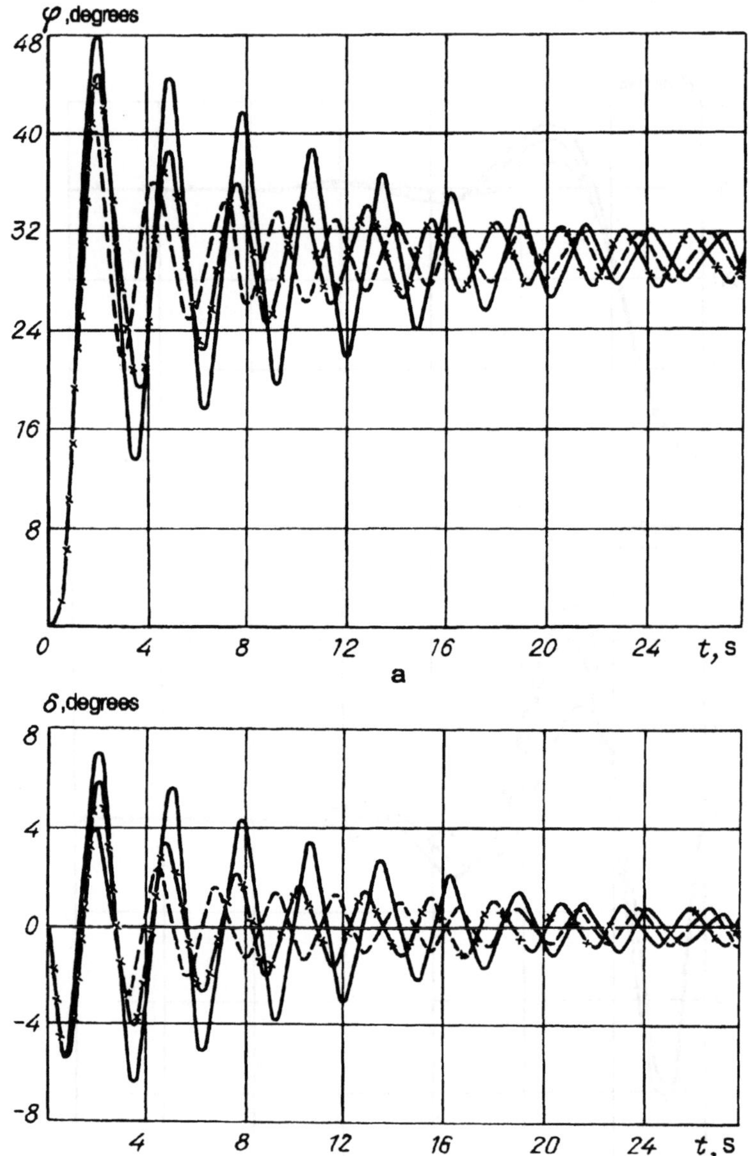

Fig. 6.14. SC turn in roll for a specified angle within the domain of instability 'in the small' and stability 'in the great'; φ is the roll angle; δ is the control deflection angle; ——— Model (6.74); – × – Model (6.75); – – – Model (6.76)

Let us in conclusion consider the problem of SC turn in roll for a specified angle with the same three modifications of mathematical model for two drive time constant values.

1. $T = 0.2$ s ($T < T_1$) corresponding to the system stable 'in the small'.

2. $T = 0.55$ s ($T > T_2$) corresponding to the system unstable ' in the small' and stable 'in the great'.

The integration results for $T = 0.2$ s are presented in Fig. 6.13. It is clear that transient processes described by mathematical Models (6.74) and (6.75) differ insignificantly both in duration and magnitude of overcontrol. This results from the fact, that the difference of damping forces caused by liquid oscillations, occurring when we change from one model to another, is insignificant, compared to the angular stabilization controller effect which is damping for ($T < T_1$).

Therefore an intermediate mathematical Model (6.75) can be used for closed-loop system stability investigation, when the system is within the domain of stability 'in the small'. At the same time, usage of conventional Model (6.76) may cause groundless optimism.

The integration results for $T = 0.55$ s are presented in Fig. 6.14. Within the domain of instability 'in the small' the angular stabilization controller can be considered as a source of 'negative damping'. When liquid oscillations reach some specific amplitude, a balance is established of this damping and the positive damping provided by in-tank devices, i.e. a stable limit cycle takes place. In this situation the difference of rib and liquid interaction description in different models reveals more explicitly.

The presented results are a good illustration of Model (6.74) features exposed in Chapter 4. Particularly, for high amplitudes of liquid oscillations the system described by Model (6.74) has weaker dissipation properties, than the systems corresponding to Models (6.75), (6.76). With oscillations attenuation, however, the roles of Models (6.74) and (6.75), (6.76) change gradually, and the first one displays more pronounced dissipative properties, than the others. Hence, although the processes corresponding to all the models converge to steady self-sustained oscillations, Model (6.74) provides lower amplitudes than Models (6.75) and (6.76).

The discussed examples are a convincing evidence of the fact, that when analyzing dynamics of a SC with developed in-tank devices, it is necessary to thoroughly consider fairly subtle effects connected with LV liquid vortex motion.

Chapter 7

Examples of Control Law Synthesis for an Object Described by a Vortex Model

Le doit y avoir une certaine proportion entre les actions et les desseins, si on en veut tirer tons les effects qu'elles peuvent produire.

La Rochefoucauld*

In Chapters 5 and 6 by the examples of specific engineering objects we showed the necessity to consider rather a strong influence of vorticity when solving the problems of closed-loop systems analysis and synthesis. In Chapter 5 we obtained a controlled electromagnet mathematical model representation considering skin effect in HECF core. The model has the form of a set of ordinary differential equations. This set of equations was tested for a closed-loop system with a control law of simple structure, synthesized ignoring eddy currents. A similar model was obtained for vehicles containing cavities partially filled with LV liquid, those cavities having developed damping devices inside.

But the mathematical model adequacy should be tested for a wider class of systems, primarily those with control laws synthesized considering eddy currents, and for nonlinear systems, particularly relay-type ones. By adequacy of a mathematical model we mean the possibility to construct a control law suitable for the real object. The tests should be carried out with the help of well developed methods of automatic control theory and numerical methods for modelling systems of that class. Those methods were presented in Chapter 3. Such testing will be performed in the present chapter for stabilized objects with electromagnetic actuators, for which the problem of synthesizing the structure and calculating the parameters of the control law is the especially urgent.

The main contents of the chapter is as follows:

a) magnetic levitation system control law synthesis considering skin effect in HECF core;

b) estimation of the introduced simplified mathematical model applicability limits based on complete analysis of dynamics for the closed-loop system with those control laws;

c) illustration of possibilities of analytical and numerical methods by those examples.

---

* La Rochefoucauld. Maximes et reflections morales. Precedees d'une par E. Deshanel, Paris, 1881, p. .51.

Carrying out this program allows to bring the 'idea' ('desseins' in French) and the 'action' to correspondence to some extent, the synthesize of control laws, that can be implemented in hardware and are 'adjusted' to the real specific features of the control object.

7.1. A Control Law Allowing Hardware Implementation, Based on Air Gap Sensor and Current Transducer Indications

Let us introduce the concept of a control law allowing hardware implementation [52, 67]. In Chapter 5 we considered a control law based on air gap sensor and accelerometer indications. Due to disadvantages caused by the accelerometer (law noise immunity, cross coupling of the channels), that control law may appear inefficient for objects subject to vibrational disturbances. Therefore it is expedient to consider the question of constructing a control law based on the indications s and I of the air gap sensor and current transducer only. A system with this control law can be expected to be more reliable, although it requires more complex methods of analysis and synthesis.

It is nessesary to note, that existance of 'spurious' time constants of real differentiating circuits for any control law degrades the closed-loop system quality (stability margins, frequensy response). On the other hand, an attempt to lower those time constants results in abrupt decrease of noise immunity. Therefore it is necessary to take into consideration the accuracy of differentiating circuits implementation when synthesizing control laws, namely, the mentioned 'spurious' time constants should be included in the control law structure.

From the point of view of practical control law implementation a very essential moment is the real amplifier-converter dynamic and static characteristics consideration. The amplifier-converter discussed above is constructed based on pulse-duration modulation (PDM) principle. It is a fairly complex device. Significant simplification of the control system power plant and hence reliability increase can be achived by using a relay-type control law, when the actuator is an ordinary thyristor switch. Therefore the relay-type control law is considered below in detail parallel to the analog one. Usage of the relay-type control law based on thyristors is preferable, for it allows to essentially increase the system operating voltages and upgrade all its quality indices.

By *hardware-realizable* we further mean a control law satisfying the requirements mentioned above.

Consider rather a simple control law based on the air gap sensor and current transducer indications, in the synthesis of which one of practical realizability aspects is taken into account, namely, that the real differentiating circuits are sluggish. That means that control operators $\Im_s(p)$ and $\Im_I(p)$ described by linear-fractional transfer functions must have the denominator's order not less than the numerator's one. The calculation scheme of the real control law parameters should include time constants calculation for the real differentiation circuits. Besides that, in order to provide the system noise immunity it is desirable to perform only one differentiation of the air gap sensor indications, because this channel is subject to greater noise interference than the current channel.

Let us write down the characteristic equation of System (5.25). Its adequacy is shown in Chapter 5 by an example of a simple control law. Let us use the following control law now.

$$\frac{\Delta V}{R} = \Im_s(p)\Delta S_s(p) - \Im_I(p)\Delta I_s(p)\,. \tag{7.1}$$

The characteristic equation is as follows:

$$\begin{aligned}
&[T(T_{J1} - T_J)]\,p^4 + (T + T_{J1})\,p^3 + p^2 + \left(a_{\eta\delta}c'_{\eta\eta} - a_{\eta\eta}T_{J1}\right)p - \\
&\quad - a_{\eta\eta} + a_{\eta\delta}\Im_s(p)\left[(T_{J1} - T_J)\,p + 1\right] + \\
&+ \Im_I(p)\left[T_{J1}p^3 + p^2 + \left(a_{\eta\delta}\,c'_{\eta\eta} - a_{\eta\eta}T_{J1}\right)p - a_{\eta\eta}\right] = 0.
\end{aligned} \tag{7.2}$$

Extract the polynomial multiplied by operator $\Im_I(p)$ and set it equal to zero:

$$T_{J1}\,p^3 + p^2 + \left(a_{\eta\delta}c'_{\eta\eta} - a_{\eta\eta}T_{J1}\right)p - a = 0. \tag{7.3}$$

As the calculation for specific electromagnet parameters values shows, Equation (7.3) has the following three roots: $p_1 = c^{\circ}_{\eta\eta}$ $\left(c^{\circ}_{\eta\eta} > 0\right)$ is a real root characterizing static instability of the object of control; $p_{2,3} = -\alpha^* \pm i\sigma^*$ $\left(\alpha^* > 0\right)$ is a pair of complex conjugated stable roots.

Denote $T_{J1} - T_J = T_F$ and rewrite the characteristic Equation (7.2):

$$\begin{aligned}
&TT_Fp^4 + (T + T_{J1})p^3 + p^2 + (a_{\eta\delta}c'_{\eta\eta} - a_{\eta\eta}T_{J1})\,p - a_{\eta\eta} + a_{\eta\delta}\,\Im_s(p)\,(T_F\,p + 1) + \\
&\quad + \Im_I(p)T_{J1}\left[(p - c^{\circ}_{\eta\eta})\left(p^2 + 2\alpha^*p + \alpha^{*2} + \sigma^{*2}\right)\right] = 0.
\end{aligned} \tag{7.4}$$

The necessary stability margin and frequency features of the closed-loop system can be provided by a control law with the following operators, whose structure allows for hardware implamentation:

$$\Im_s(p) = \frac{a_0 + a_1p}{1 + T_Fp}\,, \qquad \Im_I(p) = \frac{b_2\,p^2 + b_1\,p}{1 + d_1\,p + d_2\,p^2}\,, \tag{7.5}$$

where

$$d_1 = \frac{2\alpha^*}{\alpha^{*2} + \sigma^{*2}}\,, \qquad d_2 = \frac{1}{\alpha^{*2} + \sigma^{*2}}\,.$$

The required roots p_i $(i = 1, \ldots, 4)$ of the closed-loop system with control law described by (7.5) are provided by coefficients a_0, a_1, b_2, b_1 choice. As it can be seen, operator $\Im_s(p)$ is implemeted by a first-order differentiation circuit, $\Im_I(p)$ by a second-order differentiation circuit. The latter has, however, admissible noise immunity, for operator

$$W_F(p) = \frac{1}{1 + d_1 p + d_2 p^2}$$

is a good filter beginning from fairly low frequencies.

Let us consider the ideology of control Law (7.5) construction. Let the electromagnet be the object of control, input to which is current variation ΔI, and the controlled parameter is absolute displacement coordinate variation $\Delta \eta$.Let at first the question, how the current ΔI is generated, remain open. We shall write down dynamics equations for such an object. They result from mathematical Model (5.25) as a particular case, if we exclude the second equation from it and set $\Delta s = -\Delta \eta$:

$$\Delta \ddot{\eta} = a_{\eta\eta}\Delta \eta + a_{\eta\delta}(\Delta I + J), \quad T_{J1}\dot{J} + T_J\Delta \dot{I} + J = -c'_{\eta\eta}\Delta\dot{\eta}. \tag{7.6}$$

Let us consider the transfer function for the object of control described by (7.6) from ΔI to $\Delta \eta$:

$$\frac{\Delta \eta(p)}{\Delta I(p)} = \frac{a_{\eta\delta}}{T_{J1}p^3 + p^2 + (a_{\eta\delta}c'_{\eta\eta} - a_{\eta\eta}T_{J1})p - a_{\eta\eta}},$$

that can be written as follows with the denominations intriduced:

$$\frac{\Delta \eta(p)}{\Delta I(p)} = \frac{a_{\eta\delta}d_2}{T_{J1}(p - c^{\circ}_{\eta\eta})(1 + d_1p + d_2p^2)}. \tag{7.7}$$

We shall introduce a model of the object's stable part (if we imagine, that the object of control can be divided into two parts, and that the stable part can be considered separetely from the whole object). In operator form this model is written as

$$\eta_m(p) = \frac{\Delta I(p)}{1 + d_1 p + d_2 p^2}. \tag{7.8}$$

The term 'model of the object's stable part' means that the 2nd order polynomial

$$d_2 p^2 + d_1 p + 1 = 0$$

has roots in the left half-plane of variable p.

It is possible to ensure stability and required dynamic properties for the object of control described by Equations (5.25), where voltage ΔV is the control action, and the dynamics of current ΔI formation is taken into account, by generating control proportional to displacement $\Delta \eta$ and its three derivatives. Variable $\eta_m(p)$ allows to obtain the quantity including $\Delta \dot{\eta}(p)$. Expression (7.7) and (7.8) yield

$$\eta_m(p) = \frac{T_{J1}}{d_2 a_{\eta\delta}}(p - c^{\circ}_{\eta\eta})\Delta \eta(p). \tag{7.9}$$

The operator, which models the object's stable part in (7.8), having a a second-order denominator enables practical accomplishment of double differentiation of $\eta_m(p)$ signal. We can thus simulate the second and third derivatives of $\Delta \eta(p)$, that results in a control law of the form of (7.5).

Thus, the obtained contral law of the form of (7.5) is based on on-board construction of stable part model of the object, in contrast to the conventional concept of a reference model, where the model of the object of control is constructed taking into consideration all the degrees of freedom including unstable ones. As it shown in publications [24, 90] devoted to rockets and space engineering objects, conventional modelling results in high sensitivity of the system with respect to the reference model parameters. This is due to the fact that the closed-loop system includes two unstable units, the electromagnet and its reference model. It is also important that the electromagnet has pronounced static instability which is not observed with rockets and space engineering objects, that are not so critical to the reference model adjustment accuracy.

It is shown below by analyzing a stabilization system with control Law (7.5), that such an approach allows to obtain quite satisfactiry characteristics of the closed-loop system.

Calculation of parameters of control Law (7.5) providing the necessary properties of the closed-loop system is performed as follows. At the initial approximation roots p_j $(j = 1, \ldots, 4)$ are specified. Setting the coefficients of the equation

$$\prod_{j=1}^{4} (p - p_j) = 0,$$

equal to the coefficients of the characteristic Equation (7.4) with control Law (7.5) we obtain the values of a_0, a_1, b_2, b_1. Stability margins can be significantly increased if, when calculating the contral aw parameters, we replace $c^{\circ}_{\eta\eta}$ by the quantity

$$c_{\eta\eta} = c^{\circ}_{\eta\eta}\,(1 + \Delta\widetilde{c}_{\eta\eta}), \quad \Delta\widetilde{c}_{\eta\eta} > 0.$$

Then by interactive variation of control law parameters relative to the initial approximation we can improve the closed-loop system frequency properties and increase stability margins.

To illustrate the properties of the discussed control law let us consider an example of calculating a system with the electromagnet, whose parameters are described by (5.13). Choose the initial approximation for the roots as follows: $p_j = -\ 100\ \mathrm{s}^{-1}$ $(j = 1,\ldots,4)$. Transform the set of integrodifferential Equations (5.12) to the set of ordinary differential Equations (5.25) by the harmonic linearization method at frequency $\Omega = |p_j| = 100\ \mathrm{s}^{-1}$. Polynomial (7.3) roots are

$$p_1 = c^{\circ}_{\eta\eta} = 67.235\ \mathrm{s}^{-1}, \quad p_{2,3} = -\alpha^{*} \pm i\sigma^{*} = -58.774 \pm i\,20.875\ \mathrm{s}^{-1}.$$

Introduce a shift $\Delta\widetilde{c}_{\eta\eta} = 0.5$ for the calculation of control law parameters.

$$\begin{aligned} &a_0 = 110171\ \mathrm{A\cdot m^{-1}}, \quad a_1 = 15087\ \mathrm{A\cdot m^{-1}\cdot s}, \quad T_F = 0.01\ \mathrm{s},\\ &b_1 = 3.05355\ \mathrm{s}, \qquad b_2 = 9.44\cdot 10^{-3}\ \mathrm{s}^2,\\ &d_1 = 3.0229\cdot 10^{-2}\ \mathrm{s}, \quad d_2 = 2.57295\cdot 10^{-4}\ \mathrm{s}^2. \end{aligned} \tag{7.10}$$

The closed-loop system 'Electromagnet (5.25) – control Law (7.5)' characteristic equation has the following roots:

$$p_1 = -15.762\,\text{s}^{-1}, \quad p_{2,3} = -58.744 \pm i\,20.875\,\text{s}^{-1},$$
$$p_4 = -100\,\text{s}^{-1}, \quad p_5 = -214.135\,\text{s}^{-1}, \quad p_{6,7} = -85.051 \pm i\,15.067\,\text{s}^{-1}. \tag{7.11}$$

For comparison we calculated the roots of the characteristic equation for a two-mass system. Presented here are two additional roots only, corresponding to the sprung mass:

$$p_{8,9} = -1.0546 \pm i\,6.294\,\text{s}^{-1}. \tag{7.12}$$

The other roots of (7.11) varied within the scope of 1% due to consideration of the additional degree of freedom. The partial roots corresponding to the sprung mass degree of freedom, understood as roots for the "frozem" electromagnet, are:

$$\overset{\circ}{p}_{8,9} = -\frac{\varepsilon_2}{2} \pm i\left(\omega_2^2 - \left(\frac{\varepsilon_2}{2}\right)^2\right)^{1/2} = -1.2 \pm i\,6.26\,\text{s}^{-1}. \tag{7.13}$$

These are the roots for the 'frozen' electromagnet. We can see, that partial roots practically coincide with the closed-loop system roots. This corroborates the accuracy of the assumed hypothesis of 'dynamic decoupling' of the sprung mass and the electromagnet motions and the accuracy of the two-mass system reduction to the one-mass system. This feature enables separate design of the secondary suspension system and the electromagnetic levitation system for sufficiently 'soft' cushioning, that simplifies the problem significantly

Note, that the time constant T_F is a quantity of about 1/14 of T_d, where $T_d = a_1/a_0$ is the air gap differentiation circuit time constant. Therefore the differentiation circuit described by $\Im_s(p)$ is hardware-realizable with good noise immunity. The double differentiation circuit implementing the operator $\Im_I(p)$ has good noise immunity too.

In order to illustrate the considered control law features and to compare the mathematical Models (5.25) and (5.12) with the same control law allowing for eddy currents, all the computations were performed in parallel for both models. Note, that stability of the closed-loop system described by (5.12) with control Law (7.5) was determined according to stability Criterion (3.20). The characteristic equation of that system has the following roots in q plane for nominal values of the object's and controller's parameters:

$$q_{j,j+1} = \alpha_{j,j+1} + i\beta_{j,j+1}, \quad j = 1, 3, 5, \ldots, 11. \tag{7.14}$$

(a zero root is omitted here, just as when analyzing stability of the system with the control law including acceleration transducer indications). Real and imaginary parts of the roots are presented in Table 7.1.

Table 7.1

Roots of the Characteristic Equation

j	$\alpha_{j,j+1}$, $s^{-1/2}$	$\beta_{j,j+1}$, $s^{-1/2}$	j	$\alpha_{j,j+1}$, $s^{-1/2}$	$\beta_{j,j+1}$, $s^{-1/2}$
1	-0.29204	6.964	7	0.46264	16.589
3	0.519412	4.005	9	7.3992	10.056
5	1.7323	6.1227	11	-13.985	9.3911

We can see that all the roots correspond to a stable system, according to Criterion (3.20).

The following closed-loop system characteristics were also calculated.

1) Stability domain in the plane of the object's parameters $a_{\eta\eta}$ and$a_{\eta\delta}$ deviations from their nominal (calculated) values, i.e. in $\Delta a_{\eta\eta}$, $\Delta a_{\eta\delta}$.

2) Amplitude response $A_{\eta\eta c}(\omega)$ of transfer function $W_{\eta\eta c}(p)$.

3) Domain of stability in ΔT and$\Delta\, \tau_J$.

4) Stability margins in control law coefficients.

Fig. 7.1 shows amplitude response $A_{\eta\eta c}(\omega)$ and stability domain in $\Delta a_{\eta\eta}$, $\Delta a_{\eta\delta}$. Fig. 7.2 shows stability domain in ΔT, $\Delta\, \tau_J$. The calculations corresponding to Model (5.12) are drawn in solid lines and those for Model (5.25) in dash-dot lines.

Table 7.2 presents stability margins in control law coefficients. $\Delta < 0\%$ denotes stability margin percentage when mowing down in the parameter and $\Delta > 0\%$ when moving up in it. No testing for stability margins exceeding 100% was performed. Upper lines show stability margins for Model (5.12) and lower lines for Model (5.25).

Table 7.2

Stability Margins in Control Law Coefficients

Coefficients	$\Delta < 0$, %	$\Delta > 0$, %
a_0	- 97.5	> 100
	- 97.5	> 100
a_1	- 45	> 100
	- 48.5	> 100
b_1	- 38.5	93
	- 45.5	> 100
b_2	< - 100	> 100
	< - 100	> 100
d_1	- 49	69
	- 51.5	84.5
d_2	- 64.5	94
	- 64.5	95

Following are some comments regarding the results obtained. The magnetic levitation system has good quality in following low-frequency disturbances of the ferromagnetic rail and filtrates high-frequency ones (the resonance of $A_{\eta\eta c} \approx 2{,}1$ occurs at frequency $f \approx 25$ Hz, where high-level noise is hardly possible).

Stability margins in the main parameters of the electromagnet $\Delta a_{\eta\eta}$, $\Delta a_{\eta\delta}$, ΔT, $\Delta \tau_J$ cover possible operating variations, that allows to ensure levitation system

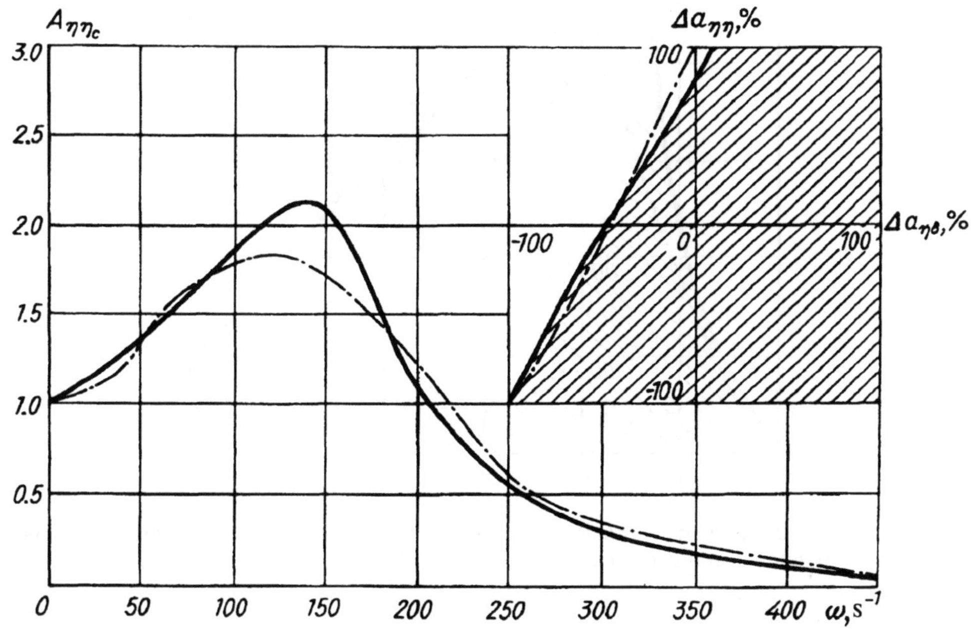

Fig. 7.1. Closed-loop system amplitude response and stability domains in coordinates $\Delta a_{\eta\eta}$, $\Delta a_{\eta\delta}$ for the control law using current in the analog modification. Stability is at the hatched side; —— Model (5.12); – · – · – Model (5.25)

stability, when the accuracy of mathematical model identification is high (good matching of the calculated and actual characteristics of the electromagnet.

Stability margins in some control law coefficients are a little lower than in the parameters of the object of control. The hardware implementation of those coefficients can, however be carried out with much higher accuracy than the electromagnet characteristics. The required accuracy of control law parameters implementation is about $\pm 10\%$, that is quite realistic.

We can conclude from the above, that, with the considered control law synthesized based on the simplified mathematical model of eddy currents in a HECF core, the stabilization system meets the necessary requirements. We should only note, that good object identification and accurate controller implementation are needed.

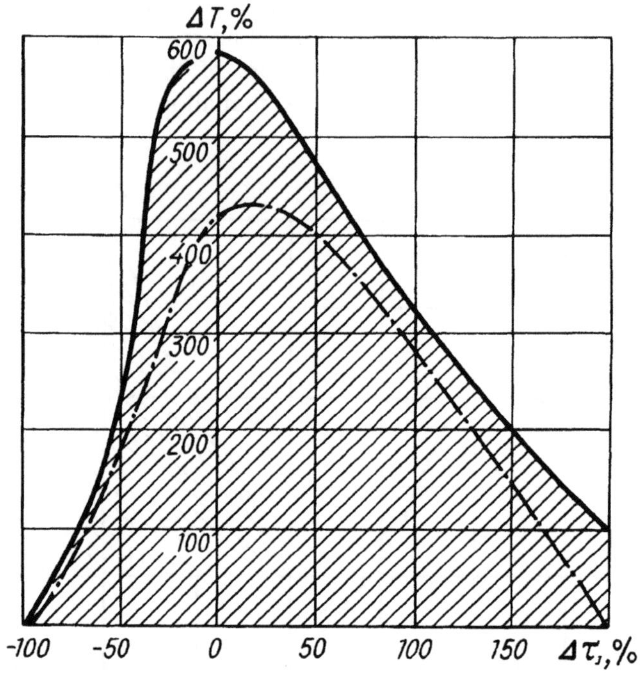

Fig. 7.2. Stability domains in coordinates ΔT, $\Delta \tau_J$ for the control law using current in the analog modification. Stability is at the hatched side; —— Model (5.12); – · – · – Model (5.25)

Practical coincidence of the closed-loop system principal characteristics obtained for the rigorous mathematical Model (5.12) and for its simplified counterpart of (5.25) corroborates the applicability of the latter for control law design.

An important characteristic of any control system is its noise immunity. By *noise immunity* we, as usual, mean the system's property to remain stable under external noise interference. Such noise may originate from the following sources:

1) noise due to the ferromagnetic rail oscillations at frequencies exceeding 5-6 Hz, which determine the programmed motion of the object;

2) noise of 'electrical' origin, among which we can outline e.g. the power source noise having the frequency of 100Hz for full-wave rectification of the supply-line voltage (the major part of this noise gets to the system via current transducer);

3) other noise getting to the system via measurement channels.

The main nonlinear factor causing loss of system stability under noise interference is the amplifier-converter nonlinearity, the static characteristic of which is presented in Fig. 5.1. The mechanism of loss of stability at the frequencies lying out of the operating range consists in the fact, that under the noise interference the total amplification coefficient of the control law decreases due to static characteristic 'saturation'. For specific level of noise the total amplification coefficient reaches some critical value, below which the system loses stability at operating frequencies. Within the operating frequency range the mechanism, how the system loses stability, is of somewhat more complex character.

Let us study the system noise immunity relative to one frequency additive noise referred to the air gap sensor output, i.e. consider the air gap sensor signal consisting of the useful signal s and noise s_n. The noise referred to the current transducer output I_n is calculated by the formula

$$I_n = s_n \left\{ \frac{[a_0^2 + (a_1\omega_n)^2]\,[(1 - d_2\omega_n^2)^2 + (d_1\omega_n)^2]}{[1 + (T_F\omega_n)^2]\,[(b_2\omega_n^2)^2 + (b_1\omega_n)^2]} \right\}^{1/2} \cdot 10^{-3}, \qquad (7.15)$$

where ω_n is the angular frequency of the noise. It is assumed in Formula (7.15), that s_n is counted in millimetres and I_n in amperes.

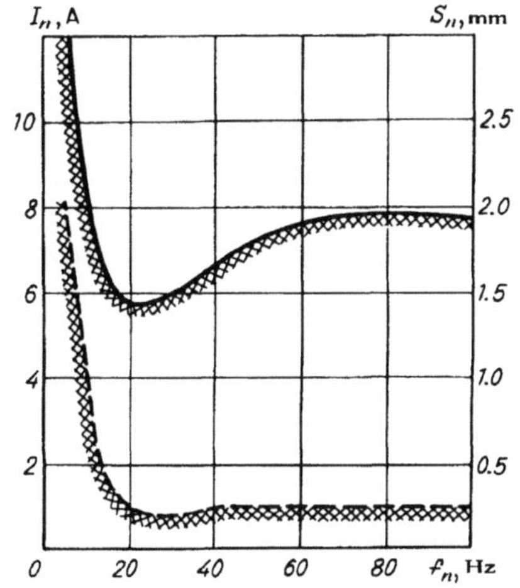

Fig. 7.3. Noise immunity curves for the analogue control law with current. Stability is at the hatched side; —— corrseponds to noise in the current measurement channel; - - - to noise in the air gap measurement channel

The calculation of noise immunity curves was performed by numerical simulation of the set of integrodifferential Equations (5.1) with control Law (7.5). By noise immunity curves we mean frequency dependence of the noise amplitude boundary value. When this value is exceeded, loss of stability occurs.

The results of the calculation are presented in Fig. 7.3. The noise immunity curves analysis shows that the most dangerous for the system is frequency f_n=10 Hz, which is the resonance frequency for the loop described by $\Im_I(p)$, where the amplitude critical value is 6 A, or 40% of the nominal current value. For all the other values noise level exceeds the actually expected values for both the air gap and current measurement channels. On the whole, we can calculate, that the considered analog control law meets all the requirements listed above.

7.2. Synthesis of a Relay-Type Control Law for the Electromagnetic Levitation System and Analysis of the Object – Controller Closed-Loop System

As it was noted above, construction of a system with a relay-type control law is caused by the necessity to simplify the control system power plant, the role of which in the analog modification is performed by the power amplifier implemented in the pulse-duration modulation scheme. Relay-type control law usage results in significant simplification of the amplifier-converter scheme [52, 67].

The system with a relay-type control law is nonlinear and can be studied with the help of methods well known and widely used in engineering practice. The most convenient method of synthesis is the approximate method of harmonic linearization due to its simplicity and obviousness. An applicability limitation of this method is the existence of filtration properties of the linear part. For the considered system this limitation is met.

The problem of synthesizing an electromagnetic levitation system relay-type control law is as follows. It should provide the required quality of tracking ferromagnetic rail disturbances in the operating frequency range. And the invariably developing self-sustained oscillation process should have frequencies lying outside the operating range and amplitudes within the limits of control accuracy. The stabilization system should have sufficient stability margins in object and controller parameters and should have good noise immunity. When synthesizing control law we assume that frequencies of self sustained oscillations are high and the main motion of the system can be considered a slowly varying process relative to those frequencies.

To explain the suggested method of synthesis let us present the necessary minimum of information following the terminology and denominations of V.A.Besekersky and Ye.P.Popov [8]. Let the control system dynamics be described by the following equation:

$$Q(p)\,x + P(p)\,F(x, px) = S(p)\,f(t). \tag{7.16}$$

Consider $f(t)$ a slowly varying function that can be assumed constant at the characteristic period of self-sustained oscillations, i.e.$f(t) = f^\circ$ =const. We shell seek the solution of Equation (7.16) in the form of

$$x = x^\circ + x_a, \quad x_a = a \sin \omega t. \tag{7.17}$$

The first terms of nonlinearity $F(x,px)$ expansion considering x° displacement can be written in the form

$$F(x, px) = F^\circ + q_s x_a + \frac{q_c}{\omega} p x_a + \textit{high harmonics}, \tag{7.18}$$

where for $\theta = \omega t$

$$F^\circ = \frac{1}{2\pi}\int_0^{2\pi} F(x^\circ + a \sin \theta, a\omega \cos \theta)\, d\theta,$$

$$q_s = \frac{1}{\pi a}\int_0^{2\pi} F(x^\circ + a \sin \theta, a\omega \cos \theta) \sin \theta\, d\theta, \tag{7.19}$$

$$q_c = \frac{1}{\pi a}\int_0^{2\pi} F(x^\circ + a \sin \theta, a\omega \cos \theta) \cos \theta\, d\theta.$$

Let us substitute Expressions (7.17) and (7.18) into (7.16) and ignore higher harmonics. This yields equation

$$Q(p)(x_a + x^\circ) + P(p)\left(F^\circ + q_s x_a + \frac{q_c}{\omega} p x_a\right) = M^\circ, \tag{7.20}$$

where $M^\circ = S(0) f^\circ$. Following the principle of dividing motions into slowly and rapidly varying components, we can reduce Equation (7.20) to two equations describing the system motion in the interval of low and high frequencies respectively:

$$Q(p)\, x^\circ + P(p)\, \Phi_l(x^\circ) = S(p) f(t), \tag{7.21}$$

$$Q(p)\, x_a + P(p)\left(q_s + \frac{q_c}{\omega} p\right) x_a = 0, \tag{7.22}$$

where $\Phi_l(x^\circ) = F^\circ$.

Function $\Phi_l(x^\circ)$ is usually proximate to a linear function in a fairly wide range of variable x° change even for essentially nonlinear function $F(x, px)$ particularly a relay-type one. This allows to make another step of simplification and assume

$$F^\circ = \Phi_l(x^\circ) \cong K_l x^\circ, \qquad K_l = \left(\frac{d\Phi_l}{dx^\circ}\right)_{x^\circ=0}. \tag{7.23}$$

Considering (7.23) Equation (7.21) is replaced by equation

$$Q(p)\, x^\circ + P(p)\, K_l x^\circ = S(p) f(t). \tag{7.24}$$

The property of $F(x,px)$ nonlinearity smoothing by the self-sustained oscillation process is called *vibrational smoothing* or *vibrational linearization.*

Generally all the three harmonic linearization coefficients of (7.19) are functions of three variables:

$$F^\circ = F^\circ(x^\circ, a, \omega), \quad q_s = q_s(x^\circ, a, \omega), \quad q_c = q_c(x^\circ, a, \omega), \tag{7.25}$$

Equations (7.22) and (7.24) are therefore connected to each other via (7.25). For self-sustained oscillations investigation with no disturbances $f(t) \equiv 0$ we can assume $x^\circ = 0$. Then coefficients $q_s(x^\circ, a, \omega) = q_s(a, \omega)$ and $q_c(x^\circ, a, \omega) = q_c(a, \omega)$ are ordinary harmonic linearization coefficients for a monofrequency process. To study stability of a periodic process being the solution of Equation (7.22) we can apply the coefficients averaging method, according to which the periodic solution stability condition coincides with negativity condition for real parts of the characteristic equation roots (see V.A.Besekersky and Ye.P.Popov [8]):

$$Q(p) + P(p)\, \kappa(a_a) = 0, \tag{7.26}$$

where

$$\kappa(a_a) = \frac{1}{2\pi}\int_0^{2\pi} \left(\frac{\partial F}{\partial x}\right)_a d\theta, \qquad \theta = \omega t. \tag{7.27}$$

Subscript a is assigned to the studied periodic mode.

We now proceed directly to presentation of the suggested method of relay-type control law synthesis. As it is clear from (7.26) the condition of the periodic solution stability coincides with stability condition for some linear system. This enables to use for a control law the law, whose linear part coincides in structure with control Law (7.5) already obtained.

For a relay-type actuator in the simplest case we can use a relay with an ideal characteristic. The limit cycle adjusted to an ideal relay characteristic is essentially dependent on the following parameters of the real switch distinguishing it from the ideal one: 1) dead band; 2) hysteresis; 3) operation lag. And the stabilization system should be coarse with respect of these parameters, because they are hardly controllable and unstable. Therefore a hysteresis relay is preferable to an ideal relay. The hysteresis loop width is chosen much greater than its dead band and natural hysteresis. The phase lag introduced by the hysteresis relay should cover the operation lag.

When these conditions are satisfied, the system becomes coarse with respect to relay characteristics straggling. Hysteresis relay usage allows to lower the limit cycle frequency in comparison with the ideal relay. It is very important, because at high frequencies reliability of the mathematical model describing electromagnet dynamics degrades, and noise level becomes comparable to limit cycle amplitudes. By lowering the frequency we manage to remain within the frequency range, in which the obtained model of the electromagnet is true.

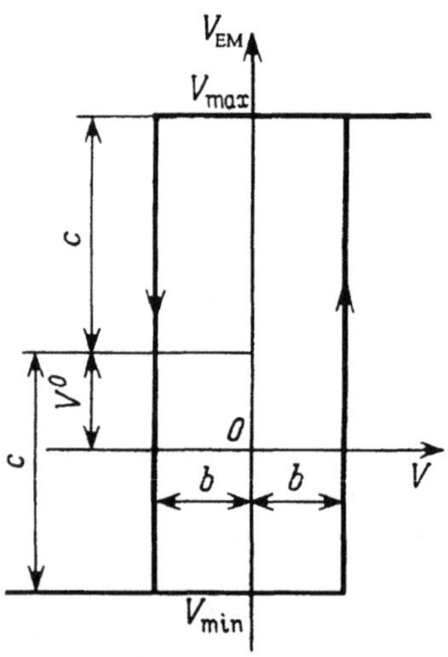

Fig. 7.4. Hysteresis relay static characteristic

The relay characteristic appearance with parameters, for which the study was performed, is shown in Fig. 7.4.

For such control law the scheme of amplifier-converter is considerably simpler. The hysteresis loop width equal to $0.2V^\circ$ was chosen based on the above considerations and for the signal, corresponding to the limit cycle, not to be suppressed by noise. The asymmetry of the characteristic in Fig. 7.4. is necessary for the constant component of the voltage V applied to the electromagnet to be equal to the nominal value V° for displaced self-sustained oscillations. In this case it is not necessary to introduce a setting to the self-sustained oscillations in a system linearized with respect to the equilibrium state, we can assume the relay characteristic symmetric.

Let us write down harmonic linearization coefficients, that will be used for further calculations:

$$q_s = \frac{4c}{\pi a_a}\left[1 - \left(\frac{b}{a_a}\right)^2\right]^{1/2} = \frac{4c}{\pi a_a}\left(1 - \lambda^2\right)^{1/2}. \tag{7.28}$$

$$q_c = -\frac{4cb}{\pi a^2} = -\frac{4c}{\pi a_a}\lambda. \tag{7.29}$$

$$K_l = \kappa = \frac{2c}{\pi a_a}\frac{1}{\left[1-\left(\frac{b}{a_a}\right)^2\right]^{1/2}} = \frac{2c}{\pi\, a_a\left(1-\lambda^2\right)^{1/2}}. \tag{7.30}$$

Here in our case for maximum acceptable voltage $V_{\max} = 80$ V we obtain $c = V_{\max} - V^{\circ} = 62.8$V, that is the maximum level of control signal in the system linearized with respect to equilibrium state; $b = 0.2\ V^{\circ} = 3.44$ V is the hysteresis loop width; a_a is the limit cycle amplitude; $\lambda = b/a_a < 1$ is a dimensionless variable having the sense of a quantity inverse to self-sustained oscillation amplitude.

To calculate the parameters of a control law having the structure similar to (7.5) let us construct the characteristic equation:

$$TT_F\,p^4 + (T + T_{J1})\,p^3 + p^2 + (a_{\eta\delta}c'_{\eta\eta} - a_{\eta\eta}T_{J1})\,p -$$
$$- a_{\eta\eta} + a_{\eta\delta}\left(q_s + \frac{q_c}{\omega_a}p\right)\frac{a_0 + a_1\,p}{1 + T_F\,p}(T_F\,p + 1) +$$
$$+\left(q_s + \frac{q_c}{\omega_a}p\right)\frac{b_2\,p^2 + b_1\,p}{1 + d_1\,p + d_2\,p^2}\left[T_{J1}\,p^3 + p^2 + (a_{\eta\delta}c'_{\eta\eta} - a_{\eta\eta}T_{J1})\,p - a_{\eta\eta}\right] = 0. \tag{7.31}$$

Having performed the necessary reductions, we shall write down Equation (7.31) using denominations introduced earlier, as follows:

$$TT_Fp^4 + (T + T_{J1})\,p^3 + p^2 + (a_{\eta\delta}c'_{\eta\eta} - a_{\eta\eta}T_{J1})\,p -$$
$$- a_{\eta\eta} + a_{\eta\delta}\left(q_s + \frac{q_c}{\omega_a}p\right)(a_0 + a_1p) +$$
$$+\left(q_s + \frac{q_c}{\omega_a}p\right)d_2T_{J1}\,(b_2\,p^2 + b_1\,p)\,(p - c^{\circ}_{\eta\eta}) = 0. \tag{7.32}$$

The reductions made mean, that stable roots of the following polynomials are excluded from consideration:

$$d_2p^2 + d_1p + 1 = 0, \quad T_Fp + 1 = 0.$$

Compose the characteristic equation to determine limit cycle stability:

$$TT_Fp^4 + (T + T_{J1})\,p^3 + p^2 + (a_{\eta\delta}c'_{\eta\eta} - a_{\eta\eta}T_{J1})\,p -$$
$$- a_{\eta\eta} + a_{\eta\delta}\,\kappa\Im_s(p)\,(T_Fp + 1) +$$
$$+ \kappa\Im_l(p)\,[T_{J1}\,p^3 + p^2 + (a_{\eta\delta}c'_{\eta\eta} - a_{\eta\eta}T_{J1})\,p - a_{\eta\eta}] = 0, \tag{7.33}$$

and compare Equations (7.33) and (7.2). They are the same, up to the multiplier κ at control law operators $\Im_s(p)$ and $\Im_l(p)$. Taking advantage of this feature, let us construct the system's limit cycle in such a way, that coefficients $a_0\kappa$, $a_1\kappa$, $b_1\kappa$, $b_2\kappa$ should be equal to those calculated while synthesizing the analog control law with parameters described by (7.10). With such technique the limit cycle is made stable automatically. Taking into account, that $K_l = \kappa$, we thus provide a repetition of the the system's frequency properties with respect to low-frequency disturbances. Such succession of analog and relay systems allows to apply the correction unit already synthesized to the relay system by changing the total amplification coefficient. This can be useful from the practical point of view.

Denote control law coefficients calculated for the analog modification of (7.10) by $\hat{a}_0$, $\hat{a}_1$, $\hat{b}_1$, $\hat{b}_2$. For a stable limit cycle the following relationships are to be fulfilled:

$$a_0\,\kappa = \frac{a_0\,2c}{\pi a_a\left(1-\lambda^2\right)^{1/2}} = \hat{a}_0, \quad a_1\,\kappa = \frac{a_1\,2c}{\pi a_a\left(1-\lambda^2\right)^{1/2}} = \hat{a}_1,$$

$$b_1\,\kappa = \frac{b_1\,2c}{\pi a_a\left(1-\lambda^2\right)^{1/2}} = \hat{b}_1, \quad b_2\,\kappa = \frac{b_2\,2c}{\pi a_a\left(1-\lambda^2\right)^{1/2}} = \hat{b}_2. \tag{7.34}$$

Let us seek the limit cycle according to the following algorithm. Number λ $(0 < \lambda < 1)$ is specified. Relationships (7.34) with specified a_0, a_1, b_1, b_2 and λ yield the following quantities

$$\frac{a_0}{a_a},\ \frac{a_1}{a_a},\ \frac{b_1}{a_a},\ \frac{b_2}{a_a}. \tag{7.35}$$

The condition of a stable limit cycle in the system for the chosen synthesizing technique is that characteristic Equation (7.32) should have a pair of complex conjugated pure imaginary roots

$$p_{ja} = \pm\, i\,\omega_a, \quad j = 1, 2. \tag{7.36}$$

Taking into account that

$$a_0 q_s = \frac{a_0}{a_a}\frac{4c}{\pi}\left(1-\lambda^2\right)^{1/2}, \quad a_0\,q_c = \frac{a_0}{a_a}\frac{4c}{\pi}\lambda \tag{7.37}$$

(and similar expressions for a_0, b_1, b_2) and that Quantities (7.35) are known for the specified λ, let us substitute $p = i\,\omega_a$ in (7.32) and set the real and imaginary parts equal to zero. Resulting is the condition of a pair of pure imaginary complex conjugated roots existence. It is written in the form of the folloing equations.

Real part of (7.32):

$$TT_F\omega_a^4 + d_2T_{J1}\,b_2\,q_c\,\omega_a^3 + (c^{\circ}_{\eta\eta}\,d_2T_{J1}b_2\,q_s - 1 - d_2T_{J1}b_1q_s)\,\omega_a^2 +$$
$$+ (c^{\circ}_{\eta\eta}\,d_2T_{J1}b_1q_c - a_{\eta\delta}\,a_1q_c)\,\omega_a + a_{\eta\delta}\,a_0\,q_s - a_{\eta\eta} = 0. \tag{7.38}$$

Imaginary part of (7.32)

$$-\ (T + T_{J1} + d_2T_{J1}b_2q_s)\,\omega_a^3 + (c^{\circ}_{\eta\eta}\, d_2T_{J1}b_2\, q_c - d_2T_{J1}b_1q_c)\,\omega_a^2 +$$
$$+\ (a_{\eta\delta}c'_{\eta\eta} - a_{\eta\eta}T_{J1} + a_{\eta\delta}a_1q_s - c^{\circ}_{\eta\eta}\, d_2T_{J1}b_1q_s)\,\omega_a + a_{\eta\delta}\, a_0\, q_c = 0. \qquad (7.39)$$

The problem of a stable limit cycle formation is reduced to finding λ value, for which Equations (7.34), (7.38) and (7.39) are satisfied simultaneously.

Let us seek the solution in the following way. For the specified λ value find Equation (7.39) roots, which are real, according to the calculations. Denote these roots by $\omega_{a1}, \omega_{a2}, \omega_{a3}$. The value of the root, for which Condition (7.38) is satisfied, is just the frequency of the possible limit cycle. Knowing λ_a and taking into account that $\lambda_a = b/a_a$, we obtain the amplitude of the signal output from the correction unit to the relay input:

$$a_a = \frac{b}{\lambda_a}\,. \qquad (7.40)$$

From (7.34) we find the values of coefficients for the control law providing the obtained limit cycle:

$$a_0 = \frac{\hat{a}_0\,\pi b\,(1 - \lambda_a^2)^{1/2}}{2c\,\lambda_a}, \qquad a_1 = \frac{\hat{a}_1\,\pi b\,(1 - \lambda_a^2)^{1/2}}{2c\,\lambda_a},$$
$$b_1 = \frac{\hat{b}_1\,\pi b\,(1 - \lambda_a^2)^{1/2}}{2c\,\lambda_a}, \qquad b_2 = \frac{\hat{b}_2\,\pi b\,(1 - \lambda_a^2)^{1/2}}{2c\,\lambda_a}. \qquad (7.41)$$

Equations (7.38), (7.39) were solved with the help of computers.

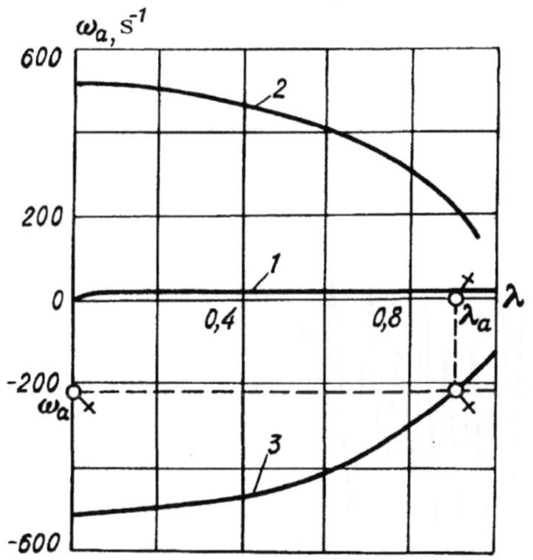

Fig. 7.5. Relay-type control law parameters determination. ω_a versus λ; 1 — ω_{a1}, 2 — ω_{a2}, 3 — ω_{a3}

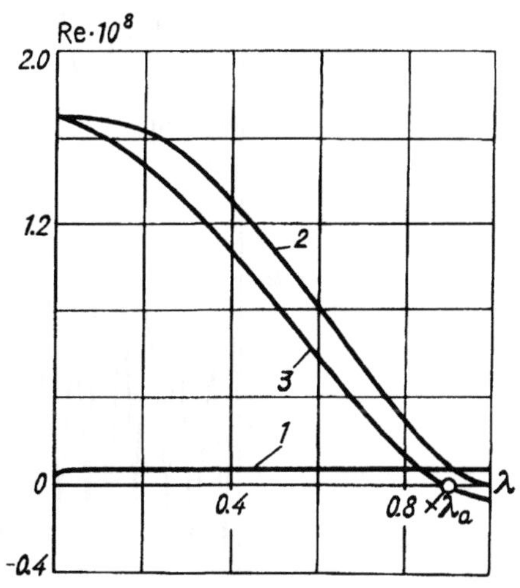

Fig. 7.6. Relay-type control law parameters determination. Re versus λ; 1 — Re_1, 2 — Re_2, 3 — Re_3

Based on the synthesis method presented above we calculated control Law (7.5) parameters for a system with a relay actuator. The actuator's characteristics are shown in Fig. 7.4. We used control law parameters calculated for the analog system described by (7.10).

Figs. 7.5 and 7.6 present a graphic form of Equations (7.34), (7.38), (7.39) computer solution. Equation (7.39) roots dependence on parameter λ is shown in Fig. 7.5; and Fig. 7.6 shows Equation (7.38) left side values Re_1, Re_2, Re_3 corresponding to roots ω_{a1}, ω_{a2}, ω_{a3} as functions of λ.
It is clear that there is only one solution corresponding to a stable limit cycle. The calculated value of the limit cycle frequency

$$\omega_a = 220\ \mathrm{s}^{-1}, \quad f_a = 36\ \mathrm{Hz}.$$

Control law coefficients, for which this limit cycle takes place, are as follows:

$$a_0 = 2160\ \mathrm{A\cdot m^{-1}}, \quad a_1 = 324\ \mathrm{A\cdot s\cdot m^{-1}},$$

$$b_1 = 0.066\ \mathrm{s}, \quad b_2 = 2.06\cdot 10^{-4}\ \mathrm{s}^2, \tag{7.42}$$

$$d_1 = 3.02289\cdot 10^{-2}\ \mathrm{s}, \quad d_2 = 2.573\cdot 10^{-4}\ \mathrm{s}^2, \quad T_F = 0.01\ \mathrm{s}.$$

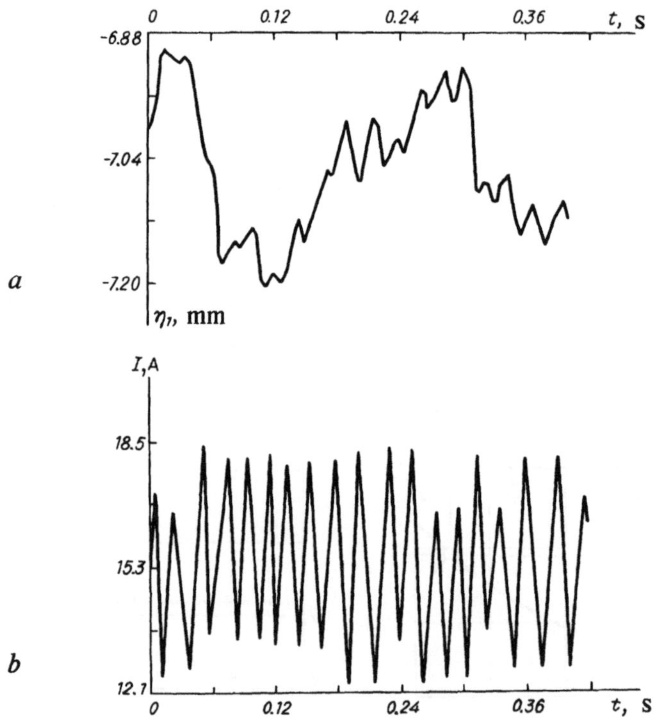

Fig. 7.7. Limit cycle in a relay-type system; *a)* time variation of the electromagnet absolute displacement coordiname η_1, mm; *b)* time variation of current in the electromagnet coil, *I*, A

As it was already mentioned, coefficients a_0, a_1, b_1, b_2 of (7.42) are proportional to coefficients, of (7.10) that enables to replace the amplifier with pulse-duration modulation by a relay actuator by means of installing a voltage divider with coefficient

$$k = \frac{a_0}{\hat{a}_0} = \frac{a_1}{\hat{a}_1} = \frac{b_1}{\hat{b}_1} = \frac{b_2}{\hat{b}_2} \approx 0.02.$$

between the correction unit and the relay. In order to check the obtained results we performed a simulation of System (5.1) with control Law (7.5) having the coefficients of (7.42) and with the relay actuator (Fig. 7.4). The simulation methodology will be described in the next article. Fig. 7.7 shows the process of reaching the limit cycle by the system starting with zero initial conditions. The following limit cycle parameters were obtained from the simulation: f_a=45 Hz, air gap oscillations amplitude s_a° = 0.04 mm for nominal air gap s° =7 mm, current oscillations amplitude I_a° = 2.5 A, for nominal current I° =15.6 A.

The calculated value of the limit cycle frequency differs from the exact one in 22%. Such a significant error is obviously due to the fact, that operator $\Im_l(p)$ is not a filter at high frequencies. Besides that, electromagnet nonlinear characteristics F, L ,c_y obtained experimentally and taken into consideration in the simulation model affect the accuracy noticeably.

But the main error is caused by the fact, that the filtration hypothesis for the considered object of control is obeyed not rigorously enough. It is easily seen from the current curve, in which the 3rd harmonic has a fairly high magnitude. The limit cycle frequency refinement taking into consideration the 3rd harmonic would yield a more accurate result. This is not necessary, however, for it is shown below, that the system synthesized according to the approximate technique results in quite satisfactory characteristics.

The limit cycle frequency obtained is, on one hand, higher than the main motion frequencies, and on the other hand, not great enough for the electromagnet dynamic characteristics instability to affect it. The limit cycle amplitude of the system principal coordinate (the air gap) is much less than deviations taking place for the required control accuracy, and at the same time it is not small enough to be buried in the noise background. The current limit cycle amplitude is also comparable in order with the possible noise.

Thus, the system obtained has quite satisfactory limit cycle parameters. In order to compare the relay-type and analog systems the following characteristics were computed:

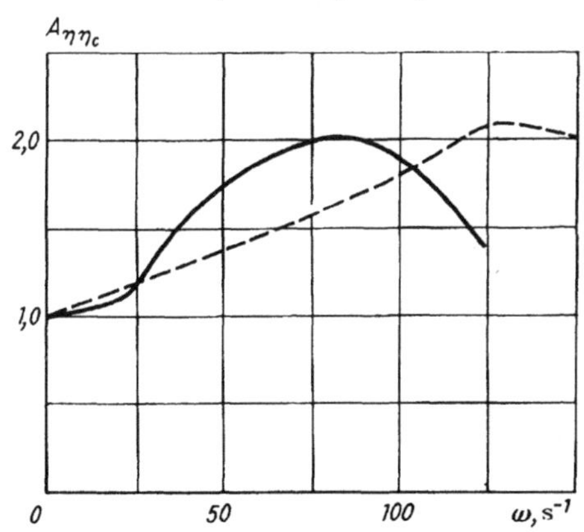

Fig. 7.8. Amplitude response for the system with relay-type control law; —— amplitude response for the relay-type system; - - - amplitude response for the analog system

1. Amplitude response $A_{\eta\eta c}(\omega)$ from ferromagnetic rail disturbance to electromagnet absolute displacement coordinate. The characteristics were computed by means of simulation. Note, that in a relay-type system perturbed motion for external harmonic disturbance is of rather complex character. Below ~20 Hz, however steady oscillations are quite noticeable at the external disturbance frequency, the limit cycle being added to those oscillations. At higher frequencies it appears impossible to extract the external disturbance harmonic, therefore frequency response was calculated up to the frequency of 20 Hz.

At each frequency the external disturbance amplitude was set about 50% of the limit value, at which the system loses stability. Amplitudes of $\eta_c(t)$ were set as follows: 0...3 Hz - 1 mm; 3...10 Hz - 0.5 mm; 10...15 Hz - 0.2 mm; 15...20 Hz - 0.1 mm.

2. Stability margins in control law coefficients.

3. Domains of noise immunity with respect to additive noise in air gap S_n and current I_n measurement channels, just as it was made for the analog system.

Studies in items 2 and 3 were also performed by means of simulation. In Fig. 7.8 the solid line shows amplitude response for the relay-type system, and the dashed line for the analog system for comparison.

As it seen, in the frequency range 0-20 Hz frequency response characteristics agree up to the accuracy of 20%. This corroborates the hypothesis of similar behaviour of the analog and the relay-type systems at law (operating) frequencies.

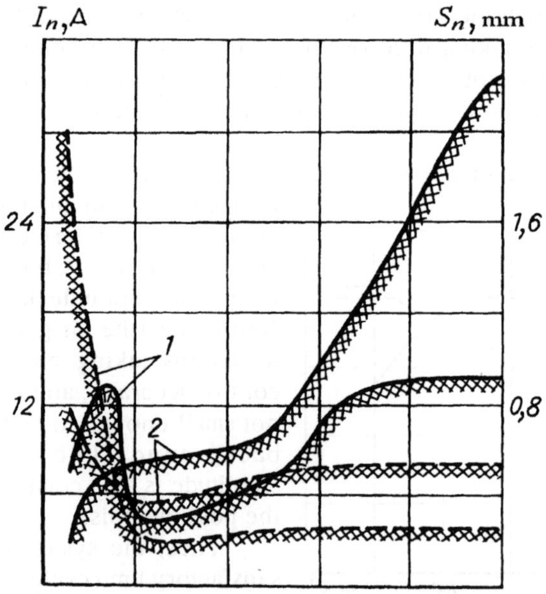

Fig. 7.9. Noise immunity curves for the relay-type system. Stability is at the hatched side; —— relay-type system; - - - analog system; 1 – noise in air gap measurement channel; 2 – noise in current measurement channel

Table 7.3 presents stability margins in control Law (7.5) coefficients with parameters specified by (7.42). Stability margins for the relay-type system control law coefficients are somewhat less than the margins for the analog system.

Table 7.3

Stability Margins in Control Law Coefficients

Coefficient	$\Delta > 0$, %	$\Delta < 0$, %
a_0	- 99	> 100
a_1	- 40	> 100
b_1	- 38	> 100
b_2	< - 100	> 100
d_1	- 46.5	84.5
d_2	- 58	87

In Fig. 7.9 the solid line shows noise immunity curves for the system with relay-type control law. The characteristics for the analog system are showed in dashed line for comparison. It is clear, that the relay-type system has a much better noise immunity than the analog one at high frequencies, where noise is most probable.

Joint analysis of dynamics of closed-loop systems with analog and relay-type control laws was performed within the scope of linearized mathematical model which was reduced to a set of ordinary differential equations by the harmonic balance method. This joint analysis corroborates, that those systems allow hardware implementation and satisfy the entire complex of technical requirements to the electromagnetic levitation system.

Let us proceed to analysis of closed-loop system specific features connected with both the mathematical description of the electromagnet in the form of a set of integrodifferential equations and nonlinear properties of the object and controller.

7.3. Mathematical Modelling of Electromagnetic Levitation System Dynamics

All the investigations presented above were performed based on the linearized model of the object of control. The obtained control laws were shown to satisfy the requirements of providing stability, quality of filtration , and noise immunity within the scope of the linear model for the object of control. But in such operating modes as 'flow-up' and (to a less extent) in the mode of control in response to disturbances stabilization system coordinates vary in the limits so wide, that the linear control model with an amplifier-converter does not guarantee, without a special testing, the control law serviceability in the variables values interval, where the above mentioned nonlinear effects are significant (see [52, 67]).

In Chapter 5 we presented the mathematical model of the controlled electromagnet with an elastoviscously suspended mass in the form of the set of nonlinear integrodifferential Equations (5.1). This set of equations has a number of features complicating its direct integration. Firstly, electromagnet characteristics F, L, α are nonlinear functions of the tabulated air gap s and variable sum $I+J$ (see Tables 4.1, 4.2, 4.3 and Figs. 4.8-4.10). Secondly, it is necessary to differentiate the tabulated function $L(s, I+J)$ and to integrate the integrand function $\alpha(s, I+J)$. Thirdly, Equations (5.1) are a set of integrodifferential equations, that should be transformed to Liapunov-Cauchy form for direct application of standard integration methods.

Differentiation and integration of tabulated functions can be avoided by change of variables. Let us introduce new variables similar to those used before (see (1.79)):

$$\Phi = L(s, I+J)(I+J). \tag{7.43}$$

$$\Psi = \alpha(s, I+J)\, J. \tag{7.44}$$

Variable Φ corresponds to magnetic flux. Expressing derivative $\dot{\Phi}$ in (5.1) using the third equation and substituting it into the fourth one, write equations (5.1) in the form

$$\ddot{\eta}_1 = -\omega_1^2(\eta_1 - \eta_2) - \varepsilon(\dot{\eta}_1 - \dot{\eta}_2) - g + \frac{1}{m_1}F,$$

$$\ddot{\eta}_2 = -\omega_2^2(\eta_2 - \eta_1) - \varepsilon_2(\dot{\eta}_2 - \eta_1) - g,$$

$$\dot{\Phi} + RI = V, \quad \int_{-\infty}^{t} \frac{d\Psi}{d\tau}\frac{d\tau}{\sqrt{1-\tau}} = -\frac{V - IR}{\alpha\gamma^{\circ}}, \tag{7.45}$$

$$s = -\eta_1 + \eta_c(t), \quad F = (s, I+J), \quad \Phi = (s, I+J),$$

$$\Psi = \alpha(s, I+J)\, J, \quad \alpha = \alpha(s, I+J).$$

The set of $L(s, I+J)$ values can be converted to the set of $\Phi(s, I+J)$ values by Formula (7.43), and function $\alpha(s, I+J)$ is recalculated by the following formula via function $c_y(s, I+J)$ specified in Table 4.2:

$$\alpha(s, I+J) = \left[\frac{c_y(s, I+J)}{\gamma}\right]^{1/2}. \tag{7.46}$$

Then function Ψ is calculated by Formula (7.44)

To transform the set of Equations (7.46) to Cauchy form it is necessary to solve the integral equation

$$\int_{-\infty}^{t} \frac{d\Psi}{d\tau}\frac{d\tau}{\sqrt{1-\tau}} = -\frac{V - IR}{\alpha\gamma^{\circ}} \tag{7.47}$$

with respect to derivative $d\Psi/d\tau$. Denote

$$\frac{d\Psi}{d\tau} = \sigma_{\Psi}(\tau), \qquad -\frac{V - IR}{\alpha\gamma^{\circ}} = f(t). \tag{7.48}$$

Following the method of numerical solution of a integral equation with a singular kernel of $(t-\tau)^{-1/2}$ form, presented in Chapter 3, and assuming that the the system has no pre-history, i.e. $\sigma_{\Psi}(\tau) \equiv 0$ for $-\infty < \tau \leq 0$, we transform Equations (7.47) and (7.48) to the form

$$\sqrt{h}\,\left\{G_\Psi(t)+\frac{4}{3}\sigma_\Psi(t)\right\}=f(t), \tag{7.49}$$

where function $G_\Psi(t)$ is calculated by Formulae (3.55) at discrete moments of time $0,h,...ih,...,nh$ and $f(t)=f(nh)$. Uniting (7.45), (7.48), and (7.49) we obtain the set of equations that can be directly integrated after transformation to Liapunov-Cauchy form:

$$\begin{aligned}
\ddot{\eta}_1 &= -\,\omega_1^2\,(\eta_1-\eta_2)-\varepsilon_1\,(\dot{\eta}_1-\dot{\eta}_2)-g+\frac{1}{m_1}F,\\
\ddot{\eta}_2 &= -\,\omega_2^2\,(\eta_2-\eta_1)-\varepsilon_2\,(\dot{\eta}_2-\dot{\eta}_1)-g,\\
&\dot{\Phi}+RI=V,\quad \dot{\Psi}(t)=\sigma_\Psi(t),\\
&\sqrt{h}\,\left\{G_\Psi(t)+\frac{4}{3}\sigma_\Psi(t)\right\}=f(t),\quad f(t)=-\frac{V-IR}{\alpha\gamma^\circ},\\
&s=-\,\eta_1+\eta_c(t),\quad F=F(s,I+J),\\
&\Phi=\Phi(s,I+J),\quad \Psi=\alpha(s,I+J)\,J,\quad \alpha=\alpha(s,I+J),
\end{aligned} \tag{7.50}$$

were $G_\Psi(t)$ is calculated by quadrature Formulae (3.55).

Let us consider in detail some specific features of the electromagnetic levitation system dynamics equations concerning account of effects associated with assumptions and limitations, within the scope of which the eddy currents mathematical model was designed. The first feature is that integral Equation (7.47) is Volterra equation of the 1st kind, for the solvability of which it is necessary (see A.F.Verlan and V.S.Sizikov [101]), that its right side and at least its first time derivative should become zero for $t=0$. On the other hand, in a number of cases it is necessary to form a right side rapidly varying from zero to a limit value, particularly Heavyside function. But rapid variation of the right side results in high-frequency components of the transient process, that contradicts the condition of infinitesimal Strouhal number specified by (1.4)

$$\mathrm{Sh}_M=\frac{\omega_M}{\Omega_M}\ll 1.$$

This condition was one of the assumptions of Chapter 1 for eddy currents mathematical model synthesis (ω_m-is the characteristic frequency of the process).

The model imperfection in the mentioned sense can be compensated by introducing the following modified function limiting time variation rate of the integral equation right side:

$$f_h(t)=\frac{t^2}{t^2+\Delta_t}\left[-\frac{V-IR}{\alpha\gamma^\circ}\right]. \tag{7.51}$$

For numerical simulation of the considered system Δ_t was selected so, that for the transient process characteristic time T_{tf} of the order of 0.1 s function $t^2/(t^2+\Delta_t)$ in the vicinity of $t=0$ should become approximately unity time of $\sim 0.1T_{tf}$, i.e. the selected value was $\Delta_t = 1 \cdot 10^{-5}\, s^2$ .Without the limitation (7.51) for Equation (7.47) the process of eddy current formation is simulated in such a way, that when e.g. 'flowing up' at the starting moment, very high eddy current values take place, that must not occur in the real system, naturally. This dummy 'bump' of eddy current corrupts the whole consequent behaviour of the system to a great extent, for, as it was shown in Chapter 3, the formed initial values of the eddy current engender a secular component of the transient process of $1/\sqrt{t}$ type, that appears to prevail over the others.

The second feature results from the fact, that skin effect mathematical model (Chapter 1) is true for great magnetic Reynolds numbers, i.e. with the assumption of Condition (1.4) satisfied

$$\mathrm{Re}_M = \mu_0 \mu \sigma \omega_M l^2 \gg 1.$$

This assumption makes the used model inapplicable to eddy current low-frequency components description, if those occur in the process of integration.

The characteristic value of the lower boundary frequency, beginning with which self-similarity in magnetic Reynolds number is conformed to, is 0.1 Hz in the considered case. Thus, the frequency range, for which the mathematical model of skin effect in HECF core is correct, overlaps by far the closed-loop system characteristic frequency range. The used model does not describe the complete vanishing of eddy current, however. Particularly, solution component of $1/\sqrt{t}$ type, that can be considered as a constant component in a comparison with the transient process characteristic time, has nothing in common with the physical essence of the problem, as it was mentioned in Chapter 3.

In order to exclude low-frequency components of the eddy current from the solution, let us introduce variable J_h connected with variable J in p plane via a unit describing a first-order low frequency filter:

$$J_h(p) = \frac{T_l p}{T_l p + 1} J(p). \tag{7.52}$$

The value of the time constant T_l of the filter described by (7.52) was chosen in such fashion as to make eddy current J vanish practically simultaneously with the end of the transient process in the closed-loop system. For instance for the considered system T_l value was chosen 0.3 s. Unless the slowly varying component of eddy current were excluded from the solution by the filter of (7.52), the system steady state would be determined by the transient process character, that is absolutely inadmissible with nonlinear system modelling.

Thus, skin effect mathematical model incompleteness in the sense of the possibility of eddy current formation (high frequencies) and vanishing (low frequencies) processes accurate description is compensated by additional Equations (7.51), (7.52) introduction. Being based on heuristic assumptions, these equations let the integration results approach the real process.

Combining Equations (7.50) with additional Equations (7.51), (7.52) with the control law equation (7.1) and the inequality describing the amplifier-converter 'saturation-type' nonlinearity, we obtain a complete set of nonlinear integrodifferential equations recognizing all the features listed above. This set of equations has the form suitable for direct numerical integration.

$$\ddot{\eta}_1 = -\omega_1^2(\eta_1 - \eta_2) - \varepsilon_1(\dot{\eta}_1 - \dot{\eta}_2) - g + \frac{1}{m_1}F,$$

$$\ddot{\eta}_2 = -\omega_2^2(\eta_2 - \eta_1) - \varepsilon(\dot{\eta}_2 - \dot{\eta}_1) - g,$$

$$\dot{\Phi} + RI = V, \quad \dot{\Psi}(t) = \sigma_\Psi(t), \quad \sqrt{h}\left\{G_\Psi(t) + \frac{4}{3}\sigma_\Psi(t)\right\} = f_h(t),$$

$$f_h(t) = \frac{t^2}{t^2 + \Delta_t}\left[-\frac{V - IR}{\alpha\gamma^\circ}\right], \quad J_h(p) = \frac{T_l p}{T_l p + 1}J(p), \quad s = -\eta_1 + \eta_c(t),$$

$$\tag{7.53}$$

$$\frac{\Delta V(p)}{R} = \Im_s(p)\Delta S(p) - \Im_{\ddot{\eta}}(p)\ddot{\eta}_1(p) - \Im_I(p)\Delta I(p),$$

$$-V_{\max} \le V \le V_{\max}, \quad F = F(s, I + J_h),$$

$$\Phi = \Phi(s, I + J_h), \quad \Psi = \alpha(s, I + J_h)J, \quad \alpha = \alpha(s, I + J_h),$$

$$\Delta V = V - V^\circ, \quad \Delta s = s - s^\circ, \quad \Delta I = I - I^\circ.$$

where $G_\Psi(t)$ is calculated by quadrature Formula (3.55).

Let us present System (7.53) integration procedure in detail. Nonlinear functions F, Φ, Ψ and α of two variables are tabulated. At intermediate points they were calculated by piecewise linear approximation in functions of two variables. Further steps compose the following algorithm. For the initial moment $t = 0$ initial conditions are set in η_1, $\dot{\eta}_1$, η_2, $\dot{\eta}_2$, I and in variables relevant to the controller equations. Initial conditions in J and J_h are automatically assumed zero. Initial conditions for Φ and Ψ are also recalculated by the formula of piecewise linear interpolation. Further, using quadrature Formulae (3.55), at each step of integration we calculate function $\sigma_\Psi(t)$ being the right side of the equation for $\dot{\Psi}$. Variable I and J are calculated at each step of integration from functions Φ and Ψ values by solving the set of two equations in two variables with the help of piecewise linear interpolation. The following steps of solution are performed in a standard fashion using well-known techniques of numerical integration. Actual calculations were carried out by Runge-Kutta method with 4-point approximation.

Thus the approach suggested enables mathematical simulation of the electromagnetic levitation system dynamics based on experimental data, that increases reliability of the results obtained.

Let us now proceed to simulation results analysis. Parallel to integration of Equation (7.53) with the analog control law we consider the relay-type system simulation. For that purpose the inequality describing saturation-type nonlinearity of the amplifier-converter static characteristic in Equations (7.53) should be replaced by an equation of the following type:

$$V_{EM} = F_R\,(V\ ), \tag{7.54}$$

where V_{EM} is the voltage applied to the electromagnet from the relay amplifier output. It can assume two values: V_{max} and V_{min}; F_R is a function describing the hysteresis relay

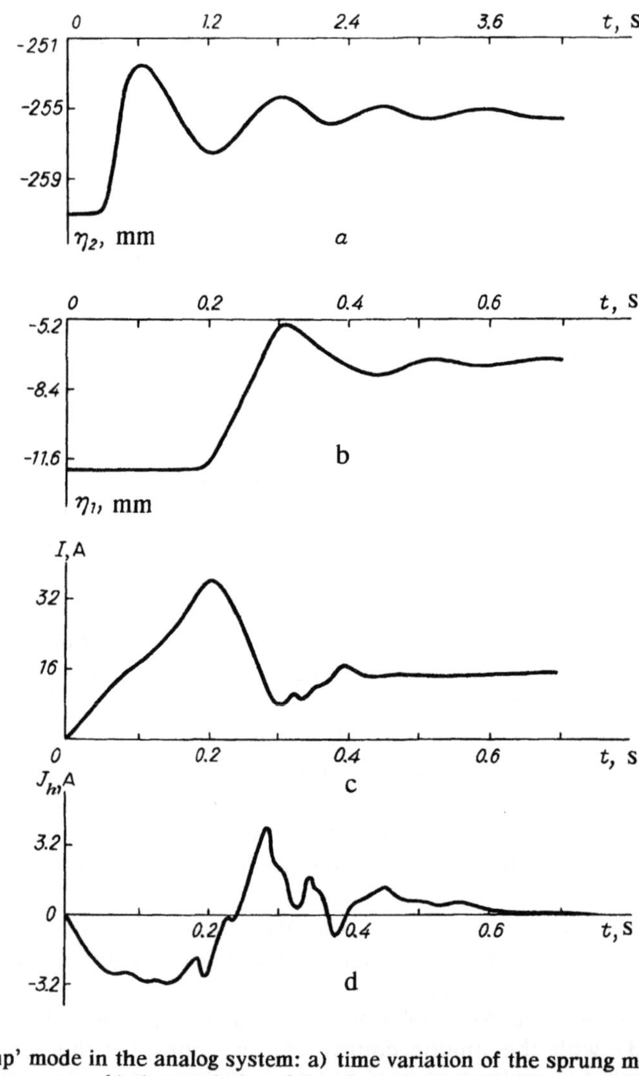

Fig. 7.10. 'Flow-up' mode in the analog system: a) time variation of the sprung mass displacement absolute coordinate η_2, mm; b) time variation of the electromagnet displacement absolute coordinate η_1, mm; c) time variation of the current I in the electromagnet coil; d) eddy current J_h,A time variation

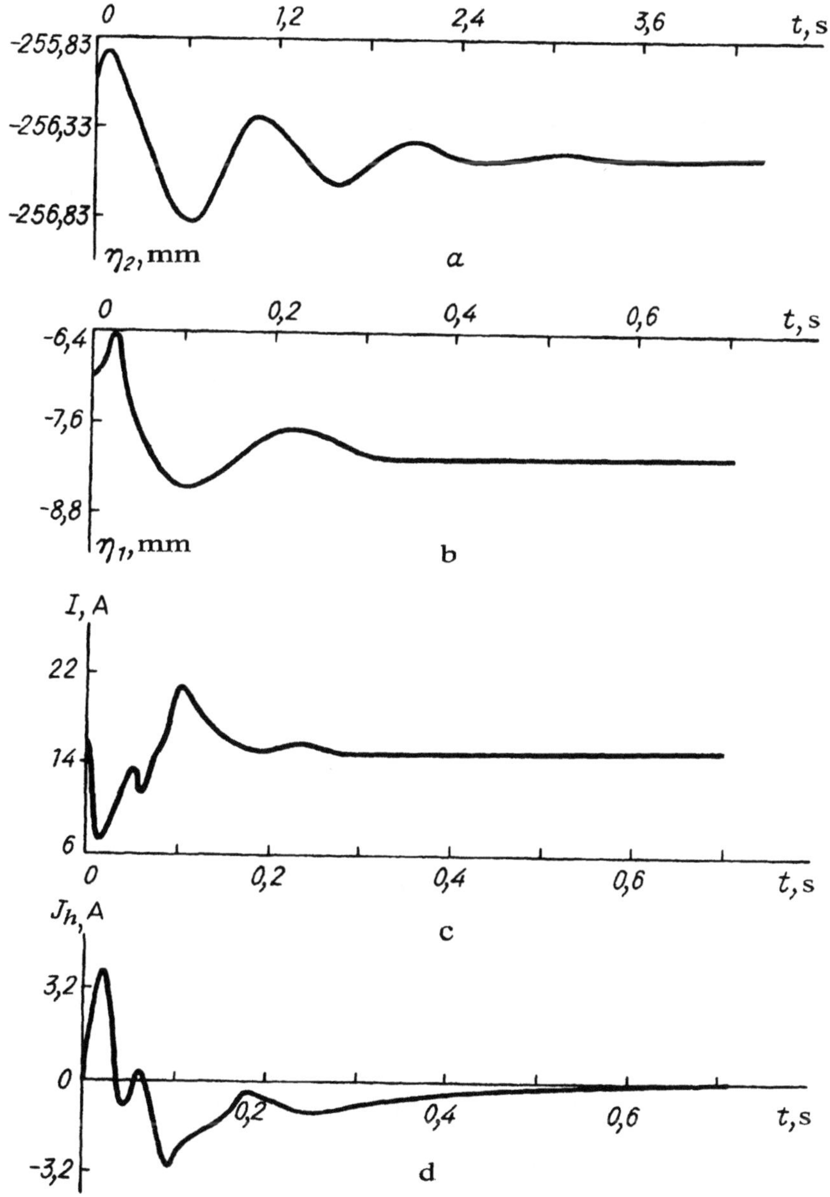

Fig.7. 11. Control mode in response to an instantaneous change of the ferromagnetic rail position in the analog system; a) time variation of the sprung mass displacement absolute coordinate η_2, mm; b) time variation of the electromagnet displacement absolute coordinate η_1, mm; c) time variation of the current I, A in the electromagnet coil; d) eddy current J_h,A time variation

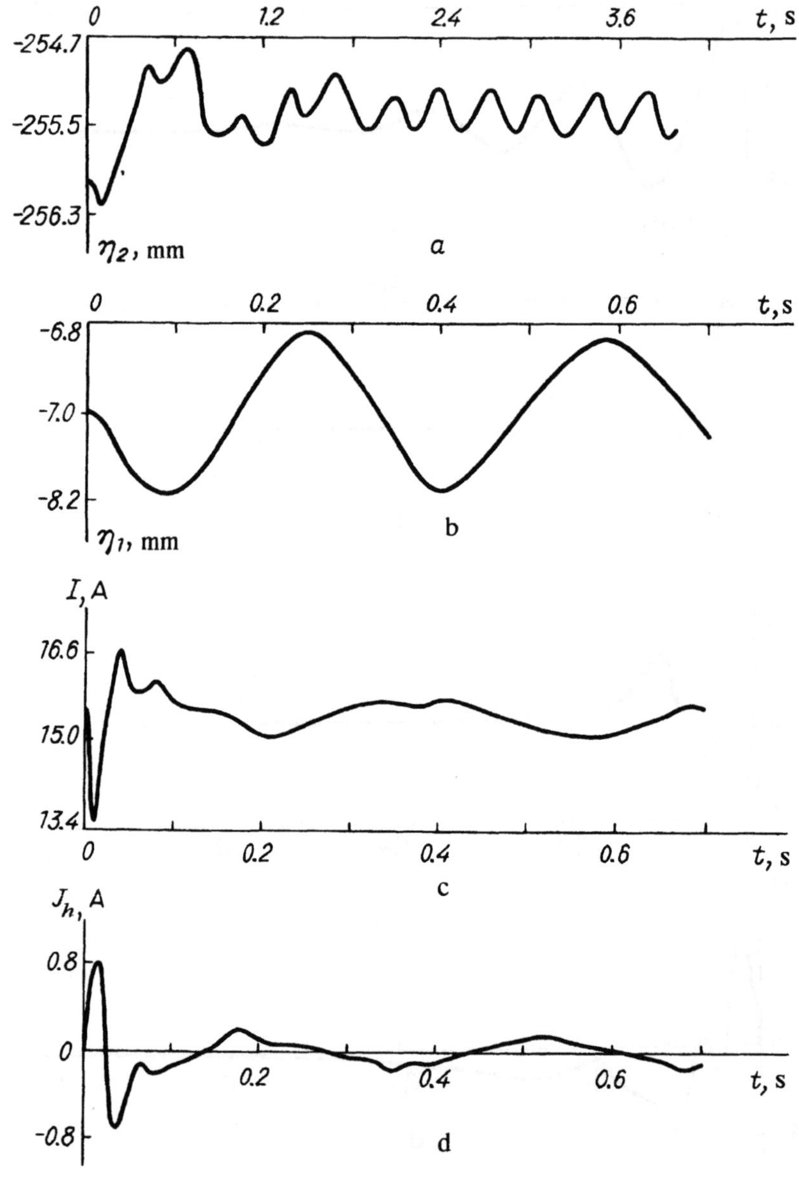

Fig. 7.12. Control mode in response to sinusoidal oscillations of the ferromagnetic rail in the analog system: a) time variation of the sprung mass displacement absolute coordinate η_2, mm; b) time variation of the electromagnet displacement absolute coordinate η_1, mm; c) time variation of the current I, A in the electromagnet coil; d) eddy current J_h,A time variation

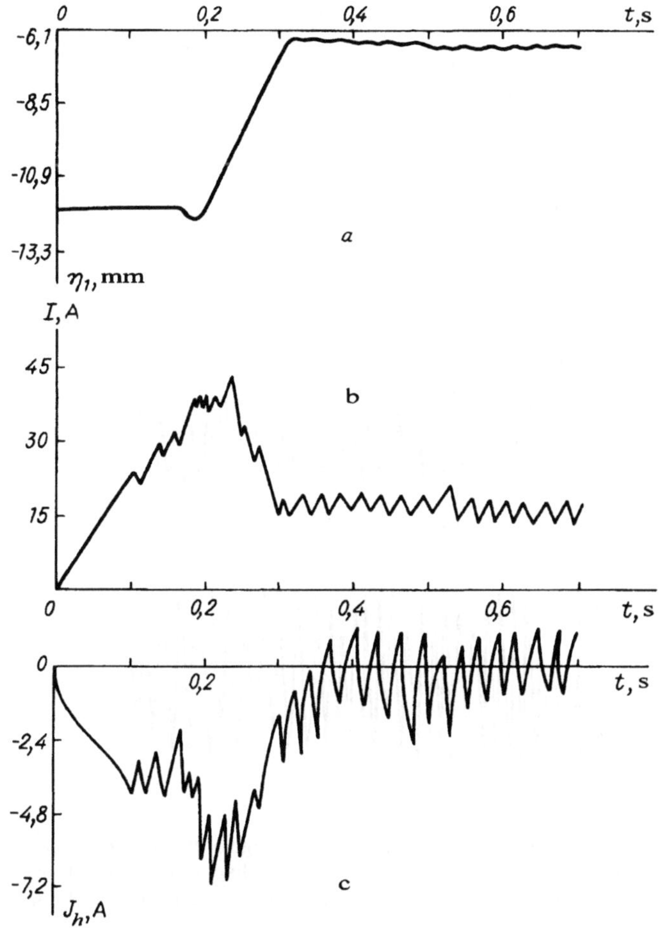

Fig. 7.13. 'Flow-up' mode in the relay-type system: a) time variation of the electromagnet displacement absolute coordinate η_1, mm; b) time variation of the current I, A in the electromagnet coil; c) eddy current J_h, A time variation

operation, the static characteristic of which is presented in Fig. 7.4. Besides that, variable V in all the equations should be replaced by V_{EM}.

The following modes of control in response to typical programmed and exciting disturbances were simulated for the analog and relay-type systems.

1. The mode of 'flow-up' from the air gap $s_0 = 12$ mm to the nominal one $s^\circ = 7$ mm. The mode of non-programmed flowing-up was simulated, that can occur when full-scale voltage is applied to the system.

2. The mode of control in response to a disturbance of an instantaneous change in the ferromagnetic rail position by 1 mm, $\eta_c(t) = 1(t)$. For the analog system

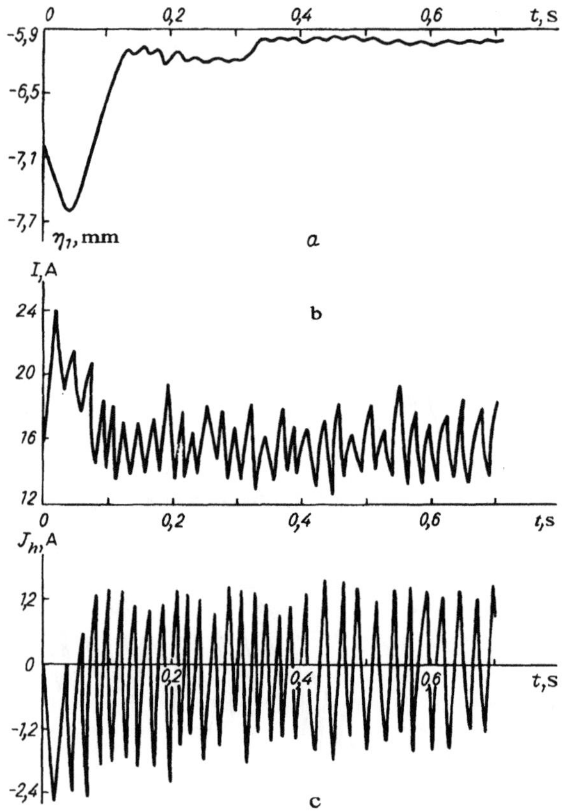

Fig. 7.14. Control mode in response to an instaneous change of the ferromagnetic rail position in the relay-type system: a) time variation of the electromagnet displacement absolute coordinate η_1, mm; b) time variation of the current I, A in the ferromagnet coil; c) eddy current J_h, A time variation

disturbance $\eta_c(t)$ was specified with a negative sign, that is equivalent to instantaneous decrease of the air gap by 1 mm, and for the relay-type system it was positive, i.e. an instantaneous increase of the air gap by 1 mm.

3. The mode of control in response to disturbances oscillations of the ferromagnetic rail harmonic in the inertial space at the frequency of 3 Hz and amplitude 1 mm.

The simulation results for the analog and relay-type systems are presented in Figs. 7.10-7.15.

The process of sprung mass motion is presented for the analog system only, because for the relay-type one this process is almost the same due to high frequency ($\omega \gg \omega_2$) filtration by the secondary suspension.

Analysis of the obtained results brings us to the following conclusions:

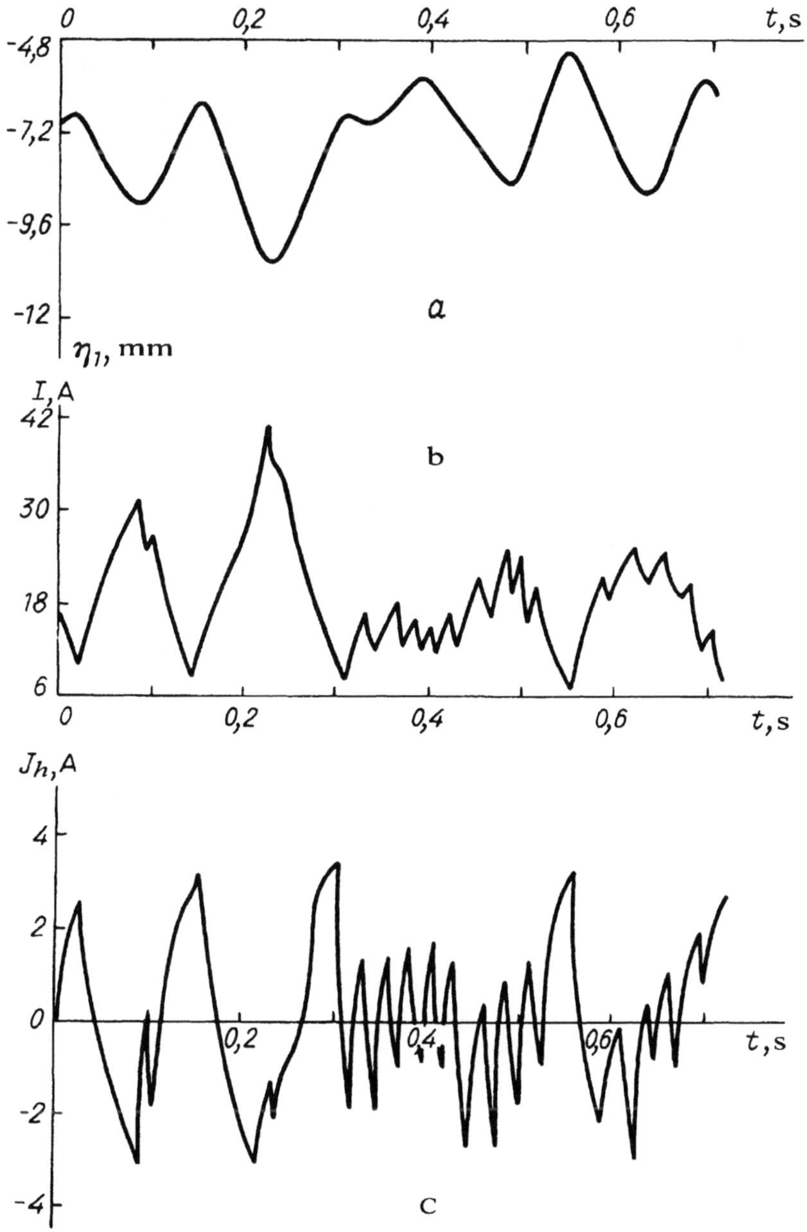

Fig. 7.15. Control mode in response to sinusoidal oscillations of the ferromagnetic rail in the relay-type system: a) time variation of the electromagnet displacement absolute coordinate η_1, mm; b) time variation of the current I, A in the ferromagnet coil; c) eddy current J_h, A time variation

1. For all the considered modes the relay-type system behaviour is practically the same as of the analog system. Self-sustained oscillations are added to the main motion of the relay-type system, that coincides with the main motion of the analog system.

2. The considered analog and relay-type control laws solve the stabilization problem both for the linear statement of the problem and when recognizing nonlinear features of the controlled object and the amplifier-converter (including eddy current in both cases).

Note in conclusion, that specific examples of control law synthesis, considered in this chapter, are not a goal in themselves. They are simply an illustration of the possibilities to construct efficient contral laws for highly complex objects, whose mathematical models allow for unsteady vortex fields.

Although we considered the electromagnetic levitation system actuator for an example of such objects, the whole methodologic part of this chapter is applicable with some reservations to synthesizing control laws for stabilisation algorithms of controlled flying vehicles, whose tanks are equipped with dampers of liquid fuel oscillations. The same refers to the methods for analyzing stability of those vehicles and mathematical modelling of their perturbed motion, taking into account vorticity of the liquid which partially fills the tanks, those methods being based on numerical solution of the relevant integrodifferential equations.

Chapter 8

Some Dynamics Problems for a Solid Body with Electrically Conductive Liquid Moving in Magnetic Field

We have said on the other side of this page, that the definition of a spirit is a power conjoined to a body; because it cannot move of its own accord, nor can it have any kind of motion in space; and if you were to say that it moves itself, this cannot be within the elements.

Leonardo da Vinci *

For the hole of the previous chapters the magnetic field induced by eddy currents in an electrically conductive ferromagnetic medium and the velocity field of vortex low-viscous liquid were considered independently. At the same time we noted analogy of those physical phenomena, which are rather far from each other. This analogy is reflected by similarity of the mathematical models describing them.

There is an ample class of problems in which the deep internal connection between eddy currents and vortex motions of liquid is displayed explicitly. These are magnetic hydrodynamics problems (see [11, 12, 13, 32, 94, 73, 49]). It is these problems to present an example of 'a power conjoined to a body' (using great Leonardo's terminology); magnetic field force lines 'frozen' into the electrically conductive liquid.

Below we consider a problem lying at the joint of magnetic hydrodynamics and solid body dynamics. It is the problem of motion for a solid body with a cavity having walls made of dielectric with $\mu = 1$, that is partially or completely filled with low-viscous electrically conductive liquid (LVEC), which can possess of ferromagnetic properties (LVECF). A solenoid with current $I(t)$ is connected with the body and specifically oriented relative to it. The unperturbed motion of the body is translational, the field of mass forces affecting the body is potential, but may be nonstationary.

Just like in Chapters 1 and 2 we shall further assign superscript $^\circ$ to quantities characterizing unperturbed values of parameters or functions. Particularly I° is the unperturbed constant value of current in the Solenoid, that induces steady magnetic

---

* The literary works of Leonardo da Vinci compiled and edited from the original manuscripts by Jean Paul Richter and Irma Richter. Oxford University Press. London – New York – Toronto, 1939, Vol. 2, p. 254

field $\mathbf{H}^{\circ}$, 'piercing' through the liquid. We shall consider this field slowly varying within the volume occupied by liquid.

Besides frozen-in magnetic' field mentioned above, the motion of liquid with high hydrodynamic and magnetic Reynolds numbers is characterized by development of wall hydrodynamic boundary layer and electromagnetic boundary layer near the boundary of the volume occupied by liquid (skin-effect). Some problems in the theory of LVEC liquid oscillations in a closed volume and magnetohydrodynamic waves (Alfven waves) are considered by D.G.Andrews and R.Hide, C.Sozou and W.M.Pickering [3, 95]. Making use of solenoidal partial solutions of magnetic hydrodynamics boundary-value problems similar to those introduced in Chapter 2 and boundary layer methods used in Chapter 1 lets us move further in solving those problems for great Reynolds numbers and construct a mathematical model of the system 'solid body - electrically conductive liquid – magnetic field' that would be suitable for applications. Derivation of this model presented in the form of a set of integrodifferential equations with singular kernels, being a superposition of the sets of equations considered in Chapters 1 and 2, is the subject of this chapter concluding the book.

8.1. Statement of the Problem. Main Assumptions

Let us suppose that the cavity inside the solid body is completely or partially filled with homogeneous incompressible LVEC liquid which may poses of ferromagnetic properties (LVECF liquid), as in R.Rosenzweig [87]. We shall further consider LVECF liquid keeping in view that LVEC liquid is its particular case. We shall consider the space of the cavity, free of liquid, occupied by gas having the properties of a low-viscous incompressible liquid with low electrical conductivity.

Let us introduce an 'absolute' coordinate system $O^{\circ}x^{\circ}y^{\circ}z^{\circ}$ with the origin in an arbitrary point, that is related to the body in its unperturbed motion, and a 'body-related' coordinate system$Oxyz$, the motion of which with respect to $O^{\circ}x^{\circ}y^{\circ}z^{\circ}$ describes perturbed motion of the body. That motion is characterized by vector $\mathbf{u}$ of point O small displacement with respect to O° and small rotation angle $\vec{\theta}$. Let us consider $O^{\circ}x^{\circ}$ axis antiparallel to unperturbed motion mass forces field potential U gradient $\mathbf{j}$, and the solenoid axis direction in $O^{\circ}x^{\circ}y^{\circ}z^{\circ}$ coordinate system specified by unit vector $\mathbf{i}_H$.

For instance, if the cavity is simply connected and the solenoid is positioned as it is shown in Fig. 8.1, $\mathbf{i}_H = \mathbf{i}_x$. If the cavity is of toroidal shape and the solenoid is positioned as in Fig. 8.2, $\mathbf{i}_H = \mathbf{i}_\theta$, where $\mathbf{i}_\theta$ is the unit vector of cylindric coordinate system $Oxr\theta$. In the first case we shall consider field $\mathbf{H}^{\circ}$ nearly longitudinal along Ox axis, and in the second case nearly circulation along circular arc r = const.

As perturbed motion character is connected, let us use hypotheses of Chapter 2. The equations of perturbed motion for body-liquid system have the same structure in this case, as the corresponding equations of Chapter 2, but the role of vortex-forming

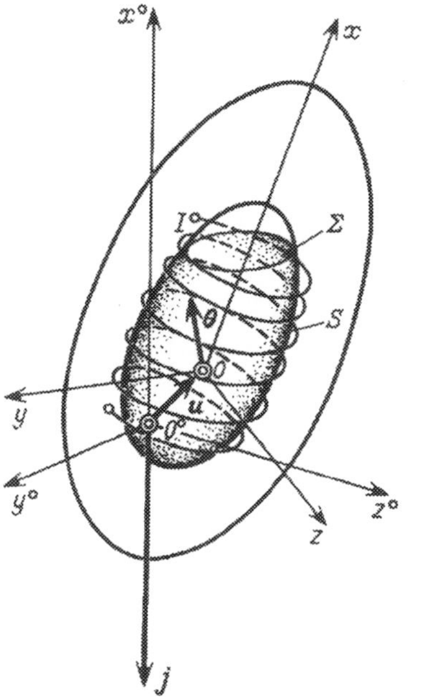

Fig. 8.1. Solid body with simply connected cavity partially filled with electrically conductive liquid in axial magnetic field. Coordinate systems and main symbols

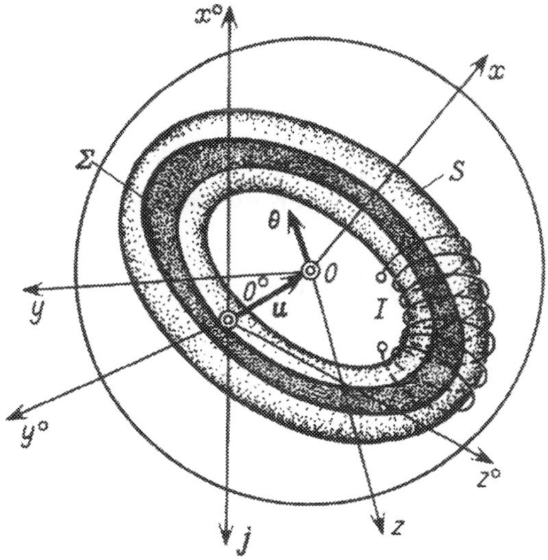

Fig. 8.2. Solid body with doubly connected domain partially filled with electrically conductive liquid in circulation magnetic field. Coordinate systems and main symbols

forces is now played by the ponderomotive forces caused by relative motion of electrically conductive liquid in external magnetic field (see A.F.Bogorodsky [11]).

A new factor in this problem is the magnetic field (analogous to the field of Chapter 1) induced by eddy currents engendered by the liquid perturbed motion. When describing this field we shall follow axiomatics and formalism of Chapter 1. Joint usage of hypotheses of Chapters 1 and 2 enables us to provide for specific magnetohydrodynamic effects by considering only relative motion of liquid, assuming $\mathbf{u} \equiv 0, \vec{\theta} \equiv 0$, and then to use the results of Chapter 2 in order to obtain the complete set of perturbed motion equations.

Let us use the following denominations for parameters characterizing liquid:

ν_F for kinematic viscosity coefficient;

ν_M for magnetic viscosity coefficient;

$\mu_0 \mu$ for magnetic permeability (μ_0 is the same for vacuum or a gas filling the free space above liquid);

σ for electrical conductance;

σ_F for surface stretching coefficient;

ρ for mass density;

p for pressure;

$\mathbf{v}^*$ for the field of relative velocities considering liquid viscosity for $\mathbf{u} \equiv 0, \vec{\theta} \equiv 0$. Let us, further, introduce the following symbols:

ω_F the characteristic frequency of liquid oscillations in its perturbed motion;

l the characteristic size of the cavity;

$\mathbf{H}^{\circ}$ magnetic field strength in unperturbed motion in the space occupied by liquid.

Dimensionless characteristic parameters are the following ones:

$\mathrm{Re} = \dfrac{\omega_F l^2}{\nu_F}$ hydrodynamic Reynolds number ;

$\mathrm{Bo} = \rho j \dfrac{l^2}{\sigma_F}$ Bond number;

$\mathrm{Re}_M = \dfrac{\omega_M l^2}{\nu_M}$ magnetic Reynolds number ;

$\mathrm{Sh}_M = \dfrac{\omega_M}{\Omega_M}$ magnetic Strouhal number ;

$\mathrm{Al} = \dfrac{H^{\circ 2}\mu_0\mu}{\rho l^2 \omega_M^2}$ Alfven number ;

ε maximum relative variation of magnetic field :

$$\varepsilon = \frac{|\mathbf{H} - \mathbf{H}^{\circ}|_{\max}}{H^{\circ}}, \qquad \nu_M = \frac{1}{\mu_0\mu\sigma}, \qquad \Omega_M \sim l\omega_M\left(\frac{\omega_M}{\nu_M}\right)^{1/2}. \tag{8.1}$$

Below we consider the class of system perturbed motions satisfying the following conditions:

$$\mathrm{Re} \gg 1, \qquad \mathrm{Bo} \gg 1, \qquad \mathrm{Re}_M \gg 1, \qquad \mathrm{Sh}_M \ll 1, \qquad \varepsilon \ll 1. \tag{8.2}$$

There are no special restraints for Alfen number. This means that the solution must include both the case of $\mathrm{Al} \ll 1$, when (for existing free surface of liquid) gravitational waves prevail, and the case of $\mathrm{Al} \gg 1$, when magnetohydrodynamic standing waves prevail. Intermediate values of Alfven number correspond to combined magneto-gravitational standing waves.

The main problem is to derive the mathematical model of the system 'body – LVECF liquid – magnetic field' in the form of a set of ordinary singular integro-differential equations being a generalization of equations of Chapters 1 and 2.

8.2. Magnetic Hydrodynamics Boundary-Value Problems for LVECF Liquid

Let $G^{(i)}$ symbolize the domain occupied by liquid; $G^{(e)}$ the domain external relative to $G^{(i)}$; S the wetted surface of walls; Σ unperturbed free surface of liquid. Further on we assign superscript e ($G^{(e)}$ domain) to external fields; absence of such superscript will mean internal field (domain).Magnetic field and liquid velocity field are described by Maxwell equations in quasi-stationary approximation (Chapter 1) and by Navier-Stokes equation (see L.D.Landau and Ye.M.Lifshits [47]) recognizing additional pondermotive forces (see A.F.Bogorodsky [11]).

Boundary conditions are zero tangential components of the stress vector, its normal component coinciding with pressure. Liquid velocity normal component must coincide with media interface at surface Σ. Both normal components of velocity at S surface must be zero. Magnetic field induction normal component and field strength tangential component at surface $S + \Sigma$ must be continuous. Besides that the conditions of magnetic field decaying at $|R| \to \infty$ and of absence of fields for $t = \to \infty$ are to be satisfied, where R is the radius vector of a point in $G^{(e)}$ domain.

The total of listed equations and boundary conditions composes the following boundary-value problem:

$$\frac{\partial \mathbf{v}^*}{\partial t} = \nabla \left(U - \frac{p}{\rho}\right) + \nu_F \Delta \mathbf{v}^* + \frac{1}{\rho} \operatorname{curl} \mathbf{B} \times \mathbf{H},$$

$$\frac{\partial \mathbf{H}}{\partial t} = \operatorname{curl} (\mathbf{v}^* \times \mathbf{H}) + \nu_M \Delta \mathbf{H}, \qquad \operatorname{div} \mathbf{v}^* = 0, \qquad \operatorname{div} \mathbf{H} = 0,$$

$$\operatorname{div} \mathbf{H}^{(e)} = 0, \qquad \operatorname{curl} \mathbf{H}^{(e)} = \mathbf{j}(\mathbf{R}, t), \qquad \operatorname{div} \mathbf{E} = 0, \qquad \operatorname{curl} \mathbf{E} = -\frac{\partial \mathbf{B}}{\partial t}, \tag{8.3}$$

$$(\mathbf{p}, \vec{\nu})\,|_\Sigma = -p^\circ, \qquad (\mathbf{p} \times \vec{\nu})\,|_\Sigma = 0, \qquad (\mathbf{v}^*, \vec{\nu})\,|_\Sigma = \frac{\partial \zeta}{\partial t}, \qquad \mathbf{v}^*\,|_S = 0,$$

$$(\mathbf{B}^{(e)} - \mathbf{B}, \ \vec{\nu})\,|_{S+\Sigma} = 0, \quad (\mathbf{H}^{(e)} - \mathbf{H}) \times \vec{\nu}\,|_{S+\Sigma} = 0,$$

$$\mathbf{B} = \mu_0 \mu \mathbf{H}, \quad \mathbf{B}^{(e)} = \mu_0 \mathbf{H}^{(e)},$$

$$\mathbf{H}^{(e)} \to 0, \quad \mathbf{B}^{(e)} \to 0 \ \text{ for } \ |\mathbf{R}| \to \infty, \quad \mathbf{v}^* = 0, \quad \mathbf{H} = 0 \ \text{ for } \ t = -\infty,$$

where $\vec{\nu}$ is the unit vector of outer normal to S or Σ surface, $\mathbf{v}^*$ is viscous liquid velocities field, $\mathbf{p}$ is the stress vector at the liquid free surface; p° is gas pressure; $\zeta(t)$ is perturbed free surface elevation; $\mathbf{j}(\mathbf{R}, t)$ is the current through unit area of the conductor cross-section. For $\nu_M = 0$ the second of Equations (8.3) becomes the well-known condition of magnetic field frozen into liquid (see A.F.Bogorodsky and L.M.Brekhovskikh, V.V.Goncharov [11, 13]).

We shall assume that for unperturbed motion the condition of magnetogravitational equilibrium of liquid [11] is satisfied:

$$(\nabla U, \ \operatorname{curl} \mathbf{F}^\circ) = 0, \qquad \mathbf{F}^\circ = \operatorname{curl} \mathbf{B}^\circ \times \mathbf{H}^\circ, \tag{8.4}$$

where $\mathbf{F}^\circ$ is the ponderomotive force. One can easily verify that Condition (8.4) is satisfied, e.g. in the case of cavities of revolution for both longitudinal and circulation fields.

Following Chapter 1, let us present magnetic field strength in the form

$$\mathbf{H} = \mathbf{H}_0^{(ie)} + \mathbf{h}^*, \qquad \mathbf{H}^{(e)} = \mathbf{H}_0^{(ei)}, \tag{8.5}$$

where fields $\mathbf{H}_0^{(ie)}$ and $\mathbf{H}_0^{(ei)}$ are the solutions of magnetostatics Problem (1.12) for the core without an air gap:

$$\text{curl}\ \mathbf{H}_0^{(ie)} = 0, \quad \text{div}\ \mathbf{H}_0^{(ie)} = 0, \quad \text{curl}\ \mathbf{H}_0^{(ei)} = \mathbf{j}\,(\mathbf{R}, t), \quad \text{div}\ \mathbf{H}_0^{(ei)} = 0,$$

$$(\mathbf{H}_0^{(ei)} - \mu \mathbf{H}_0^{(ie)}, \ \vec{\nu})\ |_{S+\Sigma} = 0, \tag{8.6}$$

$$(\mathbf{H}_0^{(ei)} - \mathbf{H}_0^{(ie)}) \times \vec{\nu}\ |_{S+\Sigma} = 0, \quad \mathbf{H}^{(ei)} \to 0 \ \text{ for } \ |\mathbf{R}| \to \infty,$$

and $\mathbf{h}^*$ is the field induced by eddy currents, that satisfies the boundary condition $\mathbf{h}^*\ |_{S+\Sigma} = 0$.

Having extracted stationary field $\mathbf{H}^\circ = \mathbf{h}^\circ I^\circ$; $\mathbf{H}^{(e)} = \mathbf{h}_e I^\circ$ corresponding to unperturbed value I° of the current through the solenoid, we obtain, denoting external current disturbance $\delta I = I - I^\circ$,

$$\mathbf{H} = \mathbf{h}^\circ (I^\circ + \delta I) + \mathbf{h}^*, \qquad \mathbf{H}^{(e)} = \mathbf{h}_e^\circ (I^\circ + \delta I). \tag{8.7}$$

By virtue of the last of Conditions (8.2),

$$|\mathbf{h}^\circ \delta I + \mathbf{h}^*| << H^\circ. \tag{8.8}$$

Let us transform the ponderomotive force expression in the right side of the first of Equations (8.3) and the first component in the right side of the second equation, taking into consideration Inequality (8.8) and continuity equation div $\mathbf{v}^* = 0$ and ignoring second order infinitesimal quantities (see [11]):

$$\frac{\mathbf{F}}{\mu_0 \mu} = \text{curl}\ \mathbf{H} \times \mathbf{H} \approx \text{curl}\ \mathbf{h}^* \times \mathbf{H} = (\mathbf{h}^*, \nabla)\,\mathbf{H} + (\mathbf{H}, \nabla)\,\mathbf{h}^* +$$

$$+ \mathbf{h}^* \times \text{curl}\ \mathbf{H} - \nabla\,(\mathbf{h}^*, \mathbf{H}) \approx (\mathbf{h}^*, \nabla)\,\mathbf{H}^\circ + (\mathbf{H}^\circ, \nabla)\,\mathbf{h}^* - \nabla\,(\mathbf{H}, \mathbf{h}^*), \tag{8.9}$$

$$\text{curl}\,(\mathbf{v}^* \times \mathbf{H}) = (\mathbf{H}, \nabla)\,\mathbf{v}^* - (\mathbf{v}^*, \nabla)\,\mathbf{H} + \mathbf{v}^*\,\text{div}\,\mathbf{H} - \mathbf{H}\,\text{div}\,\mathbf{v}^* \approx$$

$$\approx (\mathbf{H}^\circ, \nabla)\,\mathbf{v}^* - (\mathbf{v}^*, \nabla)\,\mathbf{H}^\circ.$$

It is obvious from (8.9) that $\mathbf{F} \equiv 0$ for $\mathbf{h}^* = \nabla \chi$, i.e. potential field $\mathbf{h}^*$ is forceless. Functions $\mathbf{v}^*$and $\mathbf{H}$ being asymptotic solutions of boundary-value Problem (8.3) for $\text{Re} \to \infty$, $\text{Re}_M \to \infty$ can be presented in the form

$$\mathbf{v}^* = \mathbf{v}_0 \vec{\Gamma}_0 + \mathbf{v} + \hat{\mathbf{v}}, \qquad \mathbf{H} - \mathbf{H}^\circ = \mathbf{h}^\circ \delta I + \mathbf{h}^*,$$

$$\mathbf{h}^* = \mathbf{h}_0 J_0 + n^\circ (\mathbf{h} + \hat{\mathbf{h}}), \qquad n^\circ = \frac{N^\circ}{2\pi}, \tag{8.10}$$

where n° is the normalizing multiplier (N° is the number of turns of the solenoid); $\mathbf{v}$, $n^\circ\,\mathbf{h}$ are vortex fields associated with relative motion of liquid for $\nu_F \to 0$; $\nu_M \to 0$; $h_0 J_0$ is the magnetic field that tends for $\mathbf{v} \to 0$ to the field induced by eddy currents in solidified liquid; $\mathbf{v}_0 \vec{\Gamma}_0$ is the field of liquid velocities, analogous to $\mathbf{h}_0 J_0$; $\hat{\mathbf{h}}$ and $\hat{\mathbf{v}}$ are boundary layer-type functions providing satisfaction of boundary conditions of (8.3) for tangential components of fields **H** and $\mathbf{v}^*$. Field $\hat{\mathbf{v}}$ has the order of $\mathrm{Re}^{-1/2}$ near surface S and the order of Re^{-1} near surface Σ (see F.L.Cherousko [15]), and field $\hat{\mathbf{h}}$ has the order of $\mathrm{Re}_M^{-1/2}$ near total $S + \Sigma$ surface (see Chapter 1).

Let us write down the boundary-value problem to which (8.3) is reduced for ideal superconductive liquid being a limiting case of LVEC liquid. Setting $\nu_F = 0$, $\nu_M = 0$ (8.3), changing boundary conditions appropriately, and including (8.6) - (8.9), we obtain the following boundary-value problem for **v** and **h** fields.:

$$\rho \frac{\partial \mathbf{v}}{\partial t} + \nabla(-\rho U + p) = n^\circ \, [-\nabla(\mathbf{B}^\circ, \mathbf{h}) + (\mathbf{B}^\circ, \nabla)\,\mathbf{h} - (\mathbf{h}, \nabla)\,\mathbf{B}^\circ\,],$$

$$\rho_M \left(\frac{\partial \mathbf{h}}{\partial t} + \frac{\mathbf{h}^\circ}{n^\circ}\frac{d\delta I}{dt}\right) = n^\circ \, [(\mathbf{B}^\circ, \nabla)\,\mathbf{v} - (\mathbf{v}, \nabla)\,\mathbf{B}^\circ\,], \quad \mathrm{div}\,\mathbf{v} = 0, \quad \mathrm{div}\,\mathbf{h} = 0,$$

$$(\mathbf{v}, \vec{\nu})\,|_S = 0, \quad (\mathbf{v}, \vec{\nu})\,|_\Sigma = \frac{\partial \xi}{\partial t}, \quad p\,|_\Sigma = p^\circ, \quad (\mathbf{h}, \vec{\nu})\,|_{S+\Sigma} = 0, \tag{8.11}$$

$$\mathbf{v} = 0, \quad \mathbf{h} = 0 \quad \text{for} \quad t = -\infty, \quad n^\circ \mathbf{B}^\circ = \rho_M \Gamma^\circ \left(\frac{\mathbf{h}^\circ}{n^\circ}\right), \quad \rho_M = \mu_0 \mu \, n^{\circ 2}.$$

For simply connected domains the elements of the set of Functions (2.37), in terms of which it makes sense to expand vector functions **v** and **h** , are orthogonal to constant within $G^{(i)}$ domain. We shall confine to the class of simply connected domains, for which $\mathbf{h}^\circ \delta I$ and $\mathbf{h}_0 J_0$ fields slightly differ from uniform ones within $G^{(i)}$ domain. That means that we can consider $\mathbf{h}^\circ$ and $\mathbf{h}_0$ functions orthogonal to **h** field within $G^{(i)}$ domain as a first approximation (we also neglect the feeble relation of $\mathbf{h}_0 J_0$ and $\mathbf{h}^\circ \delta I$ via). This makes grounds to consider the total of $\mathbf{h}^*$ and $\mathbf{v}^*$ fields ignoring $\mathbf{h}_0 J_0$ and $\mathbf{h}^\circ \delta I$ in this case. As to the hydrodynamic analogue $\mathbf{v}_0 \Gamma_0$ of $\mathbf{h}_0 J_0$ fields there is no such solution nonvortex within $G^{(i)}$ domain in the case of simply connected domains.

For doubly connected domains $G^{(i)}$ (of the type presented in Fig. 8.2) the situation is different, because $\mathbf{h}_0 J_0$ and $\mathbf{v}_0 \Gamma_0$ circulation fields appearance in $G^{(i)}$ domain becomes possible, where $\mathbf{h}_0$ and $\mathbf{v}_0$ are nontrivial particular solutions of boundary-value problems equivalent to (1.15):

$$\mathrm{div}\,\mathbf{h}_0 = 0, \quad \mathrm{curl}\,\mathbf{h}_0 = 0, \quad \mathrm{div}\,\mathbf{v}_0 = 0, \quad \mathrm{curl}\,\mathbf{v}_0 = 0,$$

$$(\mathbf{h}_0, \vec{\nu})\,|_{S+\Sigma} = 0, \qquad (\mathbf{v}_0, \vec{\nu})\,|_{S+\Sigma} = 0. \tag{8.12}$$

These fields structure is such, that at least some of dynamic relations mentioned above are not small, hence it is necessary to consider **v** and $n^\circ \mathbf{h}$ fields together with $\mathbf{v}_0 \Gamma_0$, $\mathbf{h}^\circ \delta I$ and $\mathbf{h}_0 J_0$ in the complete mathematical model.

Let us present in conclusion the expressions for $\hat{\mathbf{v}}$ and $\hat{\mathbf{h}}$ using (1.24). Near $G^{(i)}$ domain boundary they have the form

$$\hat{\mathbf{v}} \times \vec{\nu} = -\frac{q_1}{2\sqrt{\pi\nu_F}} \int_{-\infty}^{t} \frac{(\mathbf{v}_0\Gamma_0 + \mathbf{v}) \times \vec{\nu}\ |_S}{(t-\tau)\sqrt{t-\tau}} \exp\left[-\frac{q_1^2}{4\nu_F(t-\tau)}\right] d\tau,$$

$$\hat{\mathbf{h}} \times \vec{\nu} = -\frac{q_1}{2\sqrt{\pi\nu_M}} \int_{-\infty}^{t} \frac{\left(\frac{\mathbf{h}_0}{n^\circ} J_0 + \mathbf{h}\right) \times \vec{\nu}\ |_{S+\Sigma}}{(t-\tau)\sqrt{t-\tau}} \exp\left[-\frac{q_1^2}{4\nu_M(t-\tau)}\right] d\tau, \tag{8.13}$$

where q_1 is the curvilinear coordinate having the dimension of length. It is counted from the boundary surface in the direction of inner normal. For simply connected cavities of the considered class we should set in (8.13) $\mathbf{h}_0 \equiv 0$, $\mathbf{v}_0 \equiv 0$.

8.3. Liquid – Magnetic Field System Kinetic Energy. Generalized Forces and Generalized Voltages.

Further succession of operations is as follows. Both fields **v** and **h** should be expanded into series in functions being particular solutions of boundary-value Problem (8.11). It is sufficient to obtain the expression of kinetic energy of the system 'moving liquid – magnetic field'. Using at the next stage the same representations for **v** and **h** fields and correction fields of (8.13) we can derive expressions for generalized forces and generalized voltages engendered by:

a) mass actions (ponderomotive forces in the first Equation (8.9) and their counterparts in the second equation);

b) surface effects directly connected with hydrodynamic and magnetic boundary layers, i.e. with Functions (8.13).

Using those expressions and including additional factors (translation motion of the body) we can compose Lagrange - Maxwell equations for the entire closed-loop system 'solid body – LVECF liquid – magnetic field'.

Let us make use of vector Functions (2.37) satisfying within $G^{(i)}$ domain the continuity equation, the condition of normal component equal to zero at its boundary surface $S+\Sigma$, orthogonality Conditions (2.54a), the condition of orthogonality to constant. We shall also use functions of (2.9), (2.10) mutually orthogonal and orthogonal to constant within domain $G^{(i)}$. Following Chapter 2 and holding in mind that **v** field must be a sum of a solenoidal and a potential field, and **h** field must be solenoidal (see (8.11)), let us represent **v** and **h** fields in the form of the following expansions:

$$\mathbf{v} = \sum_{k=1}^{2N} \mathbf{v}_k + \sum_{n=1}^{\infty} \mathbf{v}^{(n)}, \qquad \mathbf{h} = \sum_{s=1}^{2N} \mathbf{h}_s, \tag{8.14}$$

where

$$\mathbf{v}_k = \vec{\Omega}_k \times \mathbf{R}_k - \nabla(\vec{\Omega}_k, \vec{\Psi}_k) = \sum_{j=1}^{3} \Omega_{kj}\,\mathbf{f}_{kj}\,,$$

$$\mathbf{h}_s = \mathbf{J}_s \times \mathbf{R}_s - \nabla(\mathbf{J}_s, \vec{\Psi}_s) = \sum_{i=1}^{3} J_{si}\,\mathbf{f}_{si}, \tag{8.15}$$

$$\mathbf{v}^{(n)} = \dot{s}_n \nabla \varphi_n, \quad \mathbf{f}_{ki} = \mathbf{i}_j \times \mathbf{R}_k - \nabla \vec{\Psi}_{kj}, \quad k, s = 1, 2, \ldots, 2N, \quad n = 1, 2, \ldots,$$

Here $\mathbf{R}_k$, $\vec{\Psi}_k$ are the functions introduced in Chapter 2 (see (2.39), (2.55)); $\vec{\Omega}_k(t)$ and $\mathbf{J}_s(t)$ are new, so far unknown vector functions of time, which have components Ω_{kj} and J_{si} along $Oxyz$ coordinate system axes ($\mathbf{i}_j$ are the unit vectors of that coordinate system); N is some integer number.

Let us, parallel to (8.14), introduce a more accurate representation for magnetic field $\mathbf{h}$ in the form of a series, each term of which satisfies boundary Conditions (8.3):

$$\mathbf{h} = \sum_{s=1}^{2N} (\mathbf{h}_s + \hat{\mathbf{h}}_s), \tag{8.16}$$

where boundary layer-type functions $\hat{\mathbf{h}}$ result from (8.13), if we substitute functions $\mathbf{h} \equiv \mathbf{h}_s$ of (8.15) and $\mathbf{h}_0 \equiv 0$ into the integrand expression. Expressions of kinetic energy T_F of the liquid relative motion and of magnetic field energy T_M for no disturbances of the external field and for $\mathbf{h}_0 \equiv 0$ can be brought to the form (see Chapters 1, 2)

$$T_F = \frac{1}{2}\sum_{i=1}^{3}\sum_{j=1}^{3}\sum_{k=1}^{2N} J_{ij}^{kk}\Omega_{ki}\Omega_{kj} + \frac{1}{2}\sum_{n=1}^{\infty} \mu_n \dot{s}_n^2\,,$$

$$T_M = \frac{1}{2}\sum_{i=1}^{3}\sum_{j=1}^{3}\sum_{s=1}^{2N} L_{ij}^{ss} J_{si} J_{sj}\,, \tag{8.17}$$

where J_{ij}^{kk} and L_{ij}^{ss} are elements of tensors J_k^* and L_s^* analogous to the tensor of inertia (Chapter 2) and the tensor of mutual inductance (Chapter 1); μ_n are associated masses of liquid in its wave motions. Those coefficients are expressed by the following formulae:

$$J_{ij}^{kk} = J_{ji}^{kk} = \rho \left[\int_{G^{(i)}} (\mathbf{i}_i \times \mathbf{R}_k\,, \mathbf{i}_j \times \mathbf{R}_k)\, dQ - \oint_{S+\Sigma} (\mathbf{R}_k \times \vec{\nu})_j\, \Psi_{ki}\, dS \right],$$

$$L_{ij}^{ss} = L_{ji}^{ss} = \rho_M \left[\int_{G^{(i)}} (\mathbf{i}_i \times \mathbf{R}_s\,, \mathbf{i}_j \times \mathbf{R}_s)\, dQ - \oint_{S+\Sigma} (\mathbf{R}_s \times \vec{\nu})_j\, \Psi_{si}\, dS \right], \tag{8.18}$$

$$\mu_n = \rho \int_{G^{(i)}} (\nabla \varphi_n)^2\, dQ.$$

The central moment is derivation of expressions for generalized forces and generalized voltages. Let us begin with pondermotive forces and their analogues of (8.9). We introduce denominations of F^*_{kj}, Φ^*_n, V^*_{si} for the respective generalized forces and generalized voltages. Taking into consideration that div $\mathbf{f}_{kj} = 0$, $\mathbf{h}^* \mid_{S+\Sigma} = 0$ we obtain from (8.9):

$$F^*_{kj} = \int_{G^{(i)}} ((\mathbf{B}^\circ, \nabla)\, \mathbf{h}^*, \mathbf{f}_{kj})\, dQ - \int_{G^{(i)}} (\nabla\, (\mathbf{B}^\circ, \mathbf{h}^*), \mathbf{f}_{kj})\, dQ + \int_{G^{(i)}} ((\mathbf{h}^*, \nabla)\, \mathbf{B}^\circ, \mathbf{f}_{kj})\, dQ =$$

$$= - \int_{G^{(i)}} ((\mathbf{B}^\circ, \nabla)\, \mathbf{f}_{kj}, \mathbf{h}^*)\, dQ + \oint_{S+\Sigma} (\mathbf{B}^\circ, \vec{\nu})\, (\mathbf{h}^*, \mathbf{f}_{kj})\, dS -$$

$$- \int_{G^{(i)}} \operatorname{div}\, [(\mathbf{B}^\circ, \mathbf{h}^*)\, \mathbf{f}_{kj}\,]\, dQ + \int_{G^{(i)}} (\mathbf{B}^\circ, \mathbf{h}^*)\, \operatorname{div} \mathbf{f}_{kj}\, dQ + \int_{G^{(i)}} ((\mathbf{h}^*, \nabla)\, \mathbf{B}^\circ, \mathbf{f}_{kj})\, dQ =$$

$$= - \int_{G^{(i)}} ((\mathbf{B}^\circ, \nabla)\, \mathbf{f}_{kj}, \mathbf{h}^*)\, dQ - \oint_{S+\Sigma} (\mathbf{B}^\circ, \mathbf{h}^*\,)\, (\mathbf{f}_{kj}, \vec{\nu})\, dS + \int_{G^{(i)}} ((\mathbf{h}^*, \nabla)\, \mathbf{B}^\circ, \mathbf{f}_{kj})\, dQ =$$

$$= \int_{G^{(i)}} ((\mathbf{h}^*, \nabla)\, \mathbf{B}^\circ, \mathbf{f}_{kj})\, dQ - \int_{G^{(i)}} ((\mathbf{B}^\circ, \nabla)\, \mathbf{f}_{kj}, \mathbf{h}^*)\, dQ\,,$$

and a similar expression for generalized force Φ^*_n. Generalized voltage V^*_{si} has the same structure by virtue of (8.9).

Assume, according to (8.10), $\mathbf{h}^* = n^\circ\, \mathbf{h}$ within $G^{(i)}$ domain for $\mathbf{h}_0$, $\hat{\mathbf{h}} \approx 0$. We obtain

$$F^*_{kj} = - \rho_M I^\circ \int_{G^{(i)}} \left(\left(\frac{\mathbf{h}^\circ}{n^\circ}, \nabla \right) \mathbf{f}_{kj}, \mathbf{h} \right) dQ + \rho_M I^\circ \int_{G^{(i)}} \left((\mathbf{h}, \nabla\,) \frac{\mathbf{h}^\circ}{n^\circ}, \mathbf{f}_{kj} \right) dQ,$$

$$\Phi^*_n = - \rho_M I^\circ \int_{G^{(i)}} \left(\left(\frac{\mathbf{h}^\circ}{n^\circ}, \nabla \right) \nabla\varphi_n, \mathbf{h} \right) dQ + \rho_M I^\circ \int_{G^{(i)}} \left((\mathbf{h}, \nabla\,) \frac{\mathbf{h}^\circ}{n^\circ}, \nabla\varphi_n \right) dQ,$$

$$V^*_{si} = \rho_M I^\circ \int_{G^{(i)}} \left(\left(\frac{\mathbf{h}^\circ}{n^\circ}, \nabla \right) \mathbf{v}, \mathbf{f}_{si} \right) dQ - \rho_M I^\circ \int_{G^{(i)}} \left((\mathbf{v}, \nabla\,) \frac{\mathbf{h}^\circ}{n^\circ}, \mathbf{f}_{si} \right) dQ\,. \qquad (8.19)$$

Expressions (8.19) present the first component of generalized forces and voltages. The second component is associated with surface effects, namely with tangential forces of liquid within the boundary layer interaction with the cavity walls and with Poynting vector (Chapter 1). Using the results obtained in Chapter 1 and Formulae (8.13) we can right away write down expressions for the second component F_{kj}, Φ_n, V_{si} of each generalized force and of the generalized voltage:

$$F_{kj} = - \rho \left(\frac{\nu_F}{\pi} \right)^{1/2} \int_{-\infty}^{t} \left[\int_S (\dot{\mathbf{v}}, \mathbf{f}_{kj})\, dS \right] \frac{d\tau}{\sqrt{t - \tau}}\,,$$

$$\Phi_n = -\rho \left(\frac{\nu_F}{\pi}\right)^{1/2} \int_{-\infty}^{t} \left[\int_S (\dot{\mathbf{v}}, \nabla\varphi_n)\, dS\right] \frac{d\tau}{\sqrt{t-\tau}},$$

$$V_{si} = -\rho_M \left(\frac{\nu_M}{\pi}\right)^{1/2} \int_{-\infty}^{t} \left[\oint_{S+\Sigma} (\dot{\mathbf{h}}, \mathbf{f}_{si})\, dS\right] \frac{d\tau}{\sqrt{t-\tau}}. \tag{8.20}$$

With the help of Formulae (8.17) - (8.20), according to the scheme presented above, we can compose Lagrange – Maxwell equations for the considered discrete-continual system. For a body with a doubly connected cavity (like in Fig. 8.2) it requires consideration of not only translation motion of the body, but also the external current and circulation field $\mathbf{v}_0\Gamma_0$ and $\mathbf{h}_0 J_0$ (see (8.12)).

8.4. Equations of Dynamics for a Solid Body Containing LVECF Liquid with Related Magnetic Field Presence

Let us consider a body with a simply connected cavity (Fig. 8.1). Let us place the coordinate origin O, arbitrary up to this point, to the mass centre G of the body with liquid solidified in its unperturbed state. Respectively point O° goes to G° being the image of G in unperturbed motion. We denote velocity of point G with respect to G° and angular velocity of the body in perturbed motion by $\mathbf{V}$ and $\vec{\omega}$, just as in Chapter 2. If there were no magnetic field, equations of perturbed motion of he system 'solid body – LV liquid' would have the same structure as (2.58). Here they are for convenience:

$$(m^\circ + m)\,\dot{\mathbf{V}} + \sum_{n=1}^{\infty} \vec{\lambda}_n \ddot{s}_n = \mathbf{P},$$

$$(\mathrm{J}^\circ + \mathrm{J}^{(0)}, \dot{\vec{\omega}}) + (\mathrm{J}_1^*, \dot{\vec{\Omega}}_1) + \sum_{n=1}^{\infty} [\vec{\lambda}_{Gn}\ddot{s}_n - (\vec{\lambda}_n \times \mathbf{j})\, s_n] = \mathbf{M}_G, \tag{8.21}$$

$$(\mathrm{J}_k^*, \dot{\vec{\Omega}}_k + \delta_k \dot{\vec{\omega}}) + \sum_{l=1}^{2N} \left(\mathrm{B}_{kl}, \int_{-\infty}^{t} \frac{\dot{\vec{\Omega}}_l(\tau)\, d\tau}{\sqrt{t-\tau}}\right) - \sum_{n=1}^{\infty} \vec{\gamma}_{kn} \int_{-\infty}^{t} \frac{\ddot{s}_n(\tau)\, d\tau}{\sqrt{t-\tau}} = 0,$$

$$\mu_n(\ddot{s}_n + \omega_n^2 s_n) + (\vec{\lambda}_n, \dot{\mathbf{V}}) + (\vec{\lambda}_{Gn}, \dot{\vec{\omega}}) - (\vec{\lambda}_n, \mathbf{j}) + \mu_n \sum_{m=1}^{\infty} \beta_{nm} \int_{-\infty}^{t} \frac{\ddot{s}_m(\tau)\, d\tau}{\sqrt{t-\tau}} - \sum_{k=1}^{2N} \left(\vec{\gamma}_{kn}, \int_{-\infty}^{t} \frac{\dot{\vec{\Omega}}_k(\tau)\, d\tau}{\sqrt{t-\tau}}\right) = 0,$$

$k = 1, 2, \ldots, 2N,\quad n = 1, 2, \ldots,\quad \delta_k = 0$ for $k \neq 1,\quad \delta_1 = 1,$

where m°, J° are the mass and tensor of inertia of the solid body, $\mathbf{P}$ and $\mathbf{M}_G$ are the resultant vector and resultant moment of the system of external exciting forces with respect to point G. All the coefficient of Equations (8.21), except for the coefficients at the integrals, are specified by the same Formulae (2.26), (2.52), (2.53b) as the coefficients of Equations (2.58), if we assume rib width equal to zero. Coefficients at the integrals are now constants depending on the geometry of the cavity with smooth walls only, and not nonlinear functions of the liquid relative velocity as in Equations (2.58). For Equations (8.21) $s_n \equiv 0$ convert to the equations of motion of a body with a cavity completely filled with LV liquid; for $s_n \equiv 0$ and $\vec{\Omega}_k \equiv 0$ to the equations of motion of a body with solidified liquid.

Proceeding to magnetic field influence consideration, we assume for definiteness that in our case of a simply connected cavity the solenoid is 'wound' around a circular cylinder (the cavity with liquid is inside that cylinder), whose axis is parallel to Ox, so that magnetic field $\mathbf{H}^\circ$ is longitudinal ($\mathbf{i}_H = \mathbf{i}_z = \mathbf{i}_1$). Let us substitute $\mathbf{v}$ and $\mathbf{h}$ from (8.14) into the right sides of Formulae (8.19) and compose the following expressions:

$$\mathbf{F}_k^* = \sum_{j=1}^{3} \mathbf{i}_j F_{kj}^* = -\sum_{s=1}^{2N} \sum_{i=1}^{3} J_{si} \sum_{j=1}^{3} \mathbf{i}_j \alpha_{ji}^{ks} = -\sum_{j=1}^{2N} (\mathsf{A}_{ks}, \mathbf{J}_s),$$

$$\vec{\Phi}_n^* = -\sum_{s=1}^{2N} \sum_{i=1}^{3} \vec{\alpha}_{sn} J_{si}' = -\sum_{s=1}^{3} (\vec{\alpha}_{sn}, \mathbf{J}_s), \tag{8.22}$$

$$\mathbf{V}_s^* = \sum_{i=1}^{3} \mathbf{i}_i V_{si}^* = -\sum_{k=1}^{2N} \sum_{i=1}^{3} \Omega_{ki} \sum_{j=1}^{3} \mathbf{i}_j \alpha_{ij}^{ks} + \sum_{n=1}^{\infty} \vec{\alpha}_{sn} \dot{s}_n =$$

$$= \sum_{k=1}^{2N} (\overline{\mathsf{A}}_{ks}, \vec{\Omega}_k) + \sum_{n=1}^{\infty} \vec{\alpha}_{sn} \dot{s}_n,$$

where $\vec{\Omega}_k$, $\mathbf{J}_s$, $\vec{\alpha}_{sn}$ are vectors, A_{ks} are second-rank tensors, $\overline{\mathsf{A}}_{ks}$ are tensors conjugated to A_{ks}:

$$\vec{\Omega}_k = \sum_{j=1}^{3} \mathbf{i}_j \Omega_{kj}, \quad \mathbf{J}_s = \sum_{i=1}^{s} \mathbf{i}_i J_{si}, \quad \vec{\alpha}_{sn} = \sum_{i=1}^{3} \mathbf{i}_i \alpha_{sni}, \tag{8.23}$$

$$\mathsf{A}_{ks} = \left\{\alpha_{ji}^{ks}\right\}, \quad \overline{\mathsf{A}}_{ks} = \left\{\alpha_{ij}^{ks}\right\}.$$

Vectors $\vec{\alpha}_{sn}$ components and tensors A_{ks} and $\overline{\mathsf{A}}_{ks}$ elements are specified by formulae:

$$\alpha_{ji}^{ks} = \rho_M \Gamma^\circ \int_{G^{(i)}} \left(\left(\frac{\mathbf{h}^\circ}{n^\circ}, \nabla \right) \mathbf{f}_{kj}, \mathbf{f}_{si} \right) dQ - \rho_M \Gamma^\circ \int_{G^{(i)}} (\mathbf{i}_H, \nabla) \frac{\mathbf{h}^\circ}{n^\circ} (\mathbf{i}_H, \mathbf{f}_{kj}) (\mathbf{i}_H, \mathbf{f}_{si}) \, dQ,$$

$$\alpha_{sni} = \rho_M \Gamma^\circ \int_{G^{(i)}} \left(\left(\frac{\mathbf{h}^\circ}{n^\circ}, \nabla \right) \nabla \varphi_n, \mathbf{f}_{si} \right) dQ - \rho_M \Gamma^\circ \int_{G^{(i)}} (\mathbf{i}_H, \nabla) \frac{\mathbf{h}^\circ}{n^\circ} (\mathbf{i}_H, \nabla \varphi_n) (\mathbf{i}_H, \mathbf{f}_{si}) \, dQ,$$

$$\alpha_{ji}^{ks} = \rho_M \Gamma^\circ \int\limits_{G^{(i)}} \left(\left(\frac{\mathbf{h}^\circ}{n^\circ}, \nabla \right) \mathbf{f}_{kj}, \mathbf{f}_{si} \right) dQ, \qquad h^\circ = \text{const}, \tag{8.24}$$

$$\alpha_{sni} = \rho_M \Gamma^\circ \int\limits_{G^{(i)}} \left(\left(\frac{\mathbf{h}^\circ}{n^\circ}, \nabla \right) \nabla\varphi_n, \mathbf{f}_{si} \right) dQ = \rho_M \Gamma^\circ \oint\limits_{S+\Sigma} \left(\frac{\mathbf{h}^\circ}{n^\circ}, \nabla\varphi_n \right) (\vec{\nu}, \mathbf{f}_{si})\, dS \equiv 0.$$

Similarly Formulae (8.20) yield after substitution of **v** and **h** from (8.14) and (8.15):

$$\mathbf{F}_k = -\sum_{l=1}^{2N} \left(\mathsf{B}_{kl}, \int\limits_{-\infty}^{t} \frac{\dot{\vec{\Omega}}_l(\tau)\, d\tau}{\sqrt{t-\tau}} \right) + \sum_{n=1}^{\infty} \vec{\gamma}_{kn} \int\limits_{-\infty}^{t} \frac{\ddot{s}_n(\tau)\, d\tau}{\sqrt{t-\tau}},$$

$$\vec{\Phi}_n = \sum_{k=1}^{2N} \left(\vec{\gamma}_{kn}, \int\limits_{-\infty}^{t} \frac{\dot{\vec{\Omega}}_k(\tau)\, d\tau}{\sqrt{t-\tau}} \right) - \mu_n \sum_{m=1}^{\infty} \beta_{nm} \int\limits_{-\infty}^{t} \frac{\ddot{s}_m(\tau)\, d\tau}{\sqrt{t-\tau}}, \tag{8.25}$$

$$\mathbf{V}_s = -\sum_{r=1}^{2N} \left(\mathsf{T}_{sr}, \int\limits_{-\infty}^{t} \frac{\dot{\mathbf{J}}_r(\tau)\, d\tau}{\sqrt{t-\tau}} \right),$$

where β_{nm} are scalar coefficients, $\vec{\gamma}_{kn}$ are vectors, B_{kl} and T_{sr} are second-rank tensors:

$$\vec{\gamma}_{kn} = \sum_{j=1}^{3} \mathbf{i}_j \gamma_{knj}, \quad \mathsf{B}_{kl} = \left\{ \beta_{ji}^{kl} \right\}, \qquad \mathsf{T}_{sr} = \left\{ \gamma_{ji}^{sr} \right\}. \tag{8.26}$$

Components of these vectors and elements of these tensors are expressed by the following formulae:

$$\gamma_{knj} = -\rho \left(\frac{\nu_F}{\pi} \right)^{1/2} \int\limits_S (\mathbf{f}_{kj}, \nabla\varphi_n)\, dS,$$

$$\beta_{nm} = \beta_{mn} = \frac{\rho}{\mu_n} \left(\frac{\nu_F}{\pi} \right)^{1/2} \int\limits_S (\nabla\varphi_m, \nabla\varphi_n)\, dS, \tag{8.27}$$

$$\beta_{ji}^{kl} = \beta_{ij}^{lk} = \rho \left(\frac{\nu_F}{\pi} \right)^{1/2} \int\limits_S (\mathbf{f}_{li}, \mathbf{f}_{kj})\, dS,$$

$$\gamma_{ji}^{sr} = \gamma_{ij}^{rs} = \rho_M \left(\frac{\nu_M}{\pi} \right)^{1/2} \oint\limits_{S+\Sigma} (\mathbf{f}_{ri}, \mathbf{f}_{sj})\, dS.$$

Let us join (8.21) and Lagrange - Maxwell equations for variable , use magnetic field energy Expression (8.17) to obtain the left sides of those equations, and introduce generalized forces and voltages expressed by (8.22), (8.25) into the right side of the obtained set of equations. Resulting are the following equations:

$$(m^\circ + m)\, \dot{\mathbf{V}} + \sum_{n=1}^{\infty} \vec{\lambda}_n \ddot{s}_n = \mathbf{P},$$

$$(J^\circ + J^{(0)}, \dot{\vec{\omega}}) + (J_1^*, \dot{\vec{\Omega}}_1) + \sum_{n=1}^{\infty} [\vec{\lambda}_{Gn}\ddot{s}_n - (\vec{\lambda}_n \times \mathbf{j})\, s_n] = \mathbf{M}_G,$$

$$(J_k^*, \dot{\vec{\Omega}}_k + \delta_k\dot{\vec{\omega}}) + \sum_{s=1}^{2N} (A_{ks}, \mathbf{J}_s) +$$

$$+ \sum_{l=1}^{2N} \left(B_{kl}, \int_{-\infty}^{t} \frac{\dot{\vec{\Omega}}_l(\tau)\, d\tau}{\sqrt{t-\tau}} \right) - \sum_{n=1}^{\infty} \vec{\gamma}_{kn} \int_{-\infty}^{t} \frac{\ddot{s}_n(\tau)\, d\tau}{\sqrt{t-\tau}} = 0,$$

$$(L_s^*, \dot{\mathbf{J}}_s) - \sum_{k=1}^{2N} (\overline{A}_{ks}, \vec{\Omega}_k) - \sum_{n=1}^{\infty} \vec{\alpha}_{sn}\dot{s}_n + \sum_{r=1}^{2N} \left(T_{sr}, \int_{-\infty}^{t} \frac{\dot{\mathbf{J}}_r(\tau)\, d\tau}{\sqrt{t-\tau}} \right) = 0, \tag{8.28}$$

$$\mu_n (\ddot{s}_n + \omega_n^2 s_n) + (\vec{\lambda}_n, \dot{\mathbf{V}}) + (\vec{\lambda}_{Gn}, \dot{\vec{\omega}}) - (\vec{\lambda}_n, \mathbf{j}) + \sum_{s=1}^{2N} (\vec{\alpha}_{sn}, \mathbf{J}_s) +$$

$$+ \mu_n \sum_{m=1}^{\infty} \beta_{nm} \int_{-\infty}^{t} \frac{\ddot{s}_m(\tau)\, d\tau}{\sqrt{t-\tau}} - \sum_{k=1}^{2N} \left(\vec{\gamma}_{kn}, \int_{-\infty}^{t} \frac{\dot{\vec{\Omega}}_k(\tau)\, d\tau}{\sqrt{t-\tau}} \right) = 0,$$

$$s, k = 1, 2, \ldots, 2N, \quad n = 1, 2, \ldots, \quad \delta_k = 0 \text{ for } k \neq 1, \quad \delta_1 = 1.$$

This is the set of equations describing perturbed motion of a solid body with a singly connected cavity containing LVECF liquid, in the presence of body-related steady magnetic field.

The coefficients of these equations are specified by Formulae (8.18), (8.24), (8.27) and (2.26), (2.52), (2.53b) in which the rib width should be considered zero.

Let us emphasize an important feature of Equations (8.28), namely that the terms engaged in the coupling of the two groups of equations, (8.21) and the new equations describing magnetic field variations, are of gyroscopic character. That means that corresponding matrices of coefficients are antisymmetric. In conjunction with symmetry of the matrices of inertial and dissipative forces and their analogues in the equations describing magnetic field variation provides for conformity with the laws of conservation.

Thus the set of Equations (8.28) can be considered a mathematical a model adequate in a certain sense to the considered discrete-continual system. This model is automatically transferred to the case of a cavity with inner ribs. If we ignore hydrodynamic boundary layer effects in comparison with much more powerful dynamic effects due to vortex formation at the ribs edges, as it was done in Chapter 2, the formal change to the case of a cavity with inner ribs is made by changing the appropriate coefficients of Equations (8.28) to (2.52), (2.53a), and (2.53b).

For $s_n \equiv 0$ Equations (8.28) yield the equations of perturbed motion of a solid body with a cavity completely filled with HECF liquid:

$$(m^\circ + m)\, \dot{\mathbf{V}} = \mathbf{P}, \qquad (J^\circ + J^{(0)}, \dot{\vec{\omega}}) + (J_1^*, \dot{\vec{\Omega}}_1) = \mathbf{M}_G,$$

$$(\mathsf{J}_k^*, \dot{\vec{\Omega}}_k + \delta_k \dot{\vec{\omega}}) + \sum_{s=1}^{2N} (\mathsf{A}_{ks}, \mathbf{J}_s) + \sum_{l=1}^{2N} \left(\mathsf{B}_{kl}, \int_{-\infty}^{t} \frac{\dot{\vec{\Omega}}_l(\tau)\, d\tau}{\sqrt{t-\tau}}\right) = 0, \tag{8.29}$$

$$(\mathsf{L}_s^*, \dot{\mathbf{J}}_s) - \sum_{k=1}^{2N} (\overline{\mathsf{A}}_{ks}, \dot{\vec{\Omega}}_k) + \sum_{r=1}^{2N} \left(\mathsf{T}_{sr}, \int_{-\infty}^{t} \frac{\dot{\mathbf{J}}_r(\tau)\, d\tau}{\sqrt{t-\tau}}\right) = 0,$$

$$s, k = 1, 2, \ldots, 2N, \quad \delta_k = 0 \quad \text{for} \quad k \neq 1, \quad \delta_1 = 1,$$

and for $\mathbf{V} \equiv 0$, $\vec{\omega} \equiv 0$ the equations of free magnetogravitational oscillations of LVECF liquid:

$$(\mathsf{J}_k^*, \dot{\vec{\Omega}}_k) + \sum_{s=1}^{2N} (\mathsf{A}_{ks}, \mathbf{J}_s) + \sum_{l=1}^{2N} \left(\mathsf{B}_{kl}, \int_{-\infty}^{t} \frac{\dot{\vec{\Omega}}_l(\tau)\, d\tau}{\sqrt{t-\tau}}\right) - \sum_{n=1}^{\infty} \vec{\gamma}_{kn} \int_{-\infty}^{t} \frac{\ddot{s}_n(\tau)\, d\tau}{\sqrt{t-\tau}} = 0,$$

$$(\mathsf{L}_s^*, \dot{\mathbf{J}}_s) - \sum_{k=1}^{2N} (\overline{\mathsf{A}}_{ks}, \dot{\vec{\Omega}}_k) - \sum_{n=1}^{\infty} \vec{\alpha}_{sn} \dot{s}_n + \sum_{r=1}^{2N} \left(\mathsf{T}_{sr}, \int_{-\infty}^{t} \frac{\dot{\mathbf{J}}_r(\tau)\, d\tau}{\sqrt{t-\tau}}\right) = 0, \tag{8.30}$$

$$\mu_n(\ddot{s}_n + \omega_n^2 s_n) + \sum_{s=1}^{2N} (\vec{\alpha}_{sn}, \mathbf{J}_s) + {}$$
$$+ \mu_n \sum_{m=1}^{\infty} \beta_{nm} \int_{-\infty}^{t} \frac{\ddot{s}_m(\tau)\, d\tau}{\sqrt{t-\tau}} - \sum_{k=1}^{2N} \left(\vec{\gamma}_{kn}, \int_{-\infty}^{t} \frac{\dot{\vec{\Omega}}_k(\tau)\, d\tau}{\sqrt{t-\tau}}\right) = 0,$$

$$s, k = 1, 2, \ldots, 2N, \quad n = 1, 2, \ldots$$

For $\mathbf{J}_s \equiv 0$ (8.28) changes to the set of equations of perturbed motion of a solid body with a cavity containing vortex LV liquid, (8.21), and for $\mathsf{Re} = \infty$, $\mathsf{Re}_M = \infty$ to the following set of equations corresponding to ideal superconductive liquid:

$$(m^\circ + m)\, \dot{\mathbf{V}} + \sum_{n=1}^{\infty} \vec{\lambda}_n \ddot{s}_n = \mathbf{P},$$

$$(\mathsf{J}^\circ + \mathsf{J}^{(0)}, \dot{\vec{\omega}}) + (\mathsf{J}_1^*, \dot{\vec{\Omega}}_1) + \sum_{n=1}^{\infty} [\vec{\lambda}_{Gn} \ddot{s}_n - (\vec{\lambda}_n \times \mathbf{j})\, s_n] = \mathbf{M}_G, \tag{8.31}$$

$$(\mathsf{J}_k^*, \dot{\vec{\Omega}}_k + \delta_k \dot{\vec{\omega}}) + \sum_{s=1}^{2N} (\mathsf{A}_{ks}, \mathbf{J}_s) = 0,$$

$$(\mathsf{L}_s^*, \dot{\mathbf{J}}_s) - \sum_{k=1}^{2N} (\overline{\mathsf{A}}_{ks}, \dot{\vec{\Omega}}_k) - \sum_{n=1}^{\infty} \vec{\alpha}_{sn} \dot{s}_n = 0,$$

$$\mu_n(\ddot{s}_n + \omega_n^2 s_n) + (\vec{\lambda}_n, \dot{\mathbf{V}}) + (\vec{\lambda}_{Gn}, \dot{\vec{\omega}}) - (\vec{\lambda}_n, \mathbf{j}) + \sum_{s=1}^{2N} (\vec{\alpha}_{sn}, \mathbf{J}_s) = 0,$$

$$s, k = 1, 2, \ldots, 2N, \quad n = 1, 2, \ldots, \quad \delta_k = 0 \quad \text{for} \quad k \neq 1, \quad \delta_1 = 1.$$

The structure of Equations (8.28) brings us to the idea of an equivalent gyroscopic system existence in which the role of ponderomotive forces is played by Coriolis forces.

The most natural aspirnt to the role of such system is a solid body rotating about longitudinal axis Ox, which preserves its constant position in the space, that is equivalent to its coincidence with $O^\circ x^\circ$ axis performing translations motion in the inertial space of coordinate system $O^\circ x^\circ y^\circ z^\circ$. Relevant equations can be obtained as particular case of Equations (2.137). A still more complete analogy should be sought in equations corresponding to a completely filled cavity, since wave motions of liquid have no counterparts in the form of magnetic field free oscillations in the considered low-frequency range.

For rotation around a motionless axis and for a completely filled cavity $\mathbf{V} \equiv 0$, $\vec{\omega} \equiv 0$, $s_n \equiv 0$ and we obtain from (2.137):

$$
\begin{aligned}
&(\mathsf{J}_1^*, \dot{\vec{\Omega}}_1) = \mathbf{M}_G, \\
&(\mathsf{J}_k^*, \dot{\vec{\Omega}}_k) + \sum_{l=1}^{2N} (\mathsf{D}_{kl}, \vec{\Omega}_l) + \sum_{l=1}^{2N} \left(\mathsf{B}_{kl}, \int_{-\infty}^{t} \frac{\dot{\vec{\Omega}}_l(\tau)\, d\tau}{\sqrt{t-\tau}} \right) = 0, \\
&k = 1, 2, \ldots, 2N.
\end{aligned}
\tag{8.32}
$$

It is these equations that should compared to Equations (8.29) in detail. Let us limit ourselves to the first N functions $\Omega_k(t)$ and $\mathbf{J}_s(t)$ in those, and assume $\mathbf{V} \equiv 0$, $\vec{\omega} \equiv 0$. This yields the following set of equations:

$$
\begin{aligned}
&(\mathsf{J}_1^*, \dot{\vec{\Omega}}_1) = \mathbf{M}_G, \\
&(\mathsf{J}_k^*, \dot{\vec{\Omega}}_k) + \sum_{s=1}^{2N} (\mathsf{A}_{ks}, \mathbf{J}_s) + \sum_{l=1}^{2N} \left(\mathsf{B}_{kl}, \int_{-\infty}^{t} \frac{\dot{\vec{\Omega}}_l(\tau)\, d\tau}{\sqrt{t-\tau}} \right) = 0, \\
&(\mathsf{L}_s^*, \dot{\mathbf{J}}_s) - \sum_{k=1}^{2N} (\overline{\mathsf{A}}_{ks}, \vec{\Omega}_k) + \sum_{r=1}^{2N} \left(\mathsf{T}_{sr}, \int_{-\infty}^{t} \frac{\dot{\mathbf{J}}_r(\tau)\, d\tau}{\sqrt{t-\tau}} \right) = 0, \\
&s, k = 1, 2, \ldots, N.
\end{aligned}
\tag{8.33}
$$

Suppose the body and the cavity have two symmetry planes, the resultant moment $\mathbf{M}_G$ of the system of external forces being perpendicular to one of them. In this case the set of functions describing vortex motions of liquid breaks into independent subsets of N each. It is obvious that if we keep denominations $\vec{\Omega}_k$ ($k = 1, 2, \ldots, N$) for the generalized angular velocities corresponding to one symmetry plane (to which moment $\mathbf{M}_G$ is orthogonal), and denote the generalized velocities corresponding to the other symmetry plane by $\mathbf{J}_s$ ($s = 1, 2, \ldots, N$), Equations (8.32) gain the structure of Equations (8.33). In this sense we can speak of an analogy between the body with a cavity completely filled with LV liquid, that rotates steadily around a motionless axis and the nonrotating body in the case of steady magnetic field related to the body and LVECF or LVEC liquid.

8.5. Equations of Dynamics for a Solid Body with a Cavity of Revolution Having Narrow Internal Ribs

Let us consider for an illustration a axisymmetric cavity with narrow internal ribs made of nonmagnetic dielectric material, with the assumption of the cavity and the body having a common symmetry axis Ox ; the field is longitudinal, $\mathbf{i}_H = \mathbf{i}_1$ (see Fig. 8.1). We shall consider two cases:

8.5.1. A cavity with radial ribs. Let us use Equations (8.28), (2.71) (for $x_G = 0$), (2.72) and write down equations of motion of the body in $O^\circ x^\circ z^\circ$ plane and its rotation around Ox axis in the case of radial ribs (k, s = 1, 2, ..., N).

1. Motion in $O^\circ x^\circ z^\circ$ plane:

$$(m^\circ + m)\,\dot{V} + \sum_{n=1}^{\infty} \lambda_n \ddot{s}_n = P_z,$$

$$(J^\circ + J^{(0)})\,\dot{\omega} + J_1^*\,\dot{\Omega}_1 + \sum_{n=1}^{\infty} (\lambda_{Gn} \ddot{s}_n + \lambda_n\, j\, s_n) = M_{Gy},$$

$$(8.34)$$

$$J_1^*\,(\dot{\Omega}_1 + \dot{\omega}) + \sum_{s=1}^{N} \overset{\circ}{\alpha}_{1s} \dot{I}_s + \sum_{k=1}^{N} \overset{\circ}{\beta}_{1k} \int_{-\infty}^{t} \frac{\dot{\Omega}_k(\tau)\, d\tau}{\sqrt{t-\tau}} - \sum_{n=1}^{\infty} \gamma_{n1}^* \int_{-\infty}^{t} \frac{\ddot{s}_n(\tau)\, d\tau}{\sqrt{t-\tau}} = 0,$$

$$J_k^*\,\dot{\Omega}_k + \sum_{s=1}^{N} \overset{\circ}{\alpha}_{ks} \dot{I}_s + \sum_{l=1}^{N} \overset{\circ}{\beta}_{kl} \int_{-\infty}^{t} \frac{\dot{\Omega}_l(\tau)\, d\tau}{\sqrt{t-\tau}} - \sum_{n=1}^{\infty} \gamma_{nk}^* \int_{-\infty}^{t} \frac{\ddot{s}_n(\tau)\, d\tau}{\sqrt{t-\tau}} = 0,$$

$$\Lambda_s^* \dot{I}_s - \sum_{k=1}^{N} \overset{\circ}{\alpha}_{ks} \dot{\Omega}_k - \sum_{n=1}^{\infty} \alpha_{ns}^* \ddot{s}_n + \sum_{r=1}^{N} \overset{\circ}{\gamma}_{sr} \int_{-\infty}^{t} \frac{\dot{I}_r(\tau)\, d\tau}{\sqrt{t-\tau}} = 0,$$

$$\mu_n\,(\ddot{s}_n + \omega_n^2 s_n) + \lambda_n\, \dot{V} + \lambda_{Gn} \dot{\omega} + \lambda_n j\, \psi + \sum_{s=1}^{N} \alpha_{ns}^* \dot{I}_s +$$

$$+ \mu_n \sum_{m=1}^{\infty} \beta_{nm} \int_{-\infty}^{t} \frac{\ddot{s}_m(\tau)\, d\tau}{\sqrt{t-\tau}} - \sum_{k=1}^{N} \gamma_{nk}^* \int_{-\infty}^{t} \frac{\dot{\Omega}_k(\tau)\, d\tau}{\sqrt{t-\tau}} = 0,$$

k = 2, 3, ...,N , s = 1, 2, ..., N, n = 1, 2, ...

2. Rotation around Ox axis

$$(I^\circ + I^{(0)})\,\dot{\gamma} + I_1^*\,\dot{\Gamma}_1 = M_{Gx},$$

$$\Gamma_1^* (\dot{\Gamma}_1 + \dot{\gamma}) + \sum_{s=1}^{N} \alpha_{1s} J_s + \sum_{k=1}^{N} \beta_{1k}^* \int_{-\infty}^{t} \frac{\dot{\Gamma}_k(\tau)\, d\tau}{\sqrt{t-\tau}} = 0, \tag{8.35}$$

$$\Gamma_k^* \dot{\Gamma}_k + \sum_{s=1}^{N} \alpha_{ks} J_s + \sum_{l=1}^{N} \beta_{kl}^* \int_{-\infty}^{t} \frac{\dot{\Gamma}_l(\tau)\, d\tau}{\sqrt{t-\tau}} = 0,$$

$$L_s^* \dot{J}_s - \sum_{k=1}^{N} \alpha_{ks} \Gamma_k + \sum_{r=1}^{N} \gamma_{sr}^* \int_{-\infty}^{t} \frac{\dot{J}_r(\tau)\, d\tau}{\sqrt{t-\tau}} = 0,$$

$$k = 2, 3, \dots, N, \quad s = 1, 2, \dots, N, \quad n = 1, 2, \dots$$

W e use simplified denominations of (2.70) here for generalized velocities of liquid and corresponding coefficients and similar simplified denominations for their counterparts relevant to magnetic field:

$$J_{s1} = J_s, \quad J_{s2} = I_s, \quad L_{11}^{ss} = L_s^*, \quad L_{22}^{ss} = \Lambda_s^*,$$

$$\beta_{11}^{kl} = \beta_{kl}^*, \quad \beta_{22}^{kl} = \overset{\circ}{\beta}_{kl}, \tag{8.36}$$

$$\alpha_{11}^{ss} = \alpha_{ks}, \quad \alpha_{22}^{ss} = \overset{\circ}{\alpha}_{ks}, \quad \alpha_{sn2} = \overset{*}{\alpha}_{ns},$$

$$\gamma_{11}^{sr} = \gamma_{sr}, \quad \gamma_{22}^{sr} = \overset{\circ}{\gamma}_{sr}, \quad \gamma_{kn2} = \overset{*}{\gamma}_{nk}.$$

8.5.2. A Cavity with a Ring-Shaped Rib. Axisymmetric Oscillations of Liquid. Below we consider equations describing axisymmetric perturbed motion of LVECF liquid in a cavity of revolution with a ring-shaped rib, which can also be interesting for practice. A characteristic kinematic feature of such motion is the presence of ring-shaped vortices coaxial to the cavity's longitudinal axis. The velocity field induced by them is included as a particular case in the general Expressions (2.36), (2.37) (coordinate functions of (2.39b)), and in Equations (2.57), but not in Expressions (2.94) and (2.101), which are the foundation of equations in Articles 2.7 and 2.8.

Let us consider an axisymmetric velocity field (see (2.39b))

$$\mathbf{v} = \sum_{k=1}^{N} \Omega_k (\mathbf{i}_\theta \times \mathbf{R}_k - \nabla\Psi_k) + \sum_{n=1}^{\infty} \dot{s}_n \nabla\varphi_n,$$

$$\mathbf{R}_k = [\mathbf{i}_x X_k(x) + \mathbf{i}_r R_k(r)]\, \frac{r_0}{r}, \qquad \mathbf{R}_1 = \mathbf{i}_x \frac{r_0 x}{r} + \mathbf{i}_r r_0, \tag{8.37}$$

$$\Delta\Psi_k = 0, \quad \left.\frac{\partial\Psi_k}{\partial\nu}\right|_{S+\Sigma} = (\mathbf{R}_k \times \vec{\nu}, \ \mathbf{i}_\theta), \quad \left.\frac{\partial\Psi_k}{\partial\nu}\right|_{r=0} = 0.$$

We have retained denominations Ω_k and s_n used in Equations (8.34) for generalized velocities corresponding to vortex and wave motion of liquid. But the coordinate functions used to expand the vortex fields (their number is N) correspond now to ring-shaped vortices with the field strength depending on coordinates r and x. The forms of wave motions naturally have the same symmetry and depend on the same coordinates.

Thus, velocity field for all the radial planes is the same.

Let us somewhat extend the statement of the problem in the following direction. Suppose some axisymmetric domain S_0, being a part of wetted surface S at the bottom pole vicinity, can move along an outer normal to S at the velocity $\dot{u}(r, t) = w(r)\,\dot{s}_0(t)$, where $w(r)$ and $s_0(t)$ are known functions. We shall introduce two new harmonic functions $\varphi_0(x, r)$ and $\Phi_0(x, r)$ being the solutions of the following boundary-value problems:

$$\Delta\varphi_0 = 0,$$

$$\left.\frac{\partial\varphi_0}{\partial\nu}\right|_{S_0} = w(r), \quad \left.\frac{\partial\varphi_0}{\partial\nu}\right|_{S-S_0} = 0, \quad \left.\frac{\partial\varphi_0}{\partial\nu}\right|_{\Sigma} = -\frac{1}{\Sigma}\int_{S_0} w(r)\,dS = 1, \tag{8.38}$$

$$\Delta\Phi_0 = 0,$$

$$\left.\frac{\partial\Phi_0}{\partial\nu}\right|_{S_0} = w(r) + 1, \quad \left.\frac{\partial\Phi_0}{\partial\nu}\right|_{S-S_0} = 0, \quad \left.\frac{\partial\Phi_0}{\partial\nu}\right|_{\Sigma} = 0. \tag{8.39}$$

where Σ is the free surface area of liquid. The last condition in (8.38) is the normalizing condition for functions φ_0, converting s_0 into free surface displacement according to the function $s_0 = s_0(t)$, parallel to itself.

Let us additionally introduce the following symbols:

V is the body's velocity of perturbed motion, parallel to its longitudinal axis;

$P_x(t)$ is Ox -axis component of the resultant vector for excitation forces applied to the cavity. It is specified as a function of time;

h is the liquid depth;

m° and m are the masses of the body and solidified liquid;

λ_0 and λ_n are new coefficients of inertial constraints:

$$\lambda_0 = \rho\oint_{S+\Sigma} x\frac{\partial\varphi_0}{\partial\nu}\,dS \approx \rho h\Sigma, \tag{8.40}$$

$$\lambda_n = \rho\int_{S_0}\varphi_n\frac{\partial\Phi_0}{\partial\nu}\,dS = \rho\int_{S_0}[w(r)+1]\,\varphi_n\,dS.$$

Having retained the other symbols used in Equations (8.34), we obtain the following set of equations, that can be considered as a mathematical model of the body-liquid system perturbed motion for axisymmetric oscillations of liquid.

$$(m^\circ + m)\dot{V} + \lambda_0\ddot{s}_0 = P_x(t),$$

$$J_k^*\dot{\Omega}_k + \sum_{s=1}^{N}\overset{\circ}{\alpha}_{ks}I_s + \sum_{l=1}^{N}\overset{\circ}{\beta}_{kl}\int_{-\infty}^{t}\frac{\dot{\Omega}_l(\tau)\,d\tau}{\sqrt{t-\tau}} - \sum_{n=1}^{\infty}\gamma_{nk}^*\int_{-\infty}^{t}\frac{\ddot{s}_n(\tau)\,d\tau}{\sqrt{t-\tau}} = 0,$$

$$\Lambda_s^* \dot{I}_s - \sum_{k=1}^{N} \overset{\circ}{\alpha}_{ks} \Omega_k - \sum_{n=1}^{\infty} \alpha_{ns}^* \dot{s}_n + \sum_{r=1}^{N} \overset{\circ}{\gamma}_{sr} \int_{-\infty}^{t} \frac{\dot{I}_r(\tau)\, d\tau}{\sqrt{t-\tau}} = 0, \tag{8.41}$$

$$\mu_n \left(\ddot{s}_n + \omega_n^2 s_n + \sum_{m=1}^{\infty} \beta_{nm} \int_{-\infty}^{t} \frac{\ddot{s}_m(\tau)\, d\tau}{\sqrt{t-\tau}} \right) + \lambda_n \ddot{s}_0 + \sum_{s=1}^{N} \alpha_{ns}^* I_s - \sum_{k=1}^{N} \gamma_{nk}^* \int_{-\infty}^{t} \frac{\dot{\Omega}_k(\tau)\, d\tau}{\sqrt{t-\tau}} = 0.$$

$$s, k = 1, 2, \ldots, N, \quad n = 1, 2, \ldots$$

Expressions for coefficients of these equations will be interpreted in the next article for the case of a cavity in the shape of a circular cylinder.

8.6. The Case of a Circular Cylinder-Shaped Cavity

Let us consider a cavity having the shape of a circular cylinder with flat top and bottom and a narrow ring-shaped rib near the free surface. Denote the cylinder radius by r_0, the rib distance from the liquid free surface by d, the rib width by b. We shall assume that turns of the winding occupy the whole of the cylinder surface from the bottom to the liquid free surface. Denote dimensionless depth of liquid by $\tilde{h} = \frac{h}{r_0}$ and introduce dimensionless parameters $\tilde{b}$ and $\tilde{d}$, small in comparison to unity:

$$\tilde{b} = \frac{b}{r_0} << 1, \qquad \tilde{d} = \frac{d}{r_0} << 1. \tag{8.42}$$

We shall take into account only first form of liquid wave motions ($n = 1$) and the first two forms of vortex fields oscillations ($k, s = 1, 2; N = 2$), the symmetric and antisymmetric one with respect to argument $\tilde{x}$ at the internal $-\frac{\tilde{h}}{2} \le \tilde{x} \le \frac{\tilde{h}}{2}$; $\tilde{x} = \frac{x}{r_0}$. This regards both axisymmetric and antisymmetric oscillations of liquid in the cavity with respect to Oxy plane.

Let us, as in Chapter 6, consider the excitation due to liquid wave motion to be the prevailing excitation of velocity field at the rib edge vicinity. We ignore 'cross' components corresponding to coefficients $\overset{\circ}{\beta}_{kl}$ for $k \neq l$ in the equations of perturbed motion.

At the same time we somewhat reduce the requirements to coordinate functions $\mathbf{f}_k$ used in the expansions approximating vortex fields. Namely we cancel the requirement that functions $\vec{\Psi}_k$ (2.38) to be harmonic and functions $\mathbf{f}_k$ to be orthogonal in the sense of satisfying Equation (2.54a) for $l = 1, k = 2$. That lets us obtain approximate solutions of boundary-value problems in potentials $\vec{\Psi}_k$, that meet boundary conditions of (2.38), but not necessarily comply with Laplace equation. We thus retain the ideology of Boubnov – Galerkin method, being the foundation for derivation of perturbed motion equations. We shall consider unperturbed magnetic field of the coil uniform all over volume Q occupied by liquid (see R.Feynman et al. [22]).

Let us confine to angular oscillations of the cylinder in $O^\circ x^\circ z^\circ$ plane ($V \equiv 0$) for antisymmetric oscillations of liquid. The exciting factor here is $\psi(t)$ (kinematic excitation) or $M_G(t)$ (dynamic exciation).
For axisymmetric oscillations of liquid we shall consider the cylinder motionless ($V \equiv 0, \psi \equiv 0$), and the exciting factor will be function $s_0(t)$ (kinematic excitation).
Function $w(r)$ will be specified as follows: $w(r) = 2\ln\tilde{r}, \tilde{r} = \dfrac{r}{r_0}$.

The set of equations including both cases can be derived from Equations (8.34) and (8.41). Taking into account (8.24), adding the terms resulting from reduction of requirements to coordinate functions and using simplified denominations we obtain the set of equations that can be considered as an approximate mathematical model for both classes liquid motions mentioned above:

$$(J^\circ + J^{(0)})\ddot{\psi} + J_1^*\dot{\Omega}_1 + \lambda_{G1}\ddot{s}_1 + \lambda_1 j s_1 = M_G , \tag{8.34}$$

$$J_1^*(\dot{\Omega}_1 + \ddot{\psi}) + J_{12}\dot{\Omega}_2 + \alpha_{12}^\circ I_2 + \sum_{l=1}^{2} \beta_{1l}^\circ \int_{-\infty}^{t} \frac{\dot{\Omega}_l(\tau)\,d\tau}{\sqrt{t-\tau}} - \gamma_{11}^* \int_{-\infty}^{t} \frac{\ddot{s}_1(\tau)\,d\tau}{\sqrt{t-\tau}} = 0,$$

$$J_2^*\dot{\Omega}_2 + J_{21}\dot{\Omega}_1 + \alpha_{21}^\circ I_1 + \sum_{l=1}^{2} \beta_{2l}^\circ \int_{-\infty}^{t} \frac{\dot{\Omega}_l(\tau)\,d\tau}{\sqrt{t-\tau}} - \gamma_{12}^* \int_{-\infty}^{t} \frac{\ddot{s}_1(\tau)\,d\tau}{\sqrt{t-\tau}} = 0,$$

$$\Lambda_1^*\dot{I}_1 + \Lambda_{12}^*\dot{I}_2 - \alpha_{21}^\circ\Omega_2 + \sum_{r=1}^{2} \gamma_{1r}^\circ \int_{-\infty}^{t} \frac{\dot{I}_r(\tau)\,d\tau}{\sqrt{t-\tau}} = 0, \tag{8.43}$$

$$\Lambda_2^*\dot{I}_2 + \Lambda_{21}^*\dot{I}_1 - \alpha_{12}^\circ\Omega_1 + \sum_{r=1}^{2} \gamma_{2r}^\circ \int_{-\infty}^{t} \frac{\dot{I}_r(\tau)\,d\tau}{\sqrt{t-\tau}} = 0,$$

$$\mu_1\left(\ddot{s}_1 + \omega_1^2 s_1 + \beta_1 \int_{-\infty}^{t} \frac{\ddot{s}_1(\tau)\,d\tau}{\sqrt{t-\tau}}\right) - \sum_{k=1}^{2} \gamma_{1k}^* \int_{-\infty}^{t} \frac{\dot{\Omega}_k(\tau)\,d\tau}{\sqrt{t-\tau}} = F.$$

The coefficients and right sides have the following structure:
B o t h   c a s e s   o f   o s c i l l a t i o n s

$$J_j^* = m r_0^2 \tilde{J}_j , \quad \Lambda_j^* = m_M r_0^2 \tilde{\Lambda}_j \quad (j = 1, 2),$$

$$J_{12} = J_{21} = m\, r_0^2 \tilde{J}_{12} , \qquad \Lambda_{12} = \Lambda_{21} = m_M\, r_0^2 \tilde{\Lambda}_{12} ,$$

$$\alpha_{ij}^\circ = \frac{2\pi}{\hbar} I^\circ m_M \tilde{\alpha}_{ij} \quad (i, j = 1, 2; \; i \neq j), \tag{8.44}$$

$$\gamma_{ij}^\circ = \gamma_{ji}^\circ = \left(\frac{\nu_M}{\pi}\right)^{1/2} m_M\, r_0\, \tilde{\gamma}_{ij}^\circ , \qquad \gamma_{ij}^* = \tilde{\gamma}_{ij}\left(|\dot{\tilde{s}}_1|\right)^{1/2},$$

$$\beta_1 = \tilde{\beta}_1\left(|\dot{\tilde{s}}_1|\right)^{1/2}, \quad (i, j = 1, 2),$$

$$m_M = \frac{\rho_M}{\rho}\, m , \quad \rho_M = \mu_0\,\mu\, n^{\circ 2} , \quad m = \pi\, r_0^3\, \hbar ,$$

$$n^\circ = \frac{N^\circ}{2\pi}, \qquad \nu_M = \frac{1}{\mu_0 \mu \sigma}, \qquad \widetilde{\dot{s}_1} = \frac{\dot{s}_1}{r_0}.$$

$$\beta_{ij}^\circ = \widetilde{\beta}_{ij}^\circ \left(| \widetilde{\dot{s}}_1 | \right)^{1/2}$$

Axisymmetric oscillations

$$\omega_1^2 = \frac{j \zeta_1}{r_0} \tanh (\zeta_1 \widetilde{h}), \qquad \mu_1 = \frac{\pi \rho r_0^3}{\zeta_1 \tanh (\zeta_1 \widetilde{h})},$$

$$J_0' (\zeta_1) = 0, \qquad \lambda_1 = \frac{2\pi \rho r_0^3 a_1}{\zeta_1 \sinh (\zeta_1 \widetilde{h})}, \tag{8.45}$$

$$F = F(t) = - \lambda_1 \ddot{s}_0 (t).$$

Antisymmetric oscillations

$$\omega_1^2 = \frac{j \xi_1}{r_0} \tanh (\xi_1 \widetilde{h}), \qquad \mu_1 = \frac{\pi \rho r_0^3 (\xi_1^2 - 1)}{\xi_1^3 \tanh (\xi_1 \widetilde{h})},$$

$$J_1' (\xi_1) = 0, \qquad \lambda_1 = \frac{\pi \rho r_0^3}{\xi_1^2}, \tag{8.46}$$

$$M_G = M_G (t), \qquad F = - \lambda_{G1} \ddot{\psi} - \lambda_1 j \psi.$$

Coordinate functions $\mathbf{f}_j$ (analogous to Functions (8.15)) and displacement potential φ_1 have the form:

Axisymmetric oscillations

Functions $\mathbf{f}_j (\widetilde{x}, \widetilde{y})$, potentials $\Psi_j (\widetilde{x}, \widetilde{r})$, and coefficients a_n, $j = 1, 2$, $\widetilde{x} = \dfrac{x}{r_0}$, $\widetilde{r} = \dfrac{r}{r_0}$, the coordinates origin being in the centre of mass for the solidified liquid are as follows

$$\mathbf{f}_j = \mathbf{i}_\theta \times \mathbf{R}_j - \nabla \Psi_j \qquad (j = 1, 2),$$

$$\mathbf{R}_1 = \mathbf{i}_x \frac{x r_0}{r} + \mathbf{i}_r r_0; \qquad \mathbf{R}_2 = \mathbf{i}_x \frac{x^2}{r} + \mathbf{i}_r r_0;$$

$$\Psi_1 = r_0^2 \left[\widetilde{x} \left(\ln \widetilde{r} - \frac{1}{2} \right) - \sum_{n=1}^{\infty} \frac{a_n \sinh(\zeta_n \widetilde{r})}{\zeta_n \cosh \left(\zeta_n \frac{\widetilde{h}}{2} \right)} \frac{J_0 (\zeta_n \widetilde{r})}{J_0 (\zeta_n)} \right], \tag{8.47}$$

$$\Psi_2 = r_0^2 \left[\widetilde{x}^2 \left(\ln \widetilde{r} + \frac{1}{2} \right) - \widetilde{x} - \widetilde{h} \sum_{n=1}^{\infty} \frac{a_n \cosh(\zeta_n \widetilde{r})}{\zeta_n \sinh \left(\zeta_n \frac{\widetilde{h}}{2} \right)} \frac{J_0 (\zeta_n \widetilde{r})}{J_0 (\zeta_n)} \right],$$

$$a_n = 2 \int_0^1 \frac{J_0 (\zeta_n \widetilde{r})}{J_0 (\zeta_n)} \widetilde{r} \ln \widetilde{r} \, d\widetilde{r} = \frac{2}{\zeta_n^2} \left[1 - \frac{1}{J_0 (\zeta_n)} \right],$$

$$J_0'(\zeta_n) = 0 \qquad (n = 1, 2, ...), \qquad 2\sum_{n=1}^{\infty} a_n = 1.$$

Functions $\varphi_1(\widetilde{x}, \widetilde{r})$, $\Psi_0(\widetilde{x}, \widetilde{r})$ are as follows, the coordinates origin being is in the centre of the bottom:

$$\varphi_1 = \frac{r_0 \cosh(\zeta_1 \widetilde{x})}{\zeta_1 \sinh(\zeta_1 \widetilde{h})} \frac{J_0(\zeta_1 \widetilde{r})}{J_0(\zeta_1)}, \tag{8.48}$$

$$\Phi_0 = 2r_0 \sum_{n=1}^{\infty} \frac{a_n \cosh \zeta_n (\widetilde{x} - \widetilde{h})}{\zeta_n \sinh(\zeta_n \widetilde{h})} \frac{J_0(\zeta_n \widetilde{r})}{J_0(\zeta_n)}.$$

Antisymmetric oscillations

Functions $\mathbf{f}_j(\widetilde{x}, \widetilde{r}, \theta)$ and potentials $\Psi_j(\widetilde{x}, \widetilde{r}, \theta)$; $\widetilde{x} = \dfrac{x}{r_0}$; $\widetilde{r} = \dfrac{r}{r_0}$, the coordinates origin is the centre of mass for the solidified liquid) are as follows:

$$\mathbf{f}_j = \mathbf{i}_y \times \mathbf{R}_j - \nabla \Psi_j \qquad (j = 1, 2),$$

$$\mathbf{R}_1 = \mathbf{i}_x x + \mathbf{i}_y y + \mathbf{i}_z z, \qquad \mathbf{R}_2 = \mathbf{i}_x \frac{x^2}{r} + \mathbf{i}_y y + \mathbf{i}_z z,$$

$$\Psi_1 = r_0^2 \left[-\widetilde{x}\widetilde{r} + 4 \sum_{n=1}^{\infty} \frac{\sinh(\xi_n \widetilde{x})}{\xi_n(\xi_n^2 - 1)\cosh\left(\xi_n \dfrac{\widetilde{h}}{2}\right)} \frac{J_0(\xi_n \widetilde{r})}{J_0(\xi_n)} \right] \sin\theta\,, \tag{8.49}$$

$$\Psi_2 = r_0^2 \left\{ -\widetilde{x}^2 \widetilde{r} + 2\widetilde{h} \sum_{n=1}^{\infty} \frac{1}{\xi_n(\xi_n^2 - 1)} \left[\frac{\cosh(\xi_n \widetilde{x})}{\sinh\left(\xi_n \dfrac{\widetilde{h}}{2}\right)} + \frac{\sinh(\xi_n \widetilde{x})}{\cosh\left(\xi_n \dfrac{\widetilde{h}}{2}\right)} \right] \frac{J_1(\xi_n \widetilde{r})}{J_1(\xi_n)} \right\},$$

$$J_1'(\xi_n) = 0 \qquad (n = 1, 2, ...), \qquad 2\sum_{n=1}^{\infty} \frac{1}{\xi_n^2 - 1} = 1.$$

and function $\varphi_1(\widetilde{x}, \widetilde{r}, \theta)$ has the form:

$$\varphi_1 = \frac{r_0 \cosh(\xi_1 \widetilde{x})}{\xi_1 \sinh(\xi_1 \widetilde{h})} \frac{J_0(\zeta_1 \widetilde{r})}{J_0(\xi_1)}. \tag{8.50}$$

The coordinates origin is in the centre of the bottom.
Functions $J_0(\zeta)$ in Expressions (8.47) and (8.48) and $J_1(\xi)$ in Expressions (8.49) and (8.50) are Bessel functions of order 0 and 1 respectively.

The Coefficients having a tilde are calculated including (8.41) - (8.43) according to quadrature Formulae (2.53a), (8.24), (8.27).

There is a certain peculiarity in calculating coefficients $\widetilde{J}_{ij}$ $(i \neq j; j = 1, 2)$ and $\widetilde{J}_2$. Since functions Ψ_2 of (8.47) and (8.49) are not harmonic, Formulae (2.53a) and (2.53b) should be changed appropriately.

Let us illustrate it by an example of antisymmetric oscillations. Having applied the 1st Green formulae and Gauss – Ostrogradsky formula to transform the volume integrals to the surface ones, we obtain:

$$r_0^5 \widetilde{J}_j = \int_Q [\,(\mathbf{i}_y \times \mathbf{R}_j)^2 - 2(\mathbf{i}_y \times \mathbf{R}_j, \nabla \Psi_j) + (\nabla \Psi_j)^2]\; dQ =$$

$$= \int_Q (\mathbf{i}_y \times \mathbf{R}_j)^2\, dQ - 2 \oint_{S+\Sigma} (\mathbf{R}_j \times \vec{\nu})_y \Psi_j\, dS + \oint_{S+\Sigma} \Psi_j \frac{\partial \Psi_j}{\partial \nu}\, dS - \int_Q \Psi_j \Delta \Psi_j\, dQ =$$

$$= \int_Q \left[(\mathbf{i}_y \times \mathbf{R}_j)^2 - \Psi_j \Delta \Psi_j\right] dQ - \oint_{S+\Sigma} \Psi_j \frac{\partial \Psi_j}{\partial \nu}\, dS,$$

$$r_0^5 \widetilde{J}_{ij} = r_0^5 \widetilde{J}_{ji} =$$

$$= \int_Q \left[(\mathbf{i}_y \times \mathbf{R}_i, \mathbf{i}_y \times \mathbf{R}_j) - (\mathbf{i}_y \times \mathbf{R}_j, \nabla \Psi_i) - (\mathbf{i}_y \times \mathbf{R}_i, \nabla \Psi_j) + (\nabla \Psi_i, \nabla \Psi_j)\right] dQ =$$

$$= \int_Q \left[(\mathbf{i}_y \times \mathbf{R}_i, \mathbf{i}_y \times \mathbf{R}_j) - \Psi_i \Delta \Psi_j\right] dQ - \oint_{S+\Sigma} \frac{\partial \Psi_i}{\partial \nu} \Psi_j\; dS =$$

$$= \int_Q \left[(\mathbf{i}_y \times \mathbf{R}_i, \mathbf{i}_y \times \mathbf{R}_j) - \Psi_j \Delta \Psi_j\right] dQ - \oint_{S+\Sigma} \frac{\partial \Psi_j}{\partial \nu} \Psi_i\; dS.$$

If we take into account, that of the two functions $\Psi_1(\widetilde{x}, \widetilde{r}, \theta)$ and $\Psi_2(\widetilde{x}, \widetilde{r}, \theta)$ only the first one is harmonic, we can derive the following formulae for $\widetilde{J}_1, \widetilde{J}_2, \widetilde{J}_{12} = \widetilde{J}_{21}$:

$$r_0^5 \widetilde{J}_1 = \int_Q (\mathbf{i}_y \times \mathbf{R}_1)^2\, dQ - \oint_{S+\Sigma} \Psi_1 \frac{\partial \Psi_1}{\partial \nu}\, dS,$$

$$r_0^5 \widetilde{J}_2 = \int_Q \left[(\mathbf{i}_y \times \mathbf{R}_2)^2 - \Psi_2 \Delta \Psi_2\right] dQ - \oint_{S+\Sigma} \Psi_2 \frac{\partial \Psi_2}{\partial \nu}\, dS,$$

$$\text{(8.51)}$$

$$r_0^5 \widetilde{J}_{12} = r_5^5 \widetilde{J}_{21} = \int_Q (\mathbf{i}_y \times \mathbf{R}_1, \mathbf{i}_y \times \mathbf{R}_2)\, dQ \;-\; \oint_{S+\Sigma} \frac{\partial \Psi_2}{\partial \nu} \Psi_1\, dS =$$

$$= \int_Q \left[(\mathbf{i}_y \times \mathbf{R}_1, \mathbf{i}_y \times \mathbf{R}_2) - \Psi_1 \Delta \Psi_2\right] dQ - \oint_{S+\Sigma} \frac{\partial \Psi_1}{\partial \nu} \Psi_2\, dS,$$

Note, that the expressions for potentials Ψ_j and φ_1 of (8.44) - (8.47) are obtained from Inequalities (8.42) without recognizing the effect of the narrow rib, therefore functions f_j and φ_1 are the same for the case of a circular cylinder equipped withradial ribs instead of a ring-shaped rib. This also refers to coefficients J_j^*, Λ_j^*, β_j°, J_{ij},. Λ_{ij} $(i, j = 1, 2)$, α_{ij}° $(i, j = 1, 2;\, i \neq j)$, μ_1, ω_1^2. As to coefficients β_1, γ_{ij}° and γ_{ij}^*

$(i, j = = 1, 2)$, they will be different for radial ribs and for a ring-shaped rib. The formulae for all the mentioned coefficients result from the appropriate quadratures. These formulae are rather clumsy, so we shall present only the two simplest ones herein:

$$\text{(a)} \qquad \widetilde{J}_{12} = \widetilde{J}_{21} = 0, \qquad \widetilde{\gamma}^{\circ}_{12} = \widetilde{\gamma}^{\circ}_{21} \approx 2a_1^2,$$
$$\tag{8.52}$$
$$\text{(b)} \qquad \widetilde{J}_{12} = \widetilde{J}_{21} = \frac{1}{2}\widetilde{J}_1, \qquad \widetilde{\gamma}^{\circ}_{12} = \widetilde{\gamma}^{\circ}_{21} = \frac{1}{2}\widetilde{\gamma}^{\circ}_{11}\,.$$

Case (a) refers to axisymmetric oscillations, case (b) to antisymmetric ones.

Let us now try to find out the mechanism of the external magnetic field influence on dynamic properties of the system 'solid body — LVECF liquid'. For a simple example we shall consider a cylindric cavity with several equidistant inner ring-shaped ribs. Let it be completely filled with liquid. We shall study the system's angular oscillations around the lateral axis passing through its centre of mass. An appropriate mathematical model can be obtained from (8.43), if we assume $s_1 \equiv 0$ and recognize the presence of several ribs (not one). From symmetry considerations, we should also set $\Omega_2 \equiv 0$; $I_1 \equiv 0$ in Equations (8.43), i. e. retain the antisymmetric harmonic of the liquid velocity solenoidal field and the symmetric harmonic of the induced solenoidal magnetic field, interacting with it.

The following set of integrodifferential equation results:

$$\begin{aligned}
&(J^{\circ} + J^{(0)})\,\ddot{\psi} + J^{*}\dot{\Omega} = M_G(t),\\
&J^{*}\,(\dot{\Omega} + \ddot{\psi}) + \alpha^{\circ} I + \beta^{\circ}\int_{-\infty}^{t}\frac{\dot{\Omega}(\tau)\,d\tau}{\sqrt{t-\tau}} = 0,\\
&\Lambda^{*}\,\dot{I} - \alpha^{\circ}\Omega + \gamma^{\circ}\int_{-\infty}^{t}\frac{\dot{I}(\tau)\,d\tau}{\sqrt{t-\tau}} = 0,
\end{aligned} \tag{8.53}$$

where simplified symbols Ω and I are introduced instead of Ω_1 and I_2 , and all subscripts, now insignificant, are omitted.

Equations (8.53) can be considered analogous to Equations (4.10) for the yaw plane, but recognizing a new physical factor, namely magnetic field, and, naturally, an additional degree of freedom, eddy current I.

Let us now make another step to clear up the situation associated with that new factor effect, namely, we shall consider the case of no ribs at all.

Having assumed $\beta^{\circ} = 0$, $J^{(0)} - J^{*} = J$ in (8.53), where J is Joukowski moment of inertia, we can eliminate variable Ω and transform those equations to the following form

$$\begin{aligned}
&(J^{\circ} + J)\,\ddot{\psi} - \alpha^{\circ}\dot{\chi} = M_G(t),\\
&\Lambda^{*}\,(\dot{\chi} + \vartheta^2\chi) + \alpha^{\circ}\dot{\psi} + \gamma^{\circ}\int_{-\infty}^{t}\frac{\ddot{\chi}(\tau)\,d\tau}{\sqrt{t-\tau}} = 0,
\end{aligned} \tag{8.54}$$

$$J^* (\Omega + \dot{\psi}) + \alpha^\circ \chi = 0,$$

here

$$\dot{\chi} = I, \quad \sigma^{\circ 2} = \frac{\alpha^{\circ 2}}{\Lambda^* J^*}, \qquad \chi_0 = -\frac{J^*}{\alpha^\circ} (\Omega_0 + \dot{\psi}_0). \tag{8.55}$$

and superscript 0 corresponds to $t = 0$. Equations (8.54) have much in common with (4.10). At the same time there is fundamental difference. The 'vortex kernel' motion connection to that of the solid body is inertial, and in Equations (8.54) it is gyroscopic.

Besides that 'vortex kernel' in the case of (8.54) has nonzero frequency of natural oscillations with respect to the solid body. This fact can be interpreted in Maxwell's terms as an effect of external magnetic field intensity lines 'elasticity' for a finite mass of liquid 'linked' to the magnetic field.

Thus the resonance phenomenon is fundamentally possible in the considered system for $M_G = M_0 \sin \omega t$, that had been out in the situation corresponding to Model (4.10).

Let us consider two limiting cases, namely when the coil's ampere-turns number tends to zero ($\alpha^\circ \to 0$) and to infinity ($\alpha^\circ \to \infty$) :

1. $\alpha^\circ \to 0, \quad \sigma^{\circ 2} \to 0,$

$$(J^\circ + J)\, \ddot{\psi} = M_G(t). \tag{8.56a}$$

2. $\alpha^\circ \to \infty, \quad \sigma^{\circ 2} \to \infty,$

$$(J^\circ + J^{(0)})\, \ddot{\psi} = M_G(t). \tag{8.56b}$$

In the first case the liquid velocity field vortex component disappears, and associated moment of inertia J Equations (8.56a) is Joukowski's moment of inertia. In the second case the whole of the liquid is involved in rotation (the moment of inertia for solidified liquid $J^{(0)}$ in Equations (8.56b)).

For $\dot{\psi} \equiv 0$ we obtain from (8.53) equations of free oscillations for the considered continuum. These oscillations have the character of standing Alfven waves with natural frequency σ° corresponding to the first form of liquid natural oscillations and to the second form of magnetic field natural oscillations. Both forms are in the considered case antisymmetric with respect to the cavity's diameter plane. Equations of similar structure are certainly obtained for axcisymmetric oscillations of liquid and magnetic field.

Let us for instance the intermediate transformations and write down the equations resulting from (8.51) and (8.53), that correspond to the first two forms of natural axisymmetric magnetohydrodynamic oscillations for completely filled cylinder cavity:

$$\Lambda_1^* (\ddot{\chi}_1 + \sigma_{21}^{\circ 2} \chi_1) + \sum_{r=1}^{2} \gamma_{1r}^\circ \int_{-\infty}^{t} \frac{\ddot{\chi}_{1r}(\tau)\, d\tau}{\sqrt{t - \tau}} = 0, \tag{8.57}$$

$$\Lambda_2^* (\ddot{\chi}_2 + \sigma_{12}^{\circ 2}\chi_2) + \sum_{r=1}^{2} \gamma_{2r}^{\circ} \int_{-\infty}^{t} \frac{\ddot{\chi}_{2r}(\tau)\, d\tau}{\sqrt{t-\tau}} = 0,$$

$$J_1^*\Omega_1 + \alpha_{12}^{\circ}\chi_2 = 0, \qquad J_2^*\Omega_2 + \alpha_{21}^{\circ}\chi_1 = 0,$$

where

$$\dot{\chi}_j = I_j\,, \qquad \sigma_{ij}^2 = \frac{\alpha_{ij}^{\circ 2}}{J_i^*\Lambda_j^*} \qquad (i, j = 1, 2), \tag{8.58}$$

$$\chi_{10} = -\frac{J_2^*}{\overset{\circ}{\alpha}_{21}}\,\Omega_{20}\,; \qquad \chi_{20} = -\frac{J_1^*}{\overset{\circ}{\alpha}_{12}}\,\Omega_{10}\,.$$

Equations of similar structure are obtained in terms of variable Ω_1, Ω_2 .

The character of the continuum's free oscillations described by these equations, will depend on coefficients J_j^*, Λ_j^*, γ_{ij}°, α_{ij}° of (8.44), (8.47) or (8.44), (8.49) and on initial conditions.

The equations obtained from (8.51) for $\dot{\psi} \equiv 0$ correspond to (8.57) in the modification of antisymmetric oscillations for $\chi_1 \equiv 0$ (the first mode of liquid velocity field and second of magnetic field).

We should note once more, that, as it clear from Equations (8.57), the liquid with frozen - in magnetic field gains the properties of an elastic medium.

All the above refers, certainly to the class of perturbed motions, corresponding to and Re >> 1, Re_M >> 1 .

Let us try to estimate the field of the equations applicability to real electromechanical systems, that is specified by Conditions (8.2). The m ost subtle moment here is the implementation of great Reynolds numbers in order to obtain pronounced skin effect. Naturally, it can be achieved either by high electrical conductivity or by high magnetic permeability.

The first opportunity of obtaining great Re_M and Al values is implemented in the problems about great volumes of liquid metals oscillating in artificial external magnetic fields. Relative magnetic permeability of liquid metals is practically equal to unity according to G.G.Branover and A.B.Zinober [12]. As to electrical conductivity, it is fairy high, e.g., according to [12], for mercury $\sigma = 1.017 \cdot 10^6\ \Omega \cdot \mathrm{m}^{-1}$. So we deal with LVEC liquids here.

Let us consider for example a magnetohydrodynamic system, the role of liquid in which is played by liquid sodium at 100° C . According to the same source $\rho = 925\ \mathrm{kg} \cdot \mathrm{m}^{-3}$, $\sigma = 1.037 \cdot 1\,\overset{\circ}{0}\ \Omega \cdot \mathrm{m}^{-1}$, $\mu = 1$, $\sigma_F = 0.206\ \mathrm{N} \cdot \mathrm{m}^{-1}$, $\nu_F = 7.4 \cdot 10^{-4}\ \mathrm{m}^2 \cdot s^{-1}$.

We set $j = g$, $l = 1$ m, $\omega_F = 3\pi\ \mathrm{s}^{-1}$ (1.5Hz) $\overset{\circ}{H} = n\overset{\circ}{I}$, $n = 10^3\ \mathrm{m}^{-1}$,$\overset{\circ}{I} = 10\mathrm{A}$ and calculate the dimensionless parameters for the case. We obtain

$$\mathrm{Re} = \frac{\omega_F l^2}{\nu_F} = \frac{3\pi \cdot 10^4}{7.4} = 1.274 \cdot 10^5,$$

$$\mathrm{Bo} = \frac{\rho g l^2}{\sigma_F} = \frac{925 \cdot 9.81}{0.206} = 0.44 \cdot 10^5,$$

$$\mathrm{Re}_M = \frac{\omega_F l^2}{\nu_M} = \omega_F l^2 \mu_0 \sigma = 3\pi \cdot 4\pi \cdot 10^{-7} \cdot 1.037 \cdot 10^7 = 1.23 \cdot 10^2,$$

$$\mathrm{Al} = \frac{\mu_0 H^{o2} \sigma}{\omega_F^2 l^2} = \frac{4\pi \cdot 10^{-7} \cdot 1.037 \cdot 10^7 \cdot 10^8}{9} \pi^2 = 1.47 \cdot 10^7.$$

As it is clear, all Equations (8.2) are satisfied with a great reserve, except for, condition $\mathrm{Re}_M >> 1$.

The second opportunity of obtaining nonsmall Re_M and Al values can be implemented in the problems about oscillations in a magnetic field of a suspension composed by a ferromagnetic powder with high magnetic permeability, suspended in a liquid with comparatively good electrical conductivity, which is still significantly lower than that of liquid metals.

Such an opportunity is caused, in the first place, by pronounced ferromagnetic properties of the considered medium. Such media, which have got a special title of ‘magnetic liquids’, area subject of increased attention lately. A number of works is devoted to studying their magnetohydrodynamic properties, particularly the monograph by R.Rosenzweig [86].

A subject of special interest is the fact, that for many ‘magnetic liquids’ equivalent viscosity is highly dependent on the external magnetic field. That dependence enables creation of controlled oscillation dampers with characteristics varying in wide limits for magnetic field variation.

8.7. Magnetic Field Influence on Solid Body – LVEC Liquid Open-Loop System Frequency Response

Let us illustrate magnetic field influence on solid body – LVEC liquid open-loop system frequency response by the example of liquid antisymmetric oscillations in a moving cavity having the shape of a circular cylinder with ring-shaped ribs. In that case the first antisymmetric mode of liquid velocity field vortex component prevails, because it is directly connected with the cavity's motion. That allows to retain only this mode and the appropriate magnetic field mode associated with it ($\Omega_2 \equiv 0, I_1 \equiv 0$ in Equations (8.43)) in all the examples considered hereinafter.

We shall further use the simplified denominations used in Equations (8.23), namely Ω, I, s instead of Ω_1, I_2, s_1 and omit subscripts at the corresponding coefficients.

Two examples are considered below: a) a cylindric cavity completely filled with LVEC liquid; b) a cylindric cavity partially filled with LVEC liquid which has free surface. In the first case there are no wave motions of liquid, i.e. $s \equiv 0$ and variable Ω corresponding to liquid velocity field vortex component completely describes this

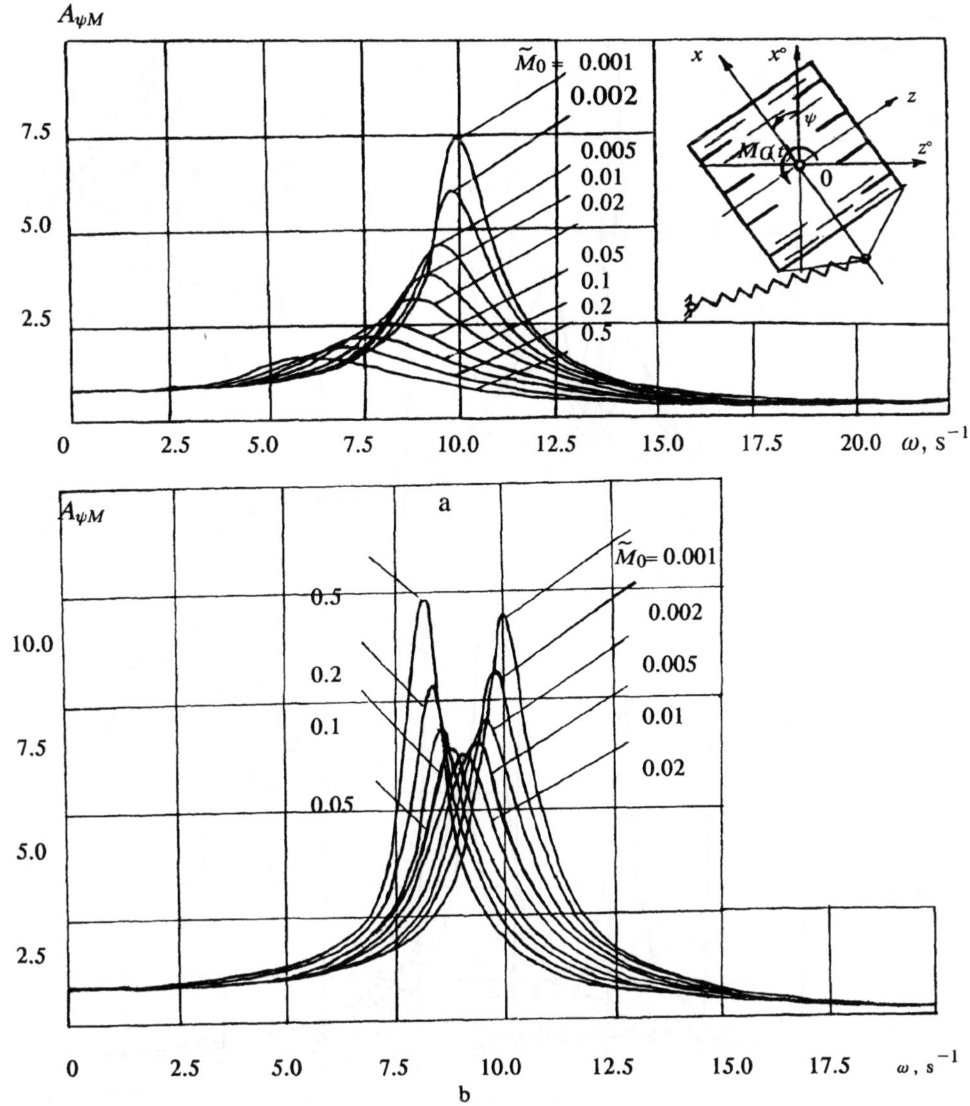

Fig. 8.3. Amplitude response for a cylindric cavity with ring-shaped ribs, completely filled with LVEC liquid with no external magnetic field: *a* – Model (8.61); *b* – Model (8.62)

component dynamics. In the second case LVEC motion dynamics is described by two variable Ω and s.

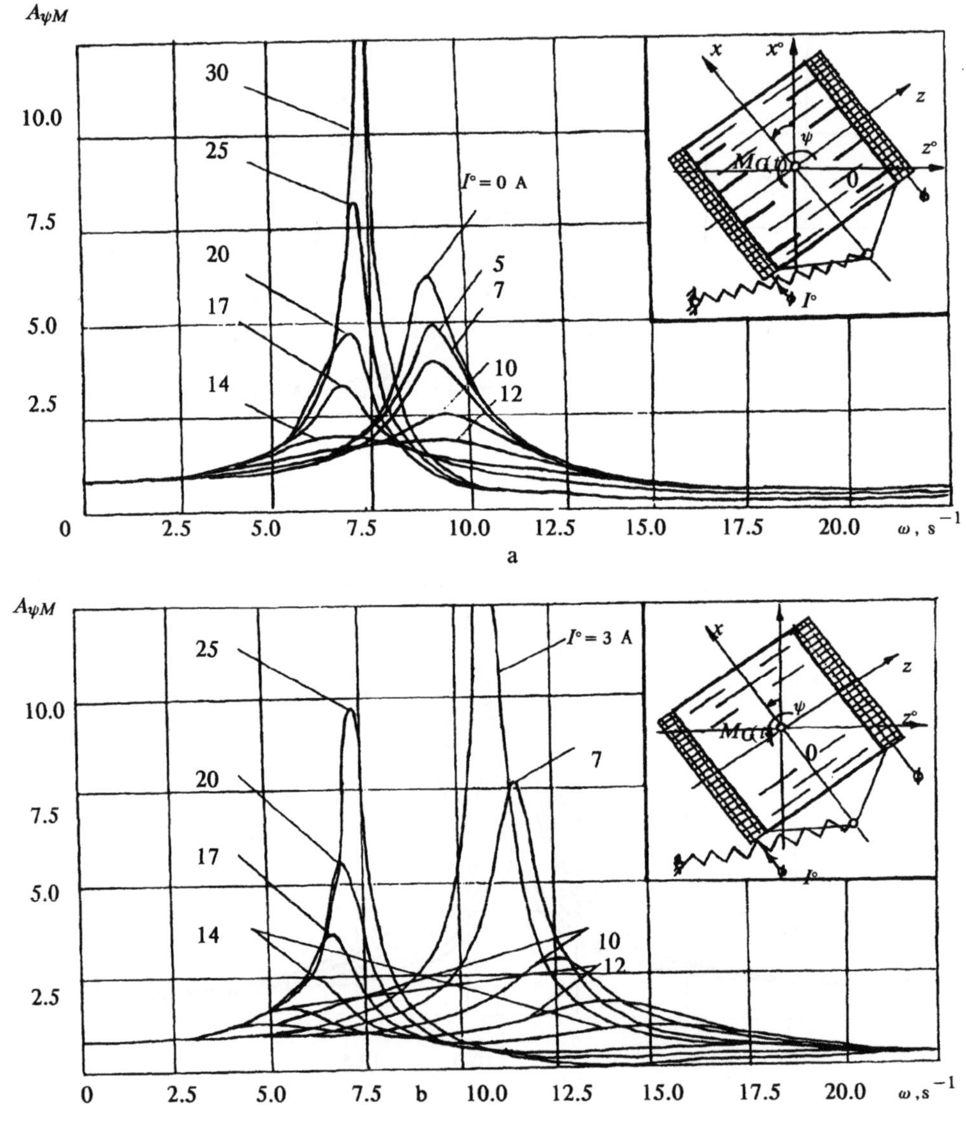

Fig. 8.4. Amplitude repose for a cylindric cavity completely filled with LVEC liquid with external magnetic field: *a* – the cavity with ring-shaped ribs; *b* – the cavity with smooth walls

With external magnetic field, for both cases it is necessary to introduce an additional variable I describing eddy current.

8.7.1. A cavity completely filled with liquid.

The scheme of the experimental plant, for which numerical analysis of magnetic field influence on open-loop system response characteristics was performed, is presented in Fig. 8.3, 8.4. The model is a cylindric cavity with radius $r_0 = 0.2$ m and relative height $\tilde{h} = h/r_0 = 1.6$ with four equidistant ribs, their relative width $\tilde{b} = b/r_0 = 0.2$. The cavity is completely filled with LVEC liquid (liquid sodium) with the following characteristics [12]:

$$\mu = 1; \qquad \rho = 925 \text{ kg} \cdot \text{m}^{-3}; \qquad \sigma = 1.037 \ \Omega \cdot \text{m}^{-1};$$

The cavity can rotate by an angle of ψ around the suspension axis running through point O.

Dynamic excitation is performed by external moment $M_G(t)$. The return spring is intended for the system to gain resonance properties. That allows a more explicit revealment of magnetic field influence on LVEC liquid's dynamic characteristics. The external magnetic field is generated by the coil having 2000 turns energized by direct current I°, which is the variable parameter in the considered case.

Let us limit ourselves to the most representative amplitude response characteristics of the open-loop system 'body – liquid' for three modifications of mathematical model for the considered system.

The first two modifications refer to the case of no magnetic field, when purely hydrodynamic effects are studied.

The first modification of the model corresponds to $\Omega = -\dot{\psi}$, so the model has the form similar to (4.7):

$$(J^\circ + J)\,\ddot{\psi} + \beta \int_{-\infty}^{t} \frac{\ddot{\psi}(\tau)\, d\tau}{\sqrt{t-\tau}} = M_G(t), \tag{8.59}$$

$$\beta = \hat{\beta}\sqrt{|\dot{\psi}|}\,.$$

When investigating frequency response characteristics, the excitation function $M_G(t)$ is described in such a way, considering the return spring influence:

$$M_G(t) = \omega_{\psi\psi}^2 (J^\circ + J)\,(\tilde{M}_0 \sin \omega t - \psi). \tag{8.60}$$

From (8.59) we obtain respectively:

$$(J^\circ + J)\,(\ddot{\psi} + \omega_{\psi\psi}^2 \psi) + \beta \int_{-\infty}^{t} \frac{\ddot{\psi}(\tau)\, d\tau}{\sqrt{t-\tau}} = \omega_{\psi\psi}^2 (J^\circ + J)\,\tilde{M}_0 \sin \omega t\,, \tag{8.61}$$

$$\beta = \hat{\beta}\sqrt{|\dot{\psi}|}\,.$$

Such form of the mathematical model representation lets us extract the partial frequency $\omega_{\psi\psi}$ of the system 'solid body – ideal liquid' oscillations and make the exciting moment $\tilde{M}_0$ have the sense of cylinder static rotation angle ($\omega = 0$) under the action of constant moment. Amplitude response $A_{\psi M}(\omega)$ value corresponding to transfer functions $W_{\psi M}(p) = \dfrac{\Psi(p)}{M(p)}$ has the sense of dynamics coefficient at the given frequency ω.

The second modification corresponds to recognition of the additional vortex degree of freedom for liquid. The appropriate mathematical model similar to (4.10) is written, including (8.60), in the form of nonlinear integrodifferential equations:

$$(J^{\circ} + J^{(0)})\,\ddot{\psi} + \omega_{\psi\psi}^{2}\,(J^{\circ} + J)\,\psi + J^{*}\dot{\Omega} = \omega_{\psi\psi}^{2}\,(J^{\circ} + J)\,\widetilde{M}_{0}\sin\omega t\,,$$

$$J^{*}\,(\dot{\Omega} + \ddot{\psi}) + \beta\int_{-\infty}^{t}\frac{\dot{\Omega}(\tau)\,d\tau}{\sqrt{t-\tau}} = 0\,, \qquad (8.62)$$

$$\beta = \hat{\beta}\sqrt{|\Omega|}\,.$$

Note, that in Models (8.61) and (8.62) one coefficient β is used. That introduces a certain error due to nonrecognition of the vortex field nonuniformity (see Chapter4).

Unlike Chapter 4, more limited problem is considered in this chapter through, therefore such assumption seems reasonable.

And finally, the third modification corresponds to introduction of a new physical factor, namely constant external magnetic field, so it takes into account magnetohydrodynamic effects. Including (8.60), the appropriate mathematical model gains the following form:

$$(J^{\circ} + J^{(0)})\,\ddot{\psi} + \omega_{\psi\psi}^{2}\,(J^{\circ} + J)\,\psi + J^{*}\dot{\Omega} = \omega_{\psi\psi}^{2}\,(J^{\circ} + J)\,\widetilde{M}_{0}\sin\omega t\,,$$

$$J^{*}(\dot{\Omega} + \ddot{\psi}) + \alpha^{\circ} I + \beta\int_{-\infty}^{t}\frac{\dot{\Omega}(\tau)\,d\tau}{\sqrt{t-\tau}} = 0\,, \qquad (8.63)$$

$$\Lambda^{*}\dot{I} - \alpha^{\circ}\Omega + \gamma^{\circ}\int_{-\infty}^{t}\frac{\dot{I}(\tau)\,d\tau}{\sqrt{t-\tau}} = 0\,,$$

$$\beta = \hat{\beta}\sqrt{|\Omega|}\,, \qquad \alpha^{\circ} = \widetilde{\alpha}^{\circ} I^{\circ}\,.$$

Equations (8.61) - (8.63) coefficients calculated using the formulae presents in Article (8.6) are shown in Table 8.1.

Table 8.1

Coefficients of Perturbed Motion Equations

J°	J^{*}	$J^{(0)}$	J
0.1 $kg \cdot m^2$	0.4105 $kg \cdot m^2$	0.7542 $kg \cdot m^2$	0.3347 $kg \cdot m^2$
$\hat{\beta}$	$\omega_{\psi\psi}$	Λ^{*}	γ°
0.5444 $H \cdot m \cdot s^{2} \cdot rad^{-3/2}$	10 s^{-1}	$0.8537 \cdot 10^{-4}$ $H \cdot m^{4}$	$3.0152 \cdot 10^{-4}$ $H \cdot m^{4} \cdot s^{-1/2}$

Coefficient $\widetilde{\alpha}^{\circ}$ has the value $0.005227\ H \cdot m^{2}$.

Note, that variable I describing eddy current in LVEC liquid has the dimension of current density $A \cdot m^{-2}$.

It is shown in Article 8.6, that the most critical from the viewpoint of self-similarity oriteria is the magnetic Reynolds number. Having assumed the cavity diameter for its characteristic size, i.e. $l = 2r_0 = 0.4$ m , we obtain

$$\mathrm{Re}_M = \frac{\omega l^2}{\nu_M} = 10 \cdot 0.4 \cdot 4\pi \cdot 10^{-7} \cdot 1.037 \cdot 10^7 \approx 20 . \tag{8.64}$$

This estimate lets us assume, that the studied magnetohydrodynamic effects in the considered model problem can be described, with some strain, though within the scope of 'great' Reynolds number asymptotics accepted above.

Amplitude response for the three modifications of mathematical models was calculated with Laplace transform and the following substitution of $p = i\omega$ (see Chapter 3). For nonlinear coefficient β determination we used iterations to obtain each point in the functions of the considered family.

The results of the calculations are shown in Fig. 8.3, 8.4.

Fig. 8.3a presents the family of amplitude response functions (parameter $\widetilde{M}_0$) for a cylindric cavity completely filled with LVEC liquid, corresponding to mathematical Model (8.61) with no magnetic field.

With parameter $\widetilde{M}_0$ growth the amplitude of cavity's oscillations increases, the damping coefficient grows, and, hence, the dynamics coefficient decreases (the resonance peak lower).

Response characteristics corresponding to Model (8.62) are presented in Fig. 8.3b. In this case their behaviour is of a more complicated character. First, with parameter $\widetilde{M}_0$ growth up to 0.02 damping grows and dynamics coefficient decreases. Beginning with that value of $\widetilde{M}_0$ the effect of 'damping saturation' occurs, that we have already observed. When $\widetilde{M}_0$ is increased further, damping decreases, and dynamics coefficient grows. From application point of view this means, that when there occur oscillation amplitudes exceeding some critical value, damping growth by construction elements development (increase in number of ribs and their relative width $\widetilde{b}$) becomes impossible. Moreover, it may cause an opposite effect, i.e. degradation of those elements efficiency. That is in qualitative agreement with the results obtained in Chapter 4.

Let us now proceed directly to investigation of magnetic field influence on response of open-loop system 'solid body – LVEC liquid'. We shall consider the situation, when liquid oscillation damping is caused by two factors: appearance of hydrodynamic vortices at construction elements edges and eddy currents induced in oscillating LVEC liquid.

The family of amplitude response functions for various values of bias current I^0 is shown in Fig. 8.4a. These functions were calculated for parameter $\widetilde{M}_0$ critical value 0.02 . For further increase of that it is impossible to increase purely hydrodynamic damping. As it is clear from the results presented, with parameter I^0 growth up to $I^0 =$

$= 12 - 14$ A it is possible to increase liquid oscillation damping by several fold: firstly, liquid velocity field vortex component amplitude Ω grows abruptly, and it is what determines hydrodynamic damping rate ($\beta = \hat{\beta}\sqrt{|\Omega|}$); secondly, an additional dissipation effect occurs, associated with eddy currents. It is obvious, that for bias current I° values corresponding to the vicinity of internal resonance at partial frequencies σ° (8.55) and $\omega_{\psi\psi}$ damping reaches its maximum. That creates a potential possibility for multiple increase of damping in a real system by an appropriate 'adjustment' of the external magnetic field.

For further growth of I° and, hence, the system's natural frequency engendered by partial frequency σ°, abrupt decrease of damping takes place, down to practically complete vanishing. Physically this means, that the external magnetic field causes as if solidification of LVEC liquid. In the limit, for $\sigma^{\circ} \to \infty$ $\Omega \to 0$, $I \to 0$ and Equations (8.63) convert to Equations (8.56b) corresponding to solidified liquid with moment of inertia $J^{(0)}$.

At the same time, the magnetic field presence makes it possible to produce a fundamentally new effect, namely liquid oscillations suppression without damping ribs. To study this possibility let us consider response functions corresponding to Equations (8.63) for $\beta \equiv 0$. The family of this functions calculated (like in Fig. 8.4a) for the critical value $\widetilde{M}_0 = 0.02$ and various I° values is presented in Fig. 8.4b. For response value $I^{\circ} = 0$ at resonance frequency $\omega_{\psi\psi}$ is infinite. With I° growth damping appears in the system, at first it is less than with coaxial ribs. For $I^{\circ} = 12 - 14\,A$ though, corresponding to internal resonance vicinity ($\sigma^{\circ} \approx \omega_{\psi\psi}$), energy dissipation practically reaches the level, maximum with coaxial ring-shape ribs. It is an evidence of the fact, that for the case of cavity with ribs dissipation due to eddy currents plays a dominating role (for the critical value of bias current I°). For further growth of I° damping decreases abruptly, that vanishes completely for $I^{\circ} \to \infty$. It should also be noted, that without ribs the effect of system's natural frequencies 'moving apart' in the vicinity of close partial frequencies ($\sigma^{\circ} \to \omega_{\psi\psi}$) is more distinctly pronounced double-hump amplitude response functions in Fig. 8.4b) The partial system corresponding to 'internal' degrees of freedom (variable Ω and I) behaves practically like a classical dynamic damper.

8.7.2. The cavity partially filled with liquid.

The scheme of the electromechanical model corresponding to this case is presented in Fig. 8.5. a cylindric cavity with one ring-shaped rib, having radius $r_0 = 0.2m$, is partially filled with liquid sodium, so there is a free surface. Relative depth of liquid $h = 1$. The ring-shape rib is situated under the free surface of liquid at the distance equal to its width (0.04 m). The system's kinematics is the same as in the previous example. LVEC liquid oscillations excitation is performed kinematically by generating oscillations of the cavity in harmonic manner $\psi(t) = \psi_0 \sin \omega t$. The external magnetic field is generated by a coil with 2000 turns energized by direct current, which embraces LVEC liquid to the whole of its depth. The direct current I° value is the variable parameter, as in the previous example.

Let us once more confine to only one antisymmetric mode of liquid velocity field vortex component and one mode of the corresponding magnetic field associated with the former one ($\Omega_2 \equiv 0$, $I_1 \equiv 0$ in Equations (8.43)), and further on let us omit subscripts at the variables and at the adjacent coefficients. We shall consider three modifications of the system's mathematical model, corresponding to the some assumptions as for completely filled cavity.

The first modification corresponds to the following mathematical model similar to (8.61):

$$\mu\,(\ddot{s} + \beta \int_{-\infty}^{t} \frac{\ddot{s}(\tau)\, d\tau}{\sqrt{t-\tau}} + \omega_s^2\, s\,) = -\,\lambda_{1j}\,\psi_0 \sin \omega t,$$

$$\beta = \hat{\beta}\left(\,|\dot{\tilde{s}}|\,\right)^{1/2}, \qquad \tilde{s} = \frac{s}{r_0}. \tag{8.65}$$

The expression for the excitation force in the right side of Equation (8.65) lacks the term $-\,\lambda_{G1}\,\ddot{\psi}$ (see Formulae (8.56) for $F(t)$). It is caused by the appropriate selection of the cavity suspension axis position from the condition $\lambda_{G1} = 0$.

The second modification corresponds to the mathematical model similar to (8.62):

$$\mu\,(\ddot{s} + \beta \int_{-\infty}^{t} \frac{\ddot{s}(\tau)\, d\tau}{\sqrt{t-\tau}} + \omega_s^2\, s\,) - \gamma^* \int_{-\infty}^{t} \frac{\dot{\Omega}(\tau)\, d\tau}{\sqrt{t-\tau}} = -\,\lambda_{1j}\,\psi_0 \sin \omega t,$$

$$J^*\dot{\Omega} + \beta^\circ \int_{-\infty}^{t} \frac{\dot{\Omega}(\tau)\, d\tau}{\sqrt{t-\tau}} - \gamma^* \int_{-\infty}^{t} \frac{\ddot{s}(\tau)\, d\tau}{\sqrt{t-\tau}} = J^*\omega^2\,\psi_0 \sin \omega t, \tag{8.66}$$

$$\beta = \hat{\beta}\left(\,|\dot{\tilde{s}}|\,\right)^{1/2}, \qquad \beta^\circ = \hat{\beta}^\circ\left(\,|\dot{\tilde{s}}|\,\right)^{1/2}, \qquad \gamma^* = \hat{\gamma}^*\left(\,|\dot{\tilde{s}}|\,\right)^{1/2}.$$

And finally, the third modification (existence of the constant external magnetic field) corresponds to the following mathematical model similar to (8.63):

$$\mu\,(\ddot{s} + \beta \int_{-\infty}^{t} \frac{\ddot{s}(\tau)\, d\tau}{\sqrt{t-\tau}} + \omega_s^2\, s\,) - \gamma^* \int_{-\infty}^{t} \frac{\dot{\Omega}(\tau)\, d\tau}{\sqrt{t-\tau}} = -\,\lambda_{1j}\,\psi_0 \sin \omega t,$$

$$J^*\dot{\Omega} + \beta^\circ \int_{-\infty}^{t} \frac{\dot{\Omega}(\tau)\, d\tau}{\sqrt{t-\tau}} - \gamma^* \int_{-\infty}^{t} \frac{\ddot{s}(\tau)\, d\tau}{\sqrt{t-\tau}} + \alpha^\circ I = J^*\omega^2\,\psi_0 \sin \omega t, \tag{8.67}$$

$$\Lambda^*\dot{I} + \gamma^\circ \int_{-\infty}^{t} \frac{\dot{I}(\tau)\, d\tau}{\sqrt{t-\tau}} - \alpha^\circ \Omega = 0,$$

$$\beta = \hat{\beta}\left(\,|\dot{\tilde{s}}|\,\right)^{1/2}, \qquad \beta^\circ = \hat{\beta}^\circ\left(\,|\dot{\tilde{s}}|\,\right)^{1/2}, \qquad \gamma^* = \hat{\gamma}^*\left(\,|\dot{\tilde{s}}|\,\right)^{1/2}, \qquad \alpha^\circ = \tilde{\alpha}^\circ r^\circ.$$

The values of coefficients included in Equations (8.65 - 8.67), calculated using the formulae presented in Article 8.6 are in Table 8.2.

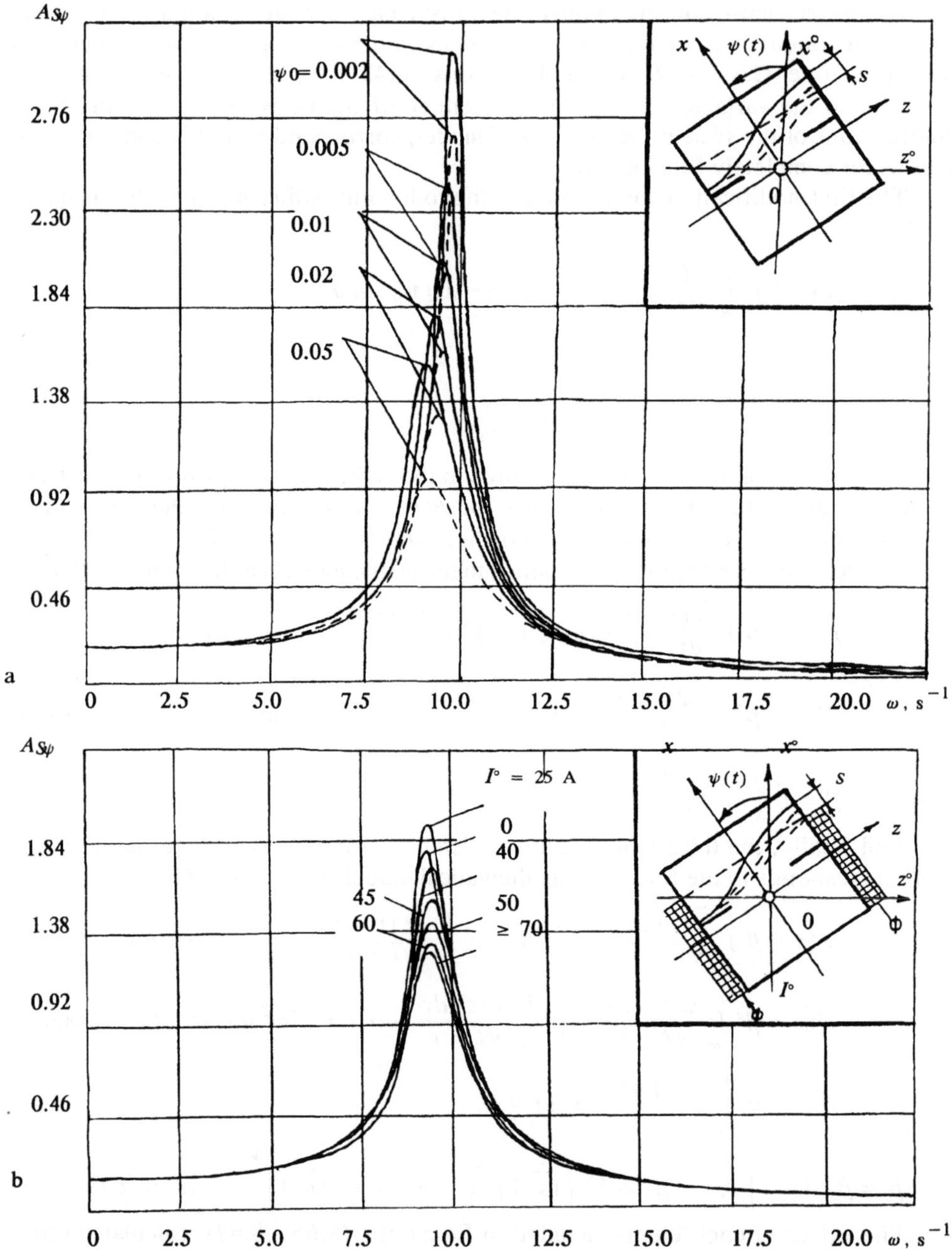

Fig. 8.5. Amplitude response for a cylindric cavity with a ring-shaped rib, partially filled with LVEC liquid: *a* – without external magnetic field; --- Model (8.65); — Model (8.66); *b* – with external magnetic field

Table 8.2

Coefficients of Perturbed Motion Equations

μ	λ_1	J^*	j	ω_S
5.06 kg	7.414 kg	0.2183 $kg \cdot m^2$	9.81 $m \cdot s^{-2}$	10 s^{-1}
$\hat{\beta}$	$\hat{\beta}^\circ$	$\hat{\gamma}^*$	γ°	Λ^*
0.3087 $N \cdot m^{-1} \cdot s^{-2}$	0.06193 $N \cdot m \cdot s^2$	-0.311 $N \cdot s^2$	$1.64 \cdot 10^{-4}$ $H \cdot m^4 \cdot s^{-1/2}$	$0.448 \cdot 10^{-4}$ $H \cdot m^4$

Coefficient $\widetilde{\alpha}^\circ$ is equal to 0.002636 $H \cdot m^2$.

As it is clear from Equations (8.65 - 8.67), coefficients β, β°, γ^* are nonlinear functions of liquid wave motions relative speed $\overset{\leftarrow}{\widetilde{s}}$, unlike Models (8.61 - 8.63), in which they depended on coordinate Ω characterizing liquid velocity field vortex component (in Model (8.61) they depend on $\dot{\psi}$). That determines a fundamentally different behaviour of the system considered in the second example for current I° variation.

Investigation of magnetic field influence on the system's dynamic properties was performed based on comparative analysis of amplitude response functions $A_{s\psi}(\omega)$ corresponding to transfer function $W_{s\psi} = S(p) / \Psi(p)$ for all the three modifications of mathematical Models (8.65 - 8.67).

Results of the calculations are presented in Fig. 8.5.

Fig. 8.5a shows the family of amplitude response functions $A_{s\psi}(\omega)$ for a cylindric cavity partially filled with LVEC liquid in the case of no external magnetic field for various amplitudes of the cavity rotation angle ψ_0. For Model (8.65) the results are shown by the dashed line, and for Model (8.66) by the solid line. The obtained results show that the additional vortical degree of freedom recognized by Model (8.66) for $\Omega = 0$ decreases energy dissipation intensity significantly, compared to Model (8.65). With the growth of ψ_0 and, hence, of liquid oscillation amplitude this effect becomes still more noticeable. The picture presents results of calculation up to $\psi_0 = 0.05$, when liquid oscillation amplitude reaches its limiting value corresponding to circular wave formation ($\widetilde{s}_0 \approx 0.2$). If we continue to increase parameter ψ_0 formally, we can achieve the effect produced in the first example, namely decrease of equivalent damping, or increase of dynamic coefficient (see Fig. 8.4b). Actual production of this effect is hardly possible, however, because of circular wave formation.

The external magnetic field influence on dynamic properties of the considered system with LVEC liquid is illustrated by Fig. 8.5b. It shows the family of amplitude response functions $A_{s\psi}(\omega)$ for various values of bias current I°. These functions were calculated using Model (8.67) for $\psi_0 = 0.05$. Parameter I° growth causes increase in negative influence of the vortical degree of freedom ($\Omega \neq 0$) on damping-amplitude response value at the resonance frequency growth, reaches its maximum value for $I^\circ = 25$ A, when internal resonance takes place at frequencies ω_S and σ° (see Formulae (8.54)). Further current increase gradually eliminates this effect, it practically vanishes for $I^\circ \geq 70$ A. It is caused by the fact, that for $I^\circ \to \infty$ vortical degrees of freedom cease to be excited, i.e. $\Omega \to 0$ and $I \to 0$.

Mathematical Model (8.66) is transformed into Model (8.65). Consequently equivalent damping grows by approximately 43 %, and dynamic coefficient decreases respectively.

Let us consider the practical aspect of using this effect. Suppose, that the liquid fluidity is the source of some closed-loop system's instability due to violation of phase stabilization conditions for the object of control. Suppression of the vortical degree of freedom by external magnetic field, generally speaking, allows to lower the limit cycle amplitude by a factor of 2, which limit cycle takes place because of the damping ribs. This is one aspect of magnetohydrodynamic effects influence on the body-liquid system stability. The other aspect is considered in a more detail in the following article.

8.8. The Possibility of Using Magnetohydrodynamic Effects to Ensure Dynamic Stability of a Spacecraft

It was shown in the previous article, that the partial system corresponding to vortical motion of liquid (variable Ω) and eddy current (variable I) behaves like a classical dynamic damper.

Let us study the possibility to use the appropriate magnetohydrodynamic effects for ensuring dynamic stability of a spacecraft with tanks partially filled with liquid fuel components. Let us use for example a spacecraft dynamically unstable at the 1st tone frequency of liquid wave motions in the upper tank, considered in Article 6.4. The SC instability is caused by the controller characteristic that does not ensure phase stabilization at this frequency. We shall try to estimate whether it is possible to cope with this instability with the help of an additional cylindric cavity completely filled with LVEC liquid and an appropriate winding which generates a coaxial magnetic field. This solution is attractive due to an opportunity to ensure object-controller closed-loop system stability without installing damping ribs in the tanks. And stability 'in the small' it is !

Let us use Equations (6.63) as the mathematical model, and assume, for there are no inner ribs in the tanks, $\vartheta_{1,2} = -\psi$ and eliminate the equations related to variables ϑ_1, ϑ_2. At the same time we add to (6.63) equations for Ω and I, associated with the cavity containing LVEC liquid which moves in the magnetic field.

Resulting is the following set of equations:

$$
\begin{aligned}
&(1 - a_{\psi\vartheta_1} - a_{\psi\vartheta_2})\ddot{\psi} + a_{\psi s_1}\ddot{s}_1 + a'_{\psi s_1}\ddot{s}_1 + \\
&\qquad\qquad + a_{\psi s_2}\ddot{s}_2 + a'_{\psi s_2}\ddot{s}_2 + a_{\psi\Omega}\dot{\Omega} = -a_{\psi\delta}\delta, \\
&(1 - a_{s_1\zeta}a_{\zeta s_1})\ddot{s}_1 + \omega_1^2 s_1 - a_{s_1\zeta}a_{\zeta s_2}\ddot{s}_2 + a_{s_1\psi}\ddot{\psi} + \\
&\qquad\qquad + (a'_{s_1\psi} - a_{s_1\zeta}a_{\zeta\psi})\psi = a_{s_1\zeta}a_{\zeta\delta}\delta,
\end{aligned}
\tag{8.68}
$$

$$(1 - a_{s2\zeta}\, a_{\zeta\, s2})\, \ddot{s}_2 + \omega_2^2\, s_2 - a_{s2\zeta}\, a_{\zeta\, s1}\, \ddot{s}_1 + a_{s1\,\psi}\, \ddot{\psi} +$$

$$+ (a'_{s2\psi} - a_{s2\zeta}\, a_{\zeta\,\psi})\, \psi = a_{s2\zeta}\, a_{\zeta\,\delta}\, \delta,$$

$$J^* (\dot{\Omega} + \ddot{\psi}) + \alpha^\circ I = 0,$$

$$\Lambda^* I^* - \alpha^\circ \Omega + \gamma^\circ \int_{-\infty}^{t} \frac{\dot{I}(\tau)\, d\tau}{\sqrt{t - \tau}} = 0,$$

$$\delta = K_\psi\, \psi + K_{\dot{\psi}}\, \dot{\psi}, \qquad \alpha^\circ = \tilde{\alpha}^\circ I^\circ.$$

The additional cavity with LVEC liquid (liquid sodium) plays the role of a dynamic damper in this case. In order to make the magnetohydrodynamic effects influence more obvious let us change some coefficients of Equations (6.63), presented in Table 6.1, and apply no special restrictions to the LVEC liquid mass and the current in the winding. New values of the coefficients and as well as those directly related to the 'dynamic damper' are presented in Table 8.3

Table 8.3.

Coefficients of Perturbed Motion Equations

$a_{\zeta\psi}$	$a_{\zeta\delta}$	$a'_{\psi S1}$	$a'_{\psi S2}$	$a_{\psi\delta}$
30.52 $m \cdot s^{-2}$	30.52 $m \cdot s^{-2}$	0.12284 $m^{-1} \cdot s^{-2}$	0.0902 $m^{-1} \cdot s^{-2}$	10.488 s^{-2}
ω_1^2	$a'_{S1\psi}$	ω_2^2	$a'_{S2\psi}$	K_ψ
28.096 s^{-2}	49.08 $m \cdot s^{-2}$	28.096 s^{-2}	47.052 $m \cdot s^{-2}$	0.03
$K_{\dot{\psi}}$	$a_{\psi\Omega}$	J^*	Λ^*	γ°
0.03 s	0.026	56645 $kg \cdot m^2$	0.4982 $H \cdot m^4$	0.699 $H \cdot m^4 \cdot s^{-1/2}$

Coefficient $\tilde{\alpha}^\circ$ is equal to 1.392 $H \cdot m^2$.

Note, that LVEC liquid mass is greatly raised up for the reasons presented above (14% of the object's total mass), so this example is for purely methodological purposes.

Linear integrodifferential Equations (8.68) solutions stability can be studied by the root locus method with parameter I° in variable $q = \sqrt{p}$ plane (see Chapter 3). The results are shown in Figs. 8.6, 8.7, 8.8.

Fig. 8.6, 8.7 shows root locus in q plane for closed-loop system (8.68) characteristic equation (the points are numbered in accordance with I° ampere values). Fig. 8.6 shows the entire locus picture, and Fig. 8.7 includes locus for the most interesting roots in a greater scale.

The following symbols are adopted:

● - the object as a solid body + angular stabilization controller;

⊕ - 'stable' root associated with the liquid motion in the lower tank;
∅ - 'unstable' root associated with the liquid motion in the upper tank;

Δ , ● - dynamic damper.

Stability domain boundary is made of two straight lines with single hatching at the stability side.

The first two roots are practically independent of I^o, the third one varies is a sophisticated manner with I^o growth, but its variation cannot be noticed in the scale of Fig. 8.6.

Roots Δ and ● are subject to maximum variations for I^o change, andΔ root is practically independent from the others, while ● is greatly connected with ∅ , The

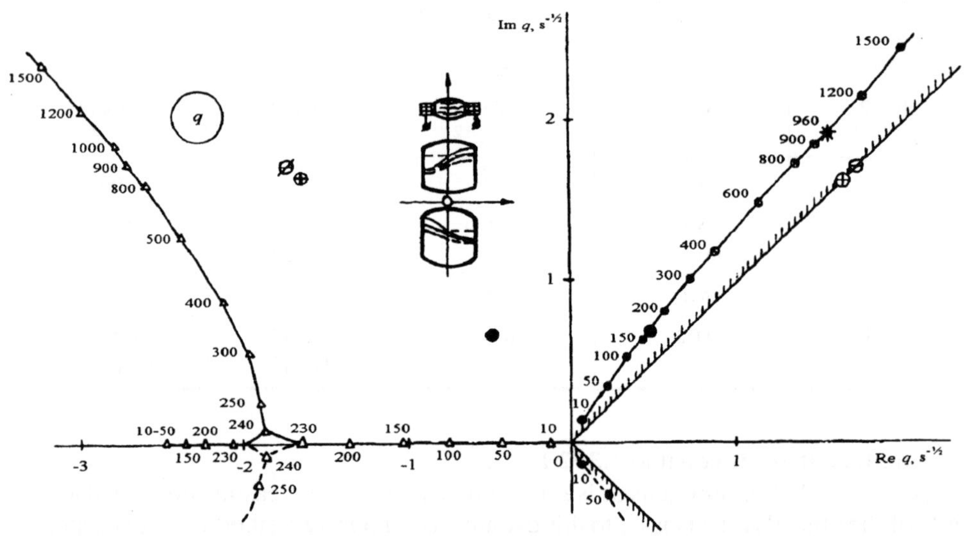

Fig.8.6ot locus in q plane for the closed-loop system solid body – liquid – controller – dynamic damper of (8.68). Parameter is bias current I^o in the outer winding: a - the general picture: ● – roots corresponding to partial system solid body – angular stabilization controller; ⊕ – 'stable' roots associated with liquid motion in the lower tank; ∅ – 'unstable' roots associated with liquid motion in the upper tank; Δ - autonomous roots corresponding to dynamic damper of oscillations; ● - roots corresponding to dynamic damper of oscillations, these roots interacting with "unsteady" root.

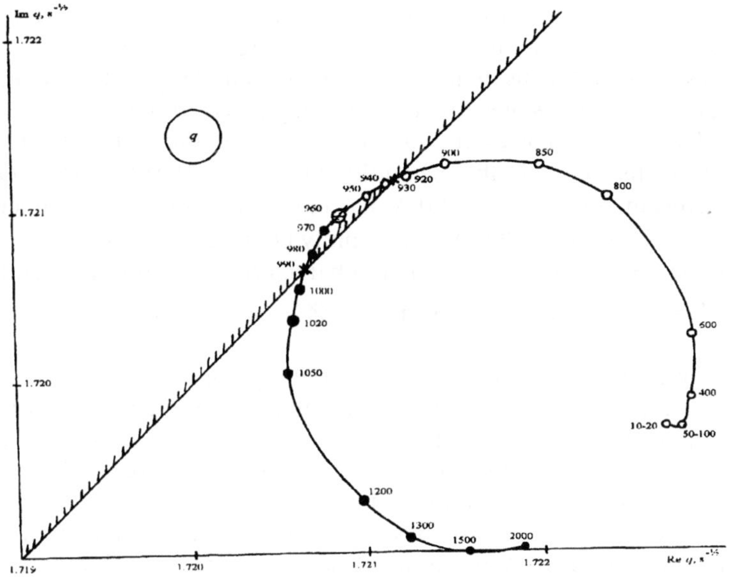

Fig. 8.7. Root locus in q plane of the "unsteady" root associated with liquid motion, showed in a larger scale.

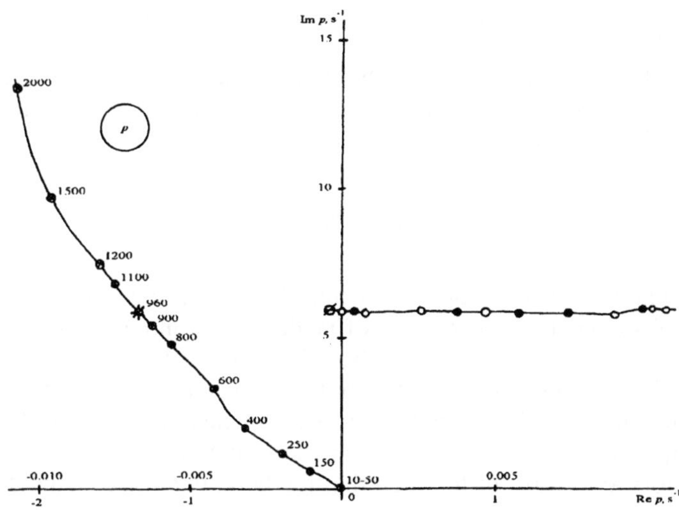

Fig 8.8. Root locus in plane p for the 'unstable' root associated with liquid motion in the upper tank : O – $I^{\circ} < 960$ A ; ● – $I^{\circ} > 960$ A ; Ø – $I^{\circ} = 960$ A ;and the root interacting with that one,

corresponding to the dynamic damper. The upper scale of the real axis is for the root associated with liquid motion in the upper tank, the lower scale is for the root corresponding to the dynamic damper

maximum convergence of the last two roots, reached at $\Gamma^{\circ} = 960$ A, corresponds to the maximum stability margin in the liquid oscillations frequency range.

Fig. 8.6 shows root locus in the upper q half-plane only, since the picture for the lower half-plane is mirror symmetric to that (conjugated roots)

Fig. 8.7 shows in a greater scale the locus for the 'unstable' root ∅. For $0 \leq \Gamma^{\circ} < 930$ A this root is in the instability domain, for $930 \text{ A} < \Gamma^{\circ} < 990$ A it is in the stability domain, and for $\Gamma^{\circ} > 990$ A it is in the instability domain once more. The values $\Gamma^{\circ} = 930$ A and $\Gamma^{\circ} = 990$ A are stability boundary values.

The following symbols for root locus characteristic domains are used in Fig. 8.7:

o — $0 \leq \Gamma^{\circ} \leq 960$ A ; ● — $\Gamma^{\circ} > 960$ A ; ∅ — $\Gamma^{\circ} = 960$ A .

Fig. 8.8 presents root locus in Laplace variable p for the 'unstable' root associated with liquid motion in the upper tank and for the root interacting with that one and corresponding to the dynamic damper. The symbols are the same as in Fig. 8.6,8.7. The real axis has two scales, the upper one is for the roots corresponding to liquid, the lower one is for those corresponding to the dynamic damper.

It is clearly, visible from Fig. 8.8, that maximum stability margin obtained for $\Gamma^{\circ} = 960$ A, when these roots are in the closest positions to each other, occurs when their imaginary parts, and hence the corresponding natural frequencies, coincide.

The analysis presented is an evidence of the l possibility to ensure stability 'in the small' for a SC dynamically unstable at the frequency of liquid oscillations without installing special mechanical dampers in the tanks, and to avoid thus the self-excited oscillation mode.

Although the distance between the hypothetical possibility and its practical implementation is rather great, may be in the future efficient engineering solutions of that problem will be found.

Conclusion

To complete the book the authors would like to express a hope, that the total of presented results concerning synthesis of mathematical models for complex objects, taking into consideration vortex fields, of control laws for those objects, and investigation of the closed-loop object – controller system dynamics may be also useful for research in dynamics of other controlled mechanical and electromechanical systems (stabilized objects) characterized by similar fundamental features.

The authors are aware of the fact, that the book is not free from drawbacks and would appreciate any critics.

Bibliography

1. Abramson, H.N. (Ed.). *The dynamic behavior of liquids in moving containers with applications to space technology* NASA SP-106. 1966, 467 p.

2. Aminov, V.R., Mikishev, G.N., Churilov, G.A. "Experimental methodology of investigating hydrodynamic characteristics of a partitioned tank rotating around the vertical axis" in *Scientific Papers Presented at the 2nd Symposium "Oscillations of Elastic Structures, Containing Liquid"*, Novosibirski Elektrotekhnicheski Institute, Novosibirsk, 1974, pp. 3-10.

3. Andrews, D.G. and Hide, R. "Hydromagnetic edge waves in a rotating stratified fluid" *J.F.M.* 1975. Vol. 72. p. 4.

4. Arens, V.D., Fedorov, S.M. and Khitrik, M.S. *Dynamics of control systems of rockets with on-board computers,* Mashinostroyeniye, Moscow, 1976, 272 p.

5. Arutiunian, N. Kh. and Kolmanovsky, V.B. *The theory of unhomogeneous bodies creeping,* Nauka, Moscow, 1983, 360 p.

6. Batchelor, G.K. *An introduction to fluid dynamics.* Cambridge University Press, Cambridge, 1970.

7. Beletsky, V.V. and Khentov, A.A. *Rotational motion of a magnetized satellite,* Nauka, Moscow, 1985, 286 p.

8. Besekersky, V.A. and Popov, Ye.P. *The theory of automatic control systems*, Nauka, Moscow, 1972, 768 p.

9. Bocharov, V.I. and Nagorsky, V.D., eds.: *High-speed surface transport with linear drive and magnetic suspension,* Transport, Moscow, 1985, 279 p.

10. Bogoriad, I.B., Druzhinin, I.A., Druzhinina E.Z. and Libin, E.I. *Introduction to dynamics of vessels with liquid,* Tomsk State University, Tomsk, 1977, 144 p.

11. Bogorodsky, A.F. *Magnetic Hydromechanics,* Kiev State University Press, Kiev, 1966.

12. Branover, G.G. and Tsinober, A.B. *Magnetic hydrodynamics of incompressible media,* Nauka, Moscow, 1970.

13. Brekhovskikh, L.M. and Goncharov, V.V. *Introduction to continuum mechanics,* Nauka, Moscow, 1982.

14. Broock, K.H., Gottzein, E., Männlein, E., and Pfefferl, J. "Control Aspects of a tracked Magnetic Levitation High Speed Vehicles" *Proc. 6th IFAC Symposium on Automatic Control in Space.* Boston/Cambridge, Massachusetts, Aug. 1975. Messerschmitt – Bölkow – Blohm GmBH, Spaceflight Division. p. 18.

15.Chernousko, F.L. *Motion of a rigid body with cavities containing a viscous liquid,* Vychislitelny Tsentr Akademii Nauk SSSR, Moscow, 1968, 230 p.

16. Chudnov, V.K. and Lomonosova, L.E. "Optimization of electromechanical systems", in *Transactions of VNIIPIhydrotruboprovod. Investigation of Dynamics for Transport Systems with Electromagnetic Levitation*, Moscow, 1988, pp. 65-74.

17. Deviatova, N.O., Nagorsky, V.D. and Scherbakov, N.I. On a multipoint EML control", in *Izvestia Vuzov. Elektromekhanika,* 1983, No. 2, pp. 43-47.

18. Dokuchaev, L.V. "On the associated moment of inertia of liquid in a partitioned cylinder rotating around its longitudinal axis", in *Izvestia AN SSSR. Mekhanika i mashinostroyeniye,* 1964, No. 2, pp. 168-171.

19. Dokuchaev, L.V. *Nonlinear dynamics of flying vehicles with deformable elements,* Mashinostroyeniye, Moscow, 1987, 232 p.

20. Fedorov, N.N , *Principles of electrodynamics,* Vysshaya Shkola, Moscow, 1980, 399 p.

21. Feshchenko, S.F., Lukovsky, I.A., Dokuchaev, L.V. and Rabinovich, B.I. *Methods for calculating associated masses of liquid in moving cavities*, Naukova dumka, Kiev, 1969, 250 p.

22. Feynman, R.P., Leighton, R.B. and Sands, M. *The Feynman lectures on physics,* Addison Wesley Publishing Company, Inc., Reading, Massachusetts, Palo Alto, London.

23. Gantmacher, F.R. *The theory of matrices.* Chelsea Publishing Company New York, 1977.

24. Goncharova, L.Ye., Kozlovskaya, A.A., Rabinovich, B.I. and Sidorov, I.M. "Composition of a control system with a reference model", in *USSR Academy of Sciences Reports,* 1973, Vol. 218, No. 5, pp. 1037-1039.

25. Gottzein, E. "Das Magnetische Rad als autonome Funktionseinheit modulares Trag-und Führsysteme für Magnetbahn" *Fahrtschr. Ber. VDI-Z.* Düsseldorf: VDI-Verlag. 1984. Reihe 8, No. 68. s. 556.

26. Gottzein, E. and Lange, B. "Magnetic Suspension Control Systems for the MBB High Speed Train" *Automatica.* 1975. Vol. 11. p. 271.

27. Gottzein, E., Miller, L. and Meisinger, R. "Magnetic Suspension Control System for High Speed Ground Transportation Vehicles" *World Electrotechnical Congress, June 21-25, 1977.* Moscow,: World Electrotechnical Congress Organizing Committee, 1977. p. 1.

28. Greenspan, H.P. *The theory of rotating fluids.* Cambridge: Cambridge University Press, 1969. p. 328.

29. Grinberg, G.A. *Selected studies of electric and magnetic phenomena*, Academic Press, USSR Academy of Sciences, Moscow-Leningrad, 1948, 727 p.

30. Ishlinsky, A.Yu. and Temchenko M.Ye. "On small oscillations of the vertical axis of a humming top having a cavity completely filled with ideal incompressible liquid", in *Prikladnaya Matematika i teoreticheskaya fizika,* 1960, No. 3, pp. 65-67.

31. Ishlinsky, A.Yu. Gorbachuk, M.L. and Temchenko M.Ye. "On rotation stability of axisymmetric solid bodies suspended on a string, having cavities filled with liquid", in *Spacecraft Dynamics and Space Research,* Mashinostroyeniye, Moscow, 1986, pp. 234-247.

32. Jepps, S.A. "Numerical models of hydromagnetic dynamics" *J.F.M.* 1975. Vol. 67. p. 4.

33. Joukowski, N.E. *On the motion of a solid having cavities filled with homogeneous dropping liquid. Selected works,* Gostekhteoretizdat, Moscow-Leningrad, 1948, Vol.1, 208 p.

34. Kalinina, A.V., Lebedev, V.G., Mytarev, A.I. and Rabinovich, B.I. "Stability investigation of controlled nonlinear systems with multiple degrees of freedom", in *Transaction of the 9th International Conference on Nonlinear Oscillations*, Naukova Dumka, Kiev, 1984, Vol.3, pp. 120-125.

35. Katsan, T.I., Lebedev, V.G. and Mytarev, A.I. "Control of an electromagnet in high-speed transport suspension systems", in *Izvestia AN SSSR. Energetika i Transport,* 1981, No. 1, pp. 108-112.

36. Keulegan, G.H. and Carpenter, L.H. "Forces on cylinders and plates in oscillating fluids" *J. National Bureau of Standards.* 1958. Vol. 60, No. 5. pp. 423-440.

37. Kobrin, A.I. and Martynenko, Yu.G. "Dynamics of conducting solid in high-frequency magnetic field", in *Soviet Physics Doklady,* 25(12), Dec. 1980, pp. 999-1000.

38. Kobrin, A.I. and Martynenko, Yu.G. "Motion of conducting Solid body about its center of mass in a slowly varying magnetic field", in *Soviet Physics Doklady,* 26(12), Dec. 1981, pp. 1134-1136.

39. Kochetov, D.A. and Kravtsova, Ye.V. "The methodology and equipment for experimental determination of control systems response characteristics", in *Transactions of the All-Union Research Institute for Electromechanics. Complex Design and Research of Electric Motors,* Moscow, 1987, Vol. 84, pp. 115-123.

40. Kochin, N.E., Kibel, I.A. and Roze, N.V. *Theoretical hydromechanics,* Interscience Publishers, New York, 1964, 577 p.

41. Kolmanovsky, V.B. and Nosov, V.R. *Stability and periodic regimes of controlled systems with aftereffect,* Nauka, Moscow, 1981, 443 p.

42. Laithwaite, E.R. *Transport without wheels.* London: Elec. Science, 1977. - 322 p.

43. Lamb, H. *Hydrodynamics,* Dover Publications, New York, 1945, 738 p.

44. Landau, L.D. and Lifshits, Ye.M. *Electrodynamics of continual media*, Pergamon, 1960.

45. Landau, L.D. and Lifshits, Ye.M. *Mechanics,* Pergamon, 1976.

46. Landau, L.D. and Lifshits, Ye.M. *The classical theory of fields*, Pergamon, 1976.

47. Landau, L.D. and Lifshits, Ye.M. *Fluid mechanics*, Pergamon, 1987.

48. Lavrentiev, M.A. and Shabat, B.V. *Problems of hydrodynamics and mathematical models,* Nauka, Moscow, 1973, 416 p.

49. Lazar Dragos. *Magnetofluid dynamics.* Abacus Press, Tunbridge Wells, Kent, England, 1975.

50. Lebedev, V.G. and Mytarev, A.I. "Dynamics Investigation of dynamics of solid with cavities containing liquid, taking into account its vorticity", in *Spacecraft Dynamics and Space Research,* Mashinostroyeniye, Moscow, 1986, pp. 121-132.

51. Lebedev, V.G., Mytarev, A.I. and Rabinovich, B.I. "A mathematical model of an electromagnet as controlled electromechanical systems actuator", in *Transaction of VNIIPIhydrotruboprovod. Transport Systems with Electromagnetic Levitation,* Moscow, 1988, pp. 6-13.

52. Lebedev, V.G., Mytarev, A.I. and Rabinovich, B.I. "On contactless electromagnetic suspension systems stability", in *Izvestia AN SSSR. Mashinovedenie,* 1988, No. 6, pp. 62-69.

53. Lebedev, V.G., Rabinovich, B.I., Ashkinazi, R.I. and Viktorov, S.V. "Experimental verification of mathematical model of eddy currents in artificial Earth satellites and spacecrafts magnetic attitude control systems actuators", *Kosmicheskie Issledovaniya,* 1984, Vol. 22, No. 6, pp. 945-947.

54. Lebedev, V.G., Mytarev, A.I., Rabinovich, B.I. and Churilov, G.A. "Experimental verification of mathematical model of spacecraft with damping systems in tanks, taking into account liquid vorticity", *Kosmicheskie Issledovaniya,* 1990, Vol. 28, No. 4, pp. 521-530.

55. Linkov, R.V. "The slow motions of conducting top in resonant interaction with an oscillating magnetic field", in *Soviet Physics. Technical Physics,* 1980, Vol. 25, No. 6, pp. 660-664.

56. Linkov, R.V. and Yurman, Yu.M. "Forces and moments on a rotating conductive shell in a magnetic field", in *Soviet Physics. Technical Physics,* 1977, Vol. 22, No. 5, pp.533-537.

57. Linkov, R.V. and Yurman, Yu.M. "Fast rotation of a conductive magnetic top in unhomogeneous alternating magnetic field", in *Soviet Physics. Technical Physics,* 1978, Vol. 23, No. 6, pp. 627-632.

58. Lurie, A.I. *Operational calculus,* Gostekhizdat, Moscow, Leningrad, 1951, 431 p.

59. Lurie, A.I. *Analytical Mechanics,* Fizmatgiz, Moscow, 1961,824 p.

60. Martynenko Yu.G. "Motion of a conducting rigid body about a fixed point in a magnetic field", in *Mechanics of Solids (Mekhanika Tverdogo Tela),* 1977, Vol. 12, No. 4, pp. 30-38.

61. Martynenko, Yu.G. *Motion of a solid body in electrical and magnetic fields,* Nauka, Moscow, 1988, 368 p.

62. Mikishev, G.N. *Experimental methods in spacecraft dynamics,* Mashinostroyeniye, 1978, 248 p.

63. Mikishev, G.N. and Rabinovich, B.I. *Dynamics of a solid with cavities partially filled with liquid,* Mashinostroyeniye, Moscow, 1968, 532 p.

64. Mikishev, G.N. and Rabinovich, B.I. *Dynamics of thin-walled structures with compartments containing a fluid,* Mashinostroyeniye, Moscow, 1971, 564 p.

65. Miles, J.W. "Ring damping of free oscillations in a circular tank" *J. Appl. Mech.* 1958. Vol. 25, No. 6. pp. 26-32.

66. Moiseyev, N.N. and Rumiantsev, V.V. *Dynamics of a body with cavities containing liquid,* Nauka, Moscow, 1965, 439 p.

67. Mytarev, A.I. "Closed-loop electromagnet-controller system dynamics and stabilization algorithms", in *Transactions of VNIIPIhydrotruboprovod . Investigation of Dynamics for Transport Systems with Electromagnetic Levitation,* Moscow, 1988, pp. 90-99.

68. Narimanov, G.S. "Motion of a rigid body with a cavity partially filled with a fluid", in *Prikladnaya Matematika i Mekhanika,* 1956, Vol. 20, No. 1, pp. 21-58.

69. Narimanov, G.S. "Motion of a symmetric gyroscope with a cavity partially filled with a fluid", in *Prikladnaya Matematika,* 1957, Vol. 21, No. 5, pp. 699-700.

70. Narimanov, G.S. *Deformable body dynamics,* Artillery Engineering Academy Publishers, Moscow, 1958, 175 p.

71. Nikitenko, Yu.A., Tushkanov, N.B., Savin, M.M., et al. "Results of experimental investigation in control systems for electromagnetic suspension and horizontal stabilization" of the three ton mass model vehicle", in *Izvestiya Vuzov. Elektromekhanika,* 1983, No. 2, pp. 57-62.

72. Okhotsimsky, D.E. "Theory of motion of a body with cavities partially filled with liquid", in *Prikladnaya Matematika i Mekhanika,* 1956, Vol. 20, No. 1, pp. 3-20.

73. Polovin, R.V. and Demutsku, V.P. *Fundamentals of magnetohydrodynamics.* Consultants Bureau, New York, 1990.

74. Rabinovich, B.I. "Equations of perturbed motion of a solid body with cylindric cavity partially filled with liquid", in *Prikladnaya Matematika i Mekhanika,* 1956, Vol. 20, No. 1, pp. 39-50.

75. Rabinovich, B.I. *Applied problems of stabilized objects stability,* Mashinostroyeniye, Moscow, 1978, 232 p.

76. Rabinovich, B.I. *Introduction to spacecraft rocket carriers dynamics,* 2nd edition, Mashinostroyeniye, Moscow, 1983, 296 p.

77. Rabinovich, B.I. "Equations for perturbed motion of a solid body containing low-viscous vorticized liquid", in *Spacecraft Dynamics and Space Research,* Mashinostroyeniye, Moscow, 1986, pp. 106-120.

78. Rabinovich, B.I. "Equations for dynamics of a solid body containing vorticized low-viscous liquid", in *Applied Problems in Dynamics and Stability of Mechanical Systems,* Institut Matematiki AN Ukr.SSR, Kiev, 1987, pp. 34-44.

79. Rabinovich, B.I. and Rogovoi, V.M. "Accounting for the viscosity of liquid fuel in investigation the motion of controlled spacecraft with liquid rocket engines", *Kosmicheskie Issledovaniya,* 1970, Vol. 8, No. 3, pp. 315-328.

80. Rabinovich, B.I. and Rogovoi, V.M. "Mathematical models of unsteady vortex currents and vortex motion of liquid in problems of spacecraft attitude control and stabilization. I", in *Kosmicheskie Issledovaniya,* 1984, Vol. 22, No. 5, pp. 683-692.

81. Rabinovich, B.I. and Rogovoi, V.M. "Mathematical models of transient eddy current and vortex motions of a fluid in satellite and spacecraft attitude control and stabilization problems. II", in *Kosmicheskie Issledovaniya,* 1984, Vol. 22, No. 6, pp. 867-874.

82. Rabinovich, B.I. and Tyurin, Yu.V. "On a recursive numerical method in conformal mapping", in *Doklady Akademii Nauk SSSR,* 1983, Vol. 272, No. 3, pp. 532-535; English translation in *Soviet Math. Dokl.,* 1983, Vol. 28, No. 2, pp. 415-418.

83. Rabinovich, B.I. and Tyurin, Yu.V. "A recursive numerical method in the conformal mapping of doubly connected regions onto a circular annulus", in *Doklady Akademii Nauk SSSR,* 1983, Vol. 272, No. 4, pp. 795-798; English translation in *Soviet Math. Dokl.,* 1983, Vol. 28, No. 2, pp. 447-450.

84. Rabinovich, B.I. and Tyurin, Yu.V. "M.A. Lavrentiev's variational principle and RT-algorithm of conformal mapping in hydrodynamics and dynamics of a solid containing liquid", in *Spacecraft Dynamics and Space Research,* Mashinostroyeniye, Moscow, 1986, pp.142-159.

85. Rabotnov, Yu.N. *Mechanics of deformable solid,* Nauka, Moscow, 1988, 744 p.

86. Rogovoi, V.M. and Cheremnykh, S.V., *Dynamic stability of spacecraft with liquid-fueled rockets,* Mashinostroyeniye, Moscow, 1975, 152 p.

87. Rosenzweig, R.E. *Ferrohydrodynamics,* Cambridge University Press, Cambridge-New York, 1985, 344 p.

88. Rvalov, R.V. and Rogovoi, V.M. "Rotational motion of a body with a cavity containing a liquid", in *Izvestiya AN SSSR. Mekhanika Tverdogo Tela,* 1972, No. 3, pp. 15-20.

89. Sidorov, I.M. and Korotaeva, I.P. "Analysis of the structural stability of mechanical systems with many degrees of freedom in the presence of a correcting device", in *Izvestiya Academii Nauk SSSR. Tekhnicheskaya Kibernetika,* 1965, No. 5, pp. 183-187.

90. Sidorov, I.M., Goncharova, L.E. and Lebedev, V.G. *Moving objects control based on algorithms with a model,* Mashinostroyeniye, Moscow, 1982, 232 p.

91. Sinha, P.K. Magnetic Suspension for low speed Vehicles // *J. Dyn. Syst. Contr.* - 1978. - V. 100. - pp. 333-342.

92. Sinha, P.K. Dynamics of magnetically suspended vehicles *Trans. Inst. Meas. and Contr.* 1979. No. 1. pp. 57-64.

93. Sobolev, S.L. "Motion of a symmetric gyroscope with a cavity filled with a liquid", in *Prikladnaya Matematika i Teoreticheskaya Fizika,* 1960, No. 3, pp. 20-25.

94. Sozou, C. and Pickering, W.M. "The development of magnetohydrodynamic flow due to an electric current discharge" *J.F.M.* 1975. Vol. 70. p. 3.

95. Sozou, C. and Pickering, W.M. "Magnetohydrodynamic flow due to the discharge of an electrical current in hemispherical container" *J.F.M.* 1976. Vol. 73. p. 4.

96. Thornton, R.D. "Magnetic Levitation and Propulsion" *IEE Trans. Magnetics.* 1975. July. - Vol. 11, No. 4. p. 981.

97. Tyurin, Yu.V. and Ashkinazi, R.I. "Magnetostatics problems solution based on RT-algorithm of conformal mapping", in *Transactions of VNIIPIhydrotruboprovod. Investigation of Dynamics for Transport Systems with Electromagnetic Levitation*, Moscow, 1988, pp. 75-81.

98. *Variational methods for the problems of oscillations of liquid and solid with liquid,* USSR Academy of Sciences Computer Centre, Moscow, 1962, 246 p.

99. Vasilieva, A.B. and Butuzov, V.F. *Asymptotic expansions of singularly perturbed equations,* Nauka, Moscow, 1973, 463 p.

100. Veinberg, D.M., Vereshchagin, V.P., Danilov-Nitusov, N.N. and Sheremet'evskii, N.N. Magnetic suspension systems in the actuating elements for attitude control of Spacecraft", in *Mechanics of Solids (Mekhanika Tverdogo Tela)*, 1981, Vol. 16, No. 3, pp. 143-148.

101. Verlan, A.F. and Sizikov, V.S. *Integral equations solution methods with computer programs*, Naukova Dumka, Kiev, 1977, 292 p.

102. Vishik, M.N. and Lusternik, L.A. "Regular singularity and boundary layer for linear differential equations with a small parameter", in *Success of Mathematical Sciences (Uspekhi matematicheskikh nauk),* 1957, Vol. 12, No. 5, pp. 3-122.

103. Van Dyke Milton. *An album of fluid motion.* The Parabolic Press Stanford, California, 1982.

Index

Mechanics

FLUID MECHANICS AND ITS APPLICATIONS

Series Editor: R. Moreau

Aims and Scope of the Series

The purpose of this series is to focus on subjects in which fluid mechanics plays a fundamental role. As well as the more traditional applications of aeronautics, hydraulics, heat and mass transfer etc., books will be published dealing with topics which are currently in a state of rapid development, such as turbulence, suspensions and multiphase fluids, super and hypersonic flows and numerical modelling techniques. It is a widely held view that it is the interdisciplinary subjects that will receive intense scientific attention, bringing them to the forefront of technological advancement. Fluids have the ability to transport matter and its properties as well as transmit force, therefore fluid mechanics is a subject that is particularly open to cross fertilisation with other sciences and disciplines of engineering. The subject of fluid mechanics will be highly relevant in domains such as chemical, metallurgical, biological and ecological engineering. This series is particularly open to such new multidisciplinary domains.

1. M. Lesieur: *Turbulence in Fluids.* 2nd rev. ed., 1990 ISBN 0-7923-0645-7
2. O. Métais and M. Lesieur (eds.): *Turbulence and Coherent Structures.* 1991 ISBN 0-7923-0646-5
3. R. Moreau: *Magnetohydrodynamics.* 1990 ISBN 0-7923-0937-5
4. E. Coustols (ed.): *Turbulence Control by Passive Means.* 1990 ISBN 0-7923-1020-9
5. A.A. Borissov (ed.): *Dynamic Structure of Detonation in Gaseous and Dispersed Media.* 1991 ISBN 0-7923-1340-2
6. K.-S. Choi (ed.): *Recent Developments in Turbulence Management.* 1991 ISBN 0-7923-1477-8
7. E.P. Evans and B. Coulbeck (eds.): *Pipeline Systems.* 1992 ISBN 0-7923-1668-1
8. B. Nau (ed.): *Fluid Sealing.* 1992 ISBN 0-7923-1669-X
9. T.K.S. Murthy (ed.): *Computational Methods in Hypersonic Aerodynamics.* 1992 ISBN 0-7923-1673-8
10. R. King (ed.): *Fluid Mechanics of Mixing.* Modelling, Operations and Experimental Techniques. 1992 ISBN 0-7923-1720-3
11. Z. Han and X. Yin: *Shock Dynamics.* 1993 ISBN 0-7923-1746-7
12. L. Svarovsky and M.T. Thew (eds.): *Hydroclones.* Analysis and Applications. 1992 ISBN 0-7923-1876-5
13. A. Lichtarowicz (ed.): *Jet Cutting Technology.* 1992 ISBN 0-7923-1979-6
14. F.T.M. Nieuwstadt (ed.): *Flow Visualization and Image Analysis.* 1993 ISBN 0-7923-1994-X
15. A.J. Saul (ed.): *Floods and Flood Management.* 1992 ISBN 0-7923-2078-6
16. D.E. Ashpis, T.B. Gatski and R. Hirsh (eds.): *Instabilities and Turbulence in Engineering Flows.* 1993 ISBN 0-7923-2161-8
17. R.S. Azad: *The Atmospheric Boundary Layer for Engineers.* 1993 ISBN 0-7923-2187-1
18. F.T.M. Nieuwstadt (ed.): *Advances in Turbulence IV.* 1993 ISBN 0-7923-2282-7
19. K.K. Prasad (ed.): *Further Developments in Turbulence Management.* 1993 ISBN 0-7923-2291-6
20. Y.A. Tatarchenko: *Shaped Crystal Growth.* 1993 ISBN 0-7923-2419-6

Kluwer Academic Publishers – Dordrecht / Boston / London

Mechanics

FLUID MECHANICS AND ITS APPLICATIONS

Series Editor: R. Moreau

21. J.P. Bonnet and M.N. Glauser (eds.): *Eddy Structure Identification in Free Turbulent Shear Flows.* 1993 ISBN 0-7923-2449-8
22. R.S. Srivastava: *Interaction of Shock Waves.* 1994 ISBN 0-7923-2920-1
23. J.R. Blake, J.M. Boulton-Stone and N.H. Thomas (eds.): *Bubble Dynamics and Interface Phenomena.* 1994 ISBN 0-7923-3008-0
24. R. Benzi (ed.): *Advances in Turbulence V.* 1994 ISBN 0-7923-3032-3
25. B.I. Rabinovich, V.G. Lebedev and A.I. Mytarev: *Vortex Processes and Solid Body Dynamics.* The Dynamic Problems of Spacecrafts and Magnetic Levitation Systems. 1994 ISBN 0-7923-3092-7
26. P.R. Voke, L. Kleiser and J.-P. Chollet (eds.): *Direct and Large-Eddy Simulation I.* Selected papers from the First ERCOFTAC Workshop on Direct and Large-Eddy Simulation. 1994 ISBN 0-7923-3106-0

Kluwer Academic Publishers – Dordrecht / Boston / London

Mechanics

From 1990, books on the subject of *mechanics* will be published under two series:
FLUID **MECHANICS AND ITS APPLICATIONS**
Series Editor: R.J. Moreau
SOLID **MECHANICS AND ITS APPLICATIONS**
Series Editor: G.M.L. Gladwell

Prior to 1990, the books listed below were published in the respective series indicated below.

MECHANICS: DYNAMICAL SYSTEMS
Editors: L. Meirovitch and G.Æ. Oravas

1. E.H. Dowell: *Aeroelasticity of Plates and Shells.* 1975 ISBN 90-286-0404-9
2. D.G.B. Edelen: *Lagrangian Mechanics of Nonconservative Nonholonomic Systems.* 1977 ISBN 90-286-0077-9
3. J.L. Junkins: *An Introduction to Optimal Estimation of Dynamical Systems.* 1978 ISBN 90-286-0067-1
4. E.H. Dowell (ed.), H.C. Curtiss Jr., R.H. Scanlan and F. Sisto: *A Modern Course in Aeroelasticity.* *Revised and enlarged edition see under Volume 11*
5. L. Meirovitch: *Computational Methods in Structural Dynamics.* 1980 ISBN 90-286-0580-0
6. B. Skalmierski and A. Tylikowski: *Stochastic Processes in Dynamics.* Revised and enlarged translation. 1982 ISBN 90-247-2686-7
7. P.C. Müller and W.O. Schiehlen: *Linear Vibrations.* A Theoretical Treatment of Multi-degree-of-freedom Vibrating Systems. 1985 ISBN 90-247-2983-1
8. Gh. Buzdugan, E. Mihăilescu and M. Radeş: *Vibration Measurement.* 1986 ISBN 90-247-3111-9
9. G.M.L. Gladwell: *Inverse Problems in Vibration.* 1987 ISBN 90-247-3408-8
10. G.I. Schuëller and M. Shinozuka: *Stochastic Methods in Structural Dynamics.* 1987 ISBN 90-247-3611-0
11. E.H. Dowell (ed.), H.C. Curtiss Jr., R.H. Scanlan and F. Sisto: *A Modern Course in Aeroelasticity.* Second revised and enlarged edition (of Volume 4). 1989 ISBN Hb 0-7923-0062-9; Pb 0-7923-0185-4
12. W. Szemplińska-Stupnicka: *The Behavior of Nonlinear Vibrating Systems.* Volume I: Fundamental Concepts and Methods: Applications to Single-Degree-of-Freedom Systems. 1990 ISBN 0-7923-0368-7
13. W. Szemplińska-Stupnicka: *The Behavior of Nonlinear Vibrating Systems.* Volume II: Advanced Concepts and Applications to Multi-Degree-of-Freedom Systems. 1990 ISBN 0-7923-0369-5
Set ISBN (Vols. 12–13) 0-7923-0370-9

MECHANICS OF STRUCTURAL SYSTEMS
Editors: J.S. Przemieniecki and G.Æ. Oravas

1. L. Frýba: *Vibration of Solids and Structures under Moving Loads.* 1970 ISBN 90-01-32420-2
2. K. Marguerre and K. Wölfel: *Mechanics of Vibration.* 1979 ISBN 90-286-0086-8

Mechanics

3. E.B. Magrab: *Vibrations of Elastic Structural Members.* 1979 ISBN 90-286-0207-0
4. R.T. Haftka and M.P. Kamat: *Elements of Structural Optimization.* 1985
   Revised and enlarged edition see under Solid Mechanics and Its Applications, Volume 1
5. J.R. Vinson and R.L. Sierakowski: *The Behavior of Structures Composed of Composite Materials.* 1986 ISBN Hb 90-247-3125-9; Pb 90-247-3578-5
6. B.E. Gatewood: *Virtual Principles in Aircraft Structures.* Volume 1: Analysis. 1989 ISBN 90-247-3754-0
7. B.E. Gatewood: *Virtual Principles in Aircraft Structures.* Volume 2: Design, Plates, Finite Elements. 1989 ISBN 90-247-3755-9
   Set (Gatewood 1 + 2) ISBN 90-247-3753-2

MECHANICS OF ELASTIC AND INELASTIC SOLIDS

Editors: S. Nemat-Nasser and G.Æ. Oravas

1. G.M.L. Gladwell: *Contact Problems in the Classical Theory of Elasticity.* 1980 ISBN Hb 90-286-0440-5; Pb 90-286-0760-9
2. G. Wempner: *Mechanics of Solids with Applications to Thin Bodies.* 1981 ISBN 90-286-0880-X
3. T. Mura: *Micromechanics of Defects in Solids.* 2nd revised edition, 1987 ISBN 90-247-3343-X
4. R.G. Payton: *Elastic Wave Propagation in Transversely Isotropic Media.* 1983 ISBN 90-247-2843-6
5. S. Nemat-Nasser, H. Abé and S. Hirakawa (eds.): *Hydraulic Fracturing and Geothermal Energy.* 1983 ISBN 90-247-2855-X
6. S. Nemat-Nasser, R.J. Asaro and G.A. Hegemier (eds.): *Theoretical Foundation for Large-scale Computations of Nonlinear Material Behavior.* 1984 ISBN 90-247-3092-9
7. N. Cristescu: *Rock Rheology.* 1988 ISBN 90-247-3660-9
8. G.I.N. Rozvany: *Structural Design via Optimality Criteria.* The Prager Approach to Structural Optimization. 1989 ISBN 90-247-3613-7

MECHANICS OF SURFACE STRUCTURES

Editors: W.A. Nash and G.Æ. Oravas

1. P. Seide: *Small Elastic Deformations of Thin Shells.* 1975 ISBN 90-286-0064-7
2. V. Panc: *Theories of Elastic Plates.* 1975 ISBN 90-286-0104-X
3. J.L. Nowinski: *Theory of Thermoelasticity with Applications.* 1978 ISBN 90-286-0457-X
4. S. Łukasiewicz: *Local Loads in Plates and Shells.* 1979 ISBN 90-286-0047-7
5. C. Fiřt: *Statics, Formfinding and Dynamics of Air-supported Membrane Structures.* 1983 ISBN 90-247-2672-7
6. Y. Kai-yuan (ed.): *Progress in Applied Mechanics.* The Chien Wei-zang Anniversary Volume. 1987 ISBN 90-247-3249-2
7. R. Negruţiu: *Elastic Analysis of Slab Structures.* 1987 ISBN 90-247-3367-7
8. J.R. Vinson: *The Behavior of Thin Walled Structures.* Beams, Plates, and Shells. 1988 ISBN Hb 90-247-3663-3; Pb 90-247-3664-1

MIX
Papier aus verantwortungsvollen Quellen
Paper from responsible sources
FSC® C105338

If you have any concerns about our products,
you can contact us on
ProductSafety@springernature.com

In case Publisher is established outside the EU,
the EU authorized representative is:
Springer Nature Customer Service Center GmbH
Europaplatz 3, 69115 Heidelberg, Germany

Printed by Libri Plureos GmbH
in Hamburg, Germany